Paleoshorelines and Prehistory: An Investigation of Method

Paleoshorelines and Prehistory: An Investigation of Method

Edited by

Lucille Lewis Johnson, Ph.D

Department of Anthropology
Vassar College
Poughkeepsie, New York

With the assistance of

Melanie Stright

CRC Press

Boca Raton Ann Arbor London

Library of Congress Cataloging-in-Publication Data

Paleoshorelines and prehistory : an investigation of method / edited by Lucille Lewis Johnson ; with the assistance of Melanie Stright.
p. cm.
Includes bibliographical references and index.
ISBN 0-8493-8855-4
1. Coastal archaeology--Methodology. 2. Indians--Antiquities.
3. America--Antiquities. I. Johnson, Lucille Lewis. II. Stright, Melanie.
GN784.P35 1991
970′.00946--dc20 91-29294
CIP

Developed by Telford Press

Direct all inquiries to CRC Press, Inc., 2000 Corporate Blvd., N.W., Boca Raton, Florida 33431.

International Standard Book Number 0-8493-8855-4

Library of Congress Card Number 91-29294

Printed in the United States of America 1 2 3 4 5 6 7 8 9 0

PREFACE

In this volume, we are looking at the state-of-the-art in coastal geoarchaeology, and it varies. The different methods that we use are due to different problems, different histories, and different backgrounds. We have much to learn from each other and, we trust, others can learn from us. While archeologists and geologists have studied coastal sites in all parts of the world, I decided, in order to increase the coherence of this volume, to focus on the Western Hemisphere (see Figure 1). The hemisphere contains a wide variety of coastlines: emergent, submergent, prograding, and variant. Thus, the insights we arrive at should be valuable to anyone interested in coastal people. Working in the Western Hemisphere also gives us the advantage of a shorter time span than would be true elsewhere in the world, so that most of us cover the entire period of human occupation of our coasts, which varies from 12,000 to 3,000 years.

When archaeologists and geologists study the same situation, there is not always agreement. However, controversy is healthy as long as it is accompanied by dialogue. In our jointly written Chapters, Winslow and Johnson, Clark and Fitzhugh, and Suigio et al. agree with one another; Watters and Donahue do not. It may be significant that the latter authors are working in a more stable situation. Might it be that less change makes that which occurs more difficult to interpret? Such controversy is healthy as long as it leads to dialogue, and in their article, Watters and Donahue outline the continuing research they hope to undertake to resolve their disagreement.

I explored many ways of organizing this volume in order to reflect the many ways in which the included Chapters reflect upon, amplify, and compliment each other. Given that the focus of all of our work is on paleoshorelines, I decided to divide the volume principally according to the stability of the coastline itself, looking first at those situations in which the land, during the period of human occupation, has been stable and second at those areas in which isostatic rebound, seismicity, or both have resulted in changing elevation of the land. Within each Section, I have divided the Chapters into those considering islands and those considering the mainland. Thus, overall, we move from the most simple situation, small, stable islands in which the human adaptation is, perforce, to the sea, to the most complex situation, a primarily agricultural occupation of a highly varying coastline.

Georgraphically, stable coasts are examined first in the Caribbean Islands (Figure 1: Chapters 1, 2), then in the northwest and southeast coasts of the hemisphere, where coastal progradation results in subaerial preservation of sites (Figure 1: Chapters 3, 4) and finally in the Gulf of Mexico (Figure 1: Chapters 5, 6) where sites have been inundated by rising Holocene seas. Unstable islands are examined in the North Pacific (Figure 1: Chapters 7, 8) while isostatic rebound raises the Northeastern mainland (Figure 1: Chapter 9) and seismicity and progradation affect the coast of Peru (Figure 1: Chapter 10).

In the archeological study of coasts, it is necessary to understand coastal processes, the settlement and land use patterns of the past human residents, and site preservation and location relative to past shorelines. Although all of the studies here concern late Pleistocene and Holocene human occupations, those authors studying stable coastlines found it important to study long-term geological history in order to understand the resource and landform distributions in their areas (Chapters 1 to 6). Under certain circumstances, longshore currents create beach ridges which significantly change the position of shorelines and also provide sand for dune formations that can materially affect the position, occupation and preservation of sites (see Chapters 1–3, 8, 10). It is also often important to look at systems of marine transport to understand the origin of sediments on the land (Chapters 1, 3, 4, 10). Many attempts have been made in the past to derive continent-wide, hemispheric, or

Figure 1. Location map—numbers are keyed to Chapters. 1. The Bahamas. 2. Barbuda, West Indies. 3. Kotzebue Sound, Alaska. 4. Brazil. 5. Gulf of Mexico. 6. Apalachee Bay, Florida. 7 and 8. Shumagin Islands, Alaska. 9. Labrador. 10. Peru.

worldwide sea level curves. However, the geologists writing in this volume recognize the need to create local sea level curves, particularly in areas with dynamic coastlines and changing oceanic currents (Chapters 3, 4, 6, 7, 9).

In order to successfully survey and find sites, and to understand the patterns revealed, it is necessary to know, or to discover, how the people related to the shorelines. All of the island dwellers discussed here focused their subsistence quest on the sea, but their patterns of residence vary considerably. In the Caribbean, the residents of Barbuda (Chapter 2) chose locations for permanent sites where they would be protected from the wind and walked from them to shore. In the Bahamas (Chapter 1), where there is no high land behind which to shelter, people lived along the beaches. To the north, the Shumagin Islanders (Chapter 8), needing even more protection from inclement weather, dug semisubterranean houses and, therefore were restricted to choosing site locations with sediments of sufficient depth and height above sea level.

On the mainland, the relationship between the people, the shore, and the sea varies a great deal and, therefore, complicates the archeologists' quest. On the Bering Sea Coast (Chapter 3), summers were spent inland while permanent winter dwellings were dug into beach ridges as close to the shore as practicable. In Newfoundland (Chapter 9), on the other hand, winter dwellings were inland while only temporary, summer settlements occurred along the shore. As in the northwest, however, these settlements were located as close to shore as possible in order to facilitate marine pursuits.

In Brazil, middens are generally found along the inner shores of lagoons, where they are not as sensitive to shoreline change. However, because of the subsistence focus on heavy-shelled mollusks, these middens were built-up close to shore and their relative positions can indicate lagoon margins, and, inferentially, sea level, at the time they were occupied. In Peru (Chapter 10), although people lived and fished along the coast and have, to some extent been studied, Moseley, Wagner, and Richardson are concerned here with the effect of coastal processes upon complex societies primarily dependent for subsistence upon irrigation agriculture.

Finally, in the Gulf of Mexico, with its wide continental shelf and relatively protected waters, the archaeologists are not even studying coastal sites, though the sites they do study are now under water. Garrison (Chapter 5) establishes the location of a late Pleistocene shoreline, but has as yet found no sites associated with it; Dunbar, Webb, and Faught (Chapter 6) are studying inland occupations currently inundated by the postglacial sea level rise. While they are moving offshore, they have not yet reached the early prehistoric shoreline and the models they use to locate sites are based on inland habitations.

The coastal processes discussed by the geologists will affect not only where people live, but also the sites they leave behind. Sites can be buried by dunes (Chapter 10) or water (Chapter 6), coastal sites can be found far inland (Chapters 4, 9), while inland sites appear far out to sea (Chapter 6). Sites can be destroyed by wave action (Chapters 7, 8) or gently covered (Chapter 6). The elevation at which a site is found can be quite distinct from that at which it was occupied (Chapters 7 to 9). Dunes can bury sites, strongly affecting our undertstanding of past settlement patterns (Chapter 10), or dunes can erode away, leaving only palimpsests of the settlements once established on their crests (Chapter 8). Finally, in a process not limited to coastlines, later inhabitants can disrupt and destroy sites, either intentionally or accidentally (Chapters 2, 8, 10).

A final focus of this volume is on the methods geologists and archaeologists have used to elucidate the history of the human habitation of coasts. All of the studies discuss methods, which range from standard boat and foot archeological surface surveys to sophisticated geophysical remote sensing. Several authors, particularly Mason, Winslow, and Moseley et al., stress the importance of air photo interpretation in understanding coastal dynamics. As mentioned above, most of the geologsits found it essential to create their own local sea level curves and discuss the methods they used to do so.

Lucille Lewis Johnson
Poughkeepsie, New York

ACKNOWLEDGEMENTS

This volume arose out of a symposium I organized for the 1988 Society for American Archaeology Meetings in Atlanta, GA. In working in the Shumagin Islands, Alaska, I had become aware of how critical it was to understand shoreline changes in order to understand past settlement patterns. From working with Margaret Winslow, I also realized that the information flow could be reversed; archaeological information could aid in the construction of geological models. I decided that it would be valuable to bring together a group of archeologists and geologists who had been working on such problems to exchange information and insights.

While I was quite aware of scholars working in the Far North and along the Pacific Coast, I did not know the Atlantic and Gulf of Mexico Coast materials. Jack Donahue suggested I call Melanie Stright, who provided names of people for me to contact, agreed to serve as a discussant at the meeting, and read and helped edit all of the Chapters as they were submitted for the volume.

After the meeting, two Chapters were added in order to complete the geographical coverage of the Americas, the first by Owen Mason and the second by Kenitiro Suguio, Louis Martin, and Jean-Marie Flexor. These people were suggested to me by Richard Jordan and Rhodes Fairbridge, and I thank them heartily.

LIST OF CONTRIBUTORS

Peter U. Clark, Ph.D.
Department of Geosciences
Oregon State University
Corvallis, Oregon

Jack Donahue, Ph.D.
Department of Geology and Planetary Sciences and Department of Anthropology
University of Pittsburgh
Pittsburgh, Pennsylvania

James S. Dunbar, B.A.
Florida Bureau of Archaeological Research
Division of Historical Resources
Department of State
Tallahassee, Florida

Michael Faughts, M.S.
Department of Anthropology
University of Arizona
Tucson, Arizona

William W. Fitzhugh, Ph.D.
Department of Anthropology
Smithsonian Institution
Washington, D.C.

Jean-Marie Flexor, Ph.D.
ON-CNPq
Departmento de Geofísica
Rio de Janeiro, Brazil

Ervan G. Garrison, Ph.D.
National Oceanic and Atmospheric Administration
Ocean and Coastal Resource Management
Washington, D.C.

Lucille Lewis Johnson, Ph.D.
Department of Anthropology
Vassar College
Poughkeepsie, New York

William F. Keegan, Ph.D.
Florida Museum of Natural History
Department of Anthropology
University of Florida
Gainesville, Florida

Louis Martin, Ph.D.
ORSTOM/ON-CNPq
Departmento de Geofísica
Rio de Janeiro, Brazil

Owen K. Mason, Ph.D.
Alaska Quaternary Center
University of Alaska Museum
University of Alaska, Fairbanks
Fairbanks, Alaska

Michael E. Moseley, Ph.D.
Department of Anthropology
University of Florida
Gainesville, Florida

James B. Richardson, III, Ph.D.
Division of Anthropology
Carnegie Museum of Natural History
Pittsburgh, Pennsylvania

Robert Stuckenrath, Ph.D.
University of Pittsburgh
Radiocarbon Laboratory
Pittsburgh, Pennsylvania

Kenitiro Suguio, Ph.D.
Instituto de Geociências
Universidade de São Paulo
São Paulo, Brazil

David Wagner
Institute for Technology Development
Space Remote Sensing Center
NSTL Station, Mississippi

David R. Watters, Ph.D.
Section of Anthropology
Carnegie Museum of Natural History
Pittsburgh, Pennsylvania

S. David Webb, Ph.D.
Department of Natural Sciences
Florida Museum of Natural History
University of Florida
Gainesville, Florida

Margaret Anne Winslow, Ph.D.
Department of Earth and Planetary Sciences
City College of New York
New York, New York

TABLE OF CONTENTS

Part II. UNSTABLE LAND

Part I

Stable Land

The chapters in this section concern tectonically stable shorelines. In the Caribbean islands, Pleistocene emergence provided the conditions which allowed the building of islands which resisted the subsequent sea level rise by making sediments available for coastal progradation during the recent past. Current information suggests only a short period of occupation although the coastal shelves have not been investigated. In Alaska, progradation has also protected settlement spaces, while in Brazil patterns of emergence and submergence are reflected both in geological and archaeological structures. Finally, the Gulf of Mexico Coast has been submerged by the postglacial sea level rise, leading to the necessity of submarine investigations.

Islands

Lucille Lewis Johnson
Department of Anthropology
Vassar College
Poughkeepsie, New York

INTRODUCTION

These first two chapters deal with the study of similar situations: tectonically stable islands with a short period of human occupation. The major difference is that in the Bahamas there are many shallow inlets and marshes whose formation and changing relationship to the open sea must be understood in order to comprehend past settlement patterns, while in Barbuda there is but one large bay whose form change must be comprehended. These chapters are also concerned with the formation of beach ridges, a focus which ties them to the chapter by Owen Mason.

William Keegan is studying the Bahamas, where the only documented prehistoric occupation was by the Lucayan Tainos from 700 to 1520 A.D. Keegan is concerned with patterns of settlement and of site survival. In small islands, particularly those with sand coasts, both normal winds and tides as well as storm action result in rapid erosion and deposition of sediments which then hamper our attempts to understand the archaeological record. Keegan shows particularly well the problems which arise when mainland settlement types are applied to small islands. On small islands of low relief, there really are no inland sites. Keegan shows that inland Lucayan sites in the Bahamas were actually coastal when they were occupied, located on the shores of inlets with subsistence focused on the sea, while many coastal sites on the other side of the island have been destroyed by erosion. Thus, geological changes in the Bahamas have influenced both where sites are and where they are not. In surveying for sites in such regions and considering the implications of where they are found, it is necessary to look everywhere since the sites may not be in the environment now that they were in when they were occupied. Additionally, careful consideration must be taken of environments in which past sites will now be missing.

The major occupation of Barbuda seems to have begun at the same time as in the Bahamas to the north, although one, probably Archaic, site has been dated to 1800 B.C. and others may be located offshore. In Barbuda there have also been rapid changes in the relationship between the sea and the land. Barbuda's karst core has been acquiring beaches during the Holocene as the rising sea level has resulted in deposition. Whereas one might expect that, as in the Bahamas and in the Bering sea region, sites of the past marine-based societies would be close to the shore, Watters' survey has shown that on Barbuda access to fresh water and protection from wind seem to be more important. Therefore, sites, particularly large village sites, are found near the highlands, where hills protect them from prevailing winds and karst and runoff lead to drinking water being available. However, it must be noted that, as in the Bahamas, the island is small enough that no location is very far from the sea.

This chapter is particularly interesting in considering how science gets done. During the initial survey, a lithified beach ridge was discovered with *Strombus* shell deposits along it. Writing up the survey, the authors realized that these shells represented an anomaly which had to be explained. However, the evidence is incomplete and equivocal, leading to competing hypotheses being developed by Watters and Donahue. The C-14 dating of these shells by Stuckenrath, which, it was hoped, would resolve this controversy, did not do so. Therefore, the authors generate plans for further investigation of this issue.

Chapter

1

Lucayan Settlement Patterns and Recent Coastal Changes in the Bahamas

William F. Keegan
Florida Museum of Natural History
Department of Anthropology
University of Florida
Gainesville, Florida

INTRODUCTION

> . . . for if the others already seen are very beautiful and green and fertile, this one is much more so . . . and in all of the island the groves are all green and the verdure like that in April in Andalusia . . . there are trees of a thousand kinds and all [with] their own kinds of fruit and all smell so that it is a marvel.

So wrote Christopher Columbus at Crooked Island on October 21, 1492 (translated in Dunn and Kelley, 1989). Were he to return today, Columbus would not recognize these islands. Plantation clearings followed by 2 centuries of slash-and-burn cultivation have reduced the vegetation to a low, xerophytic scrub (Sears and Sullivan, 1978). In addition, geological processes that were set in motion at the end of the Pleistocene have continued to modify the coastal geomorphology. In some instances, geological processes have been amplified by biotic changes (Mitchell and Keegan, 1987).

Gone too are the people who discovered Columbus—the first West Indians, who the Spanish called Lucayans (meaning people of the small islands). The Lucayans were offshoots from the Taino culture whose geopolitical center was the island of Hispaniola (Figure 1). They had a fishing and farming economy, and first entered the Bahamas from Hispaniola around 700 A.D. The Lucayans had established settlements on all 17 of the largest islands and many of the smaller islands and cays by the time of Columbus' arrival (Sears and Sullivan, 1978; Keegan, 1985).

The present chapter examines the relationship between prehistoric settlements and coastal geology on two scales: (1) long-term processes of coastline development, which relate to patterns of settlement, and (2) short-term changes that have affected the survival of abandoned sites. The objectives are to identify how a continuously changing coastal

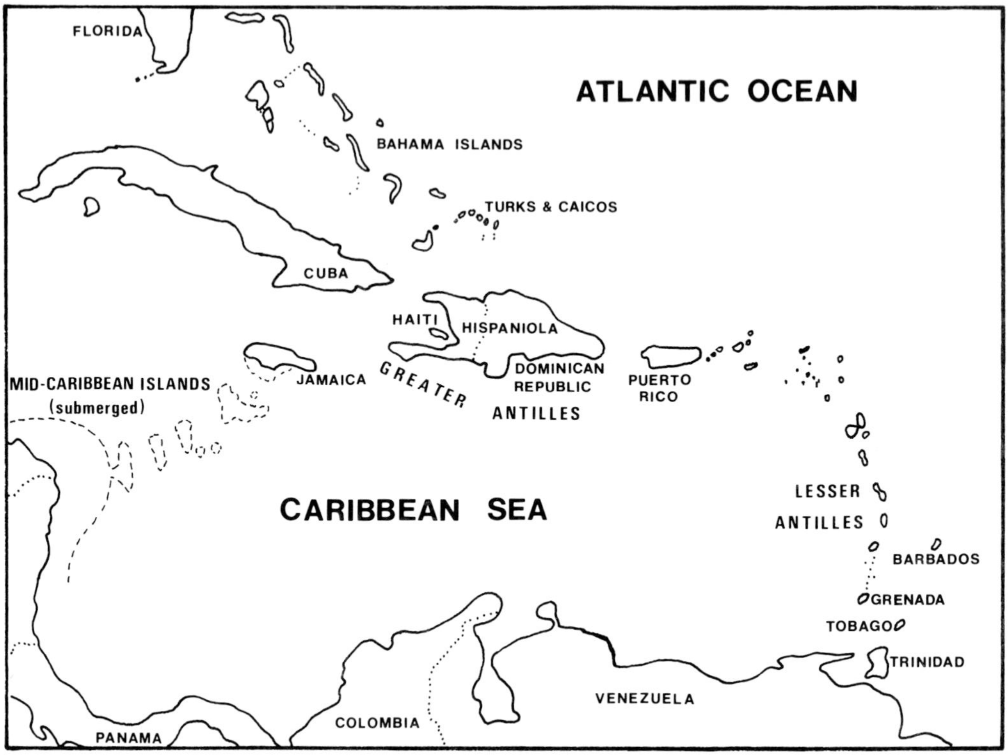

Figure 1. Map of the West Indies.

geomorphology affects interpretations of settlement patterns, and how these changes affect the survival of archaeological sites. These issues are addressed in three parts: (1) a very brief review of the geology of the Bahamas during the 200 million years prior to the arrival of humans, (2) a description of the major coastal landforms and their importance with regard to prehistoric settlement patterns, and (3) a consideration of the geologic processes that impact site survival.

CREATION OF THE BAHAMAS

The Bahama Islands developed on a tectonicly stable limestone platform created with the separation of the continents that opened the North Atlantic Ocean (Beach and Ginsburg, 1980; Garrett and Gould, 1984). For 200 million years, the Bahamas was a vast shallow bank on which at least 6000 m of marine sediments accumulated in pace with the subsidence of the Earth's crust (Adams, 1981; Dietz et al., 1970).

About 80 million years ago, deep troughs and channels developed within and between the banks, the Bahamas became separated from Cuba and Florida, and the southeastern Bahamas broke into a series of small banks surrounded by deep troughs and basins (Figure 2) (Sealey, 1985). Each bank became an independent unit, with the surrounding coral reefs acting to trap fine sediments. The banks were first exposed during the late Tertiary, which explains their borrowed and recent biota (Gillis, 1977).

The islands are products of the Pleistocene. The sea level fell by well over 120 m during the third, and by just under 120 m during the final, glacial advances (Sealey, 1985). Exposure of the expansive banks modified the local climate and produced extreme seasonal variations in temperature and rainfall. The prevailing easterly trade winds were also amplified. Very fine clays, silts, and sands from the deserts of northern Africa were blown into the Bahamas and deposited during rain showers. Wind and waves reworked these marine and terrigenous deposits into high dunes that were cemented or lithified as arcuate and linear ridges that parallel the shore at right angles to the prevailing winds (Figure 3). Almost all features with elevations greater than 3 m were formed in this way (Doran, 1955).

The lithified fossil sand dunes are known as eolianites. Eolianite ridges today form the windward skeleton of most islands. The ridges afford protection to the lee shores, which have continued to build seaward through the gradual accretion of wave-transported sediments. Following a period of quiet accretion, storm waves rework the sediments into 2-m-high dunes (Doran, 1955). The series of low dunes, separated by narrow swales, gives a corrugated texture to many lee shores. Where bays with narrow outlets, called tidal creeks, occurred, these outlets became choked by the rapidly accumulating sediments. Once closed, windblown sediments and evaporation conspired to form periodically flooded swashes which in turn gave rise to permanently dry land. Through these continuing processes of expansion, the lee shores slowly emerged from behind the protection of the high ridges, creating the characteristic pattern of a high windward coast backed by a low leeward plain.

Finally, changes in sea level have had another effect expressed in the growth of coral reefs. The significance of coral reefs is most apparent in the southern Bahamas. Mayaguana,

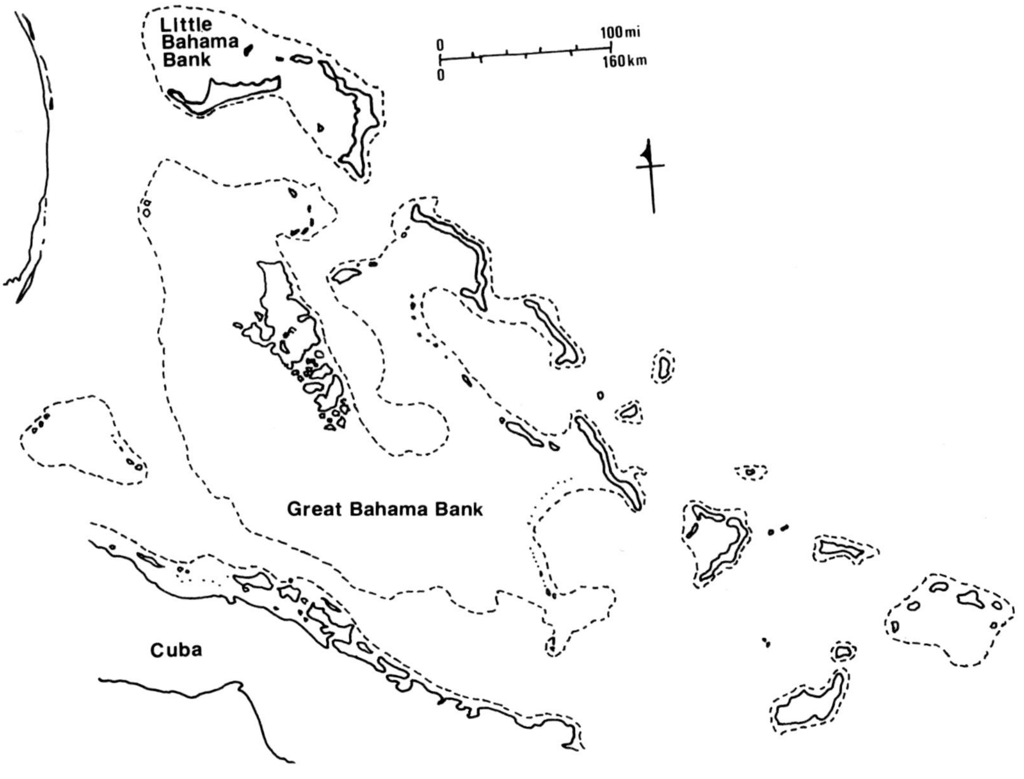

Figure 2. Map of the Bahama Banks. Dashed line marks 10 fathoms. (After Sealey, 1985: Figure 1.3.)

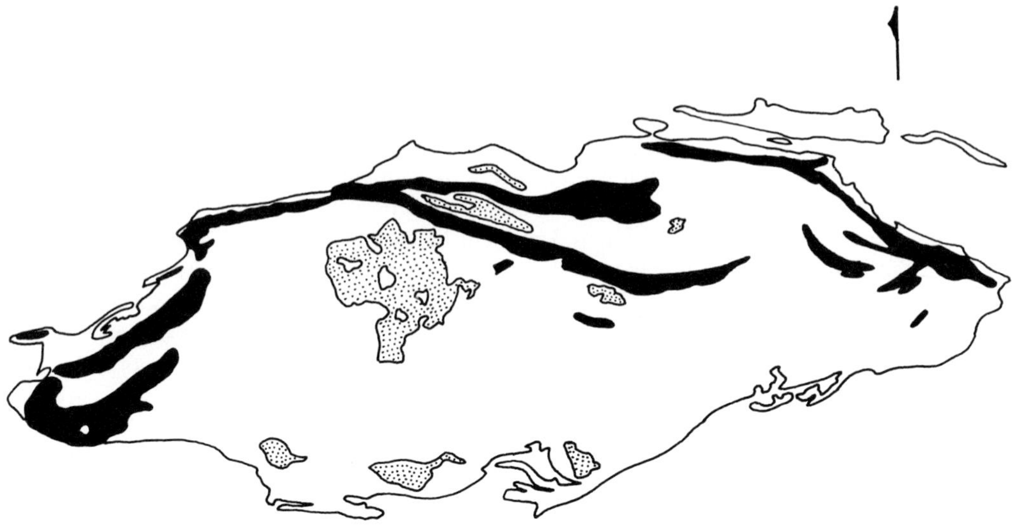

Figure 3. Ridges of New Providence Island (Nassau). Ridges are in black; lakes are dotted. (After Sealey, 1985: Figure 4.6.)

for example, has the appearance of a coral atoll—a fossil coral-reef coastline with a low interior of permanently or seasonally flooded zones and now-dry Pleistocene-age estuaries. Without its protective ring of coral, the island's poorly cemented sediments would be washed into the Atlantic.

Small fluctuations in sea level have been recorded during the Holocene. However, the pristine landscape onto which humans migrated over 1200 years ago was established early in the Holocene. Since then, coastal geomorphology has been affected by the same processes that today affect the islands.

COASTLINES AND SETTLEMENTS

A series of comprehensive archaeological and geological surveys have been conducted in the central and southern Bahamas since 1982 (Keegan, 1983a,b, 1988; Keegan and Mitchell, 1984a,b; Keegan et al., 1990). These surveys achieved 100% coverage of substantial areas of coastline (e.g., the coast of Mayaguana was completely surveyed), and every type of coastline was given equal attention, even though Lucayan settlements had previously been associated almost exclusively with sand beaches (Sears and Sullivan, 1978). The surveys covered approximately 1500 km of 8 island coastlines, and identified 123 previously unrecorded sites of 235 total for the Bahamas (Keegan, 1985).

For archaeological purposes, six coastline types have been identified: (1) eolianite, composed of cemented sediments that have eroded into dissected surfaces punctuated by sharp pinnacles, (2) fossil coral reef, composed of in-place corals in a well-cemented matrix derived from periods of higher sea level, (3) beach rock, a sandstone formed in coastal sand dunes and composed of poorly cemented marine sediments, (4) salinas, level expanses of sand or marl that are exposed coastal shallows that flood seasonally, (5) tidal creeks, the Bahamian equivalent of estuaries, called tidal creeks because they lack a freshwater influx, and (6) sand beaches, composed of unconsolidated accumulations of marine sediments. These types are discussed in turn.

Only minor changes have occurred in eolianites and fossil coral reefs during the past 500 years. They tend to have sharp dissected surfaces and steep sides at the water's edge which makes them difficult to traverse. Sites associated with these coastal formations always have a sand beach in close proximity, usually in direct contact with at least one part of the site.

Beach rock is formed beneath sand dunes through the cementation of sand, crushed shell, precipitates, and other beach materials. Its age can be estimated from color, texture, and degree of cementation. Fossil beach-rock terraces attest to higher eustatic sea levels (prior to 35,000 years ago), and extensions of beach rock into the sea attest to periods of lower relative sea level (e.g., the "Bimini Road"; see Richards, 1989). There is no evidence of tectonic uplift. Recent beach-rock formations have incorporated modern artifacts such as nails and bottle caps. Archaeological sites are common above recent beach-rock formations, but are usually absent from the vicinity of older formations.

The importance of beach rock is its use as a means for gauging the size of prehistoric beaches in places where the sand beach has eroded. For example, site AN-1 on Andros Island had artifacts scattered through mangrove roots on the shore, but no subsurface deposits of artifacts. In association with the mangroves was an exposed beach-rock shore that extended from the coastal dune into the water, a distance of 5 m (Keegan and Sealey, 1988). This beach rock was formed within a coastal dune that must have at one time extended that additional 5 m from the shore.

Salinas are level plains with scant vegetation that form a lee-shore border between marine tidal flats and the mainland (Figure 4). They are composed of sand or marl that

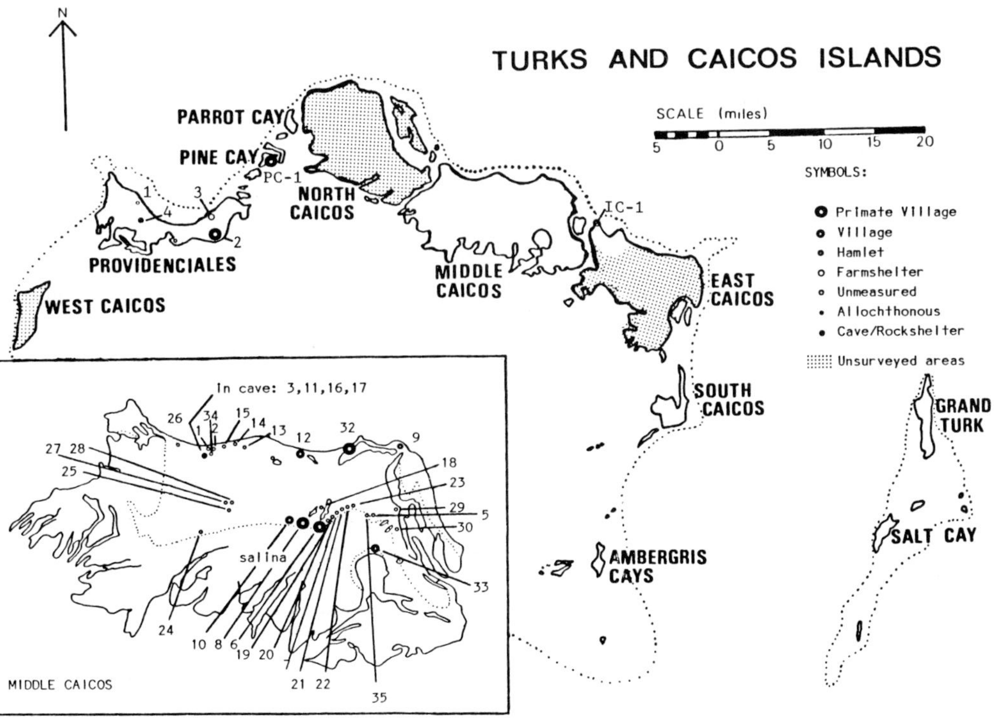

Figure 4. Lucayan sites in the Caicos Islands. The interface of salina and permanently dry land is marked by the dotted line. (After Keegan, 1985.)

was leveled by shallow-water tidal action. Salinas have since been exposed by changes in marine circulation, minor fluctuations in sea level, or the accumulation of sand. They are often flooded by heavy rains and/or extreme high tides. Salinas occur along the lee shores of Great Abaco, Andros, Crooked, and Acklins Islands, and the Caicos Islands. The Caicos salinas are featured in the present discussion.

In 1978, Sears and Sullivan proposed a two-phase sequence of colonization for the Bahama archipelago based on Sullivan's archaeological survey of the Turks and Caicos Islands (1980, 1981). They proposed an initial Antillean period (800–1000 A.D.) during which settlements were found at inland locations, followed by a Lucayan Period (1000–1500 A.D.), at which time settlements shifted to the coast. The shift from an inland to a coastal settlement pattern made sense because it followed the shift that had been identified for the preceding Saladoid colonization of the Lesser Antilles (Keegan and Diamond, 1987).

Sears and Sullivan (1978) classified all of the known open-air sites in the Bahamas as either coastal or inland. Of the 43 in the central Bahamas, 42 were classed as coastal, while 24 of 25 sites in the southern Bahamas (= Caicos Islands) were classified as inland. On the basis of this classification, they argued that the Caicos Islands were colonized first. Their efforts to demonstrate the logic of identifying the Caicos Islands as the first colonized ignored the possibility that Great Inagua, the other major southern island, may have been settled (Keegan, 1985).

A variety of new evidence has led to the rejection of the inland/coastal dichotomy (Keegan, 1985; Keegan and Mitchell, 1986), not the least problem being that the dichotomy is difficult to apply because the islands are only, on average, 5 to 8 km wide. Their classification is discussed here as an example of the need to consider coastal geomorphology when interpreting settlement patterns. Of the 24 inland sites in the Caicos Islands, 14 are aligned atop the first tract of permanently dry coastal plain that borders the salina along the south coast of Middle Caicos (Figure 4). The salina is up to 7 km wide, with a maximum elevation of 25 cm above mean high tide.

Sullivan (1980, 1981) chose to interpret these sites as inland. Alternatively, it could be argued that the sites are coastal because they are located on the first tract of permanently dry land. Moreoever, it is possible that the salina was actually a submerged tidal flat at the time the sites were in use. Small fluctuations in sea level (30 to 60 cm), the accumulation of sediments along the south coast, and changes in water circulation on the Caicos Bank have all been reported by modern observers. Salinas form a dynamic border between land and sea; it cannot be assumed that they have remained unchanged, even over a period as short as 500 years (Doran, 1955).

Of the remaining ten sites, eight were classified as inland by virtue of their location on the interior margins of coastal lakes. Mitchell's (1986a,b) studies of the long-term evolution of coasts from bays to tidal creeks to lakes has demonstrated that many coastal lakes were actually tidal creeks during the period of Lucayan settlement. In 60 cases, Mitchell has found that modern lakes that were tidal creeks 500 to 1000 years ago have evidence for Lucayan settlement, while lakes that were lakes during that period have no evidence for prehistoric settlement (Mitchell and Keegan, 1987). It is apparent that access to marine habitats is the key to settlement on tidal creeks, and that sites on tidal creeks should be considered coastal.

Mitchell's technique involves the use of shallow cores which are examined for stratigraphic changes in the ratio of silt to clay, the size and abundance of interclasts, and tidal creek vs. lacustrine biofacies (Mitchell, 1986b; Mitchell and Keegan, 1987). The technique uses depositional rates estimated from sediments dated to the past 200 years for calculating the timing of the tidal creek-to-lake transition. Mitchell's studies have proven that the windward coasts, largely ignored by archaeologists as uninhabitable, were settled by the Lucayans wherever significant tidal creeks were present.

In sum, Sears and Sullivan found a substantial number of apparently "inland" settlements on Middle Caicos and interpreted these as following a pathway that was previously identified for the Lesser Antilles, namely, a shift from inland to coastal locations. Further research has demonstrated that the settlements postdate the initial colonization of the Bahamas, and that coastal geomorphology indicates that the inland classification is misleading if not inaccurate (Keegan, 1985).

SITE IDENTIFICATION AND SITE SURVIVAL

With regard to the identification and survival of Lucayan sites, two major geologic processes are paramount. Holocene processes of coastline change can be divided into two categories: erosional and depositional (Mitchell and Keegan, 1987). Erosional processes are observed as biological and mechanical erosion of rock on the surf bench, the erosion of headlands, and the storm removal of coastal sediments (e.g., uncemented beach/dune ridges, sand bars, etc.). Depositional processes include the formation of beach/dune ridges, the bay-to-tidal creek-to-lake transition, and storm deposits (e.g., boulder ramparts). These processes have a pronounced effect on the survival of sites as well as on our ability to identify sites. Relative rises in sea level seem to have been less than the height of coastal dunes on which sites were established. In this regard, the site is closer or further, on a vertical scale, from the high-tide line, but not sufficiently close to have been inundated.

The significance of recent changes in coastal geomorphology has been highlighted by the debate surrounding Columbus' first landfall. In 1986, *National Geographic* magazine announced that Samana Cay, rather than the traditionally accepted San Salvador (Watling) Island, was Columbus' first landfall (Judge, 1986). One piece of evidence used to support that hypothesis was the identification of a 66-m-long peninsula on the south coast of Samana Cay as the "piece of land like an island, although it is not one" on which six Lucayan houses were observed. Geological investigations have demonstrated that the prograding end of this sand-spit peninsula has a growth rate of about 3 m per year. Thus, the entire peninsula may be less than 250 years old, if it grew at a constant rate (Sealey, 1987). In addition, the peninsula would not have been suited for human habitation even if it did exist 500 years ago as, even today, it is swept by waves during storms.

Archaeological surveys have demonstrated that the majority of Lucayan sites occur in direct association with sand beaches—either modern beaches or eroded beaches whose former presence is denoted by recent beach-rock formations. Sand beaches are the most susceptible to short-term changes and therefore figure prominently in the consideration of postabandonment disturbances.

Depositional processes can be described as continuous or discrete. Continuous deposition involves the long-term, quiet accumulation of sediments on a coast. For instance, it has been noted that north- and south-facing beaches tend to accumulate sediments along their coasts. Exceptions to this general rule are beaches that face extensive tidal flats or banks. In conjunction, prehistoric sites located along north- or south-facing beaches are typically buried by 40 to 60 cm of overburden. Sites in these areas are identified only when substantial soil disturbances have occurred (e.g., road cuts, farm plots, test excavations, or crab burrows).

TABLE I. Common Tree Species of the Whitelands Formation Type

Common Name	Scientific Name	Canopy Height	Possible Uses
Round Top Palm (Palmetto)	*Sabal palmetto*	5–8 m	Baskets/thatch
Cocoplum	*Chrysobalanus icaco*	3–5 m	Edible fruit
Seagrape	*Coccoloba uvifera*	5–10 m	Edible fruit
Milk Berry	*Bumelia americana*	3–6 m	Edible fruit
Buttonwood	*Conocarpus erectus*	5–12 m	
Frangipanni	*Plumeria obtusa*	3–5 m	
Bay Cedar	*Suriana maritima*	3–4 m	
Seven Year Apple	*Casasia clusiifolia*	3–5 m	
Guana Berry	*Byrsonima lucida*	3–6 m	
Joewood	*Jacquinia keyensis*	3–5 m	
Wild Onion	*Epidendron* sp.		Edible bulb?
Salt Grass	*Spartina patens*		Tea
Sea Oats	*Uniola Paniculata*		Tea
Cat's Claw	*Pithecolobium unquiscati*		Antidote for fish poisoning
Australian Pine	*Casuarina litorea*		Historic introduction
Coconut Palm	*Cocos nucifera*		Historic introduction

Areas of continuous sediment accumulation often can be recognized by their white sands and pioneering vegetation (Little et al., 1977). Together, these comprise the Whitelands formation type (Table I). A very diagnostic type of Whitelands vegetation is the *Casuarina* or Australian Pine, an historic introduction that rapidly colonizes new beaches. Archaeological surveys have repeatedly shown that Lucayan sites range from 10 to 100 m behind *Casuarina*-lined beaches (Keegan et al., 1990). This situation occurs because more-recent dunes, which are stabilized by the pines, have formed in front of sites that were at one time directly on the beach.

Discrete depositional episodes are the product of storms. An example is site AC-10 on Acklins Island where we observed a record of two major storm events, separated by periods of quiet sediment accumulation (Keegan, 1988). Evidence of these storm events proved important for interpreting the erosional disturbances at other sites on the bay. Stratigraphic evidence of depositional episodes must be carefully monitored because these sites are close to the sea.

The erosion of sediments has had a more profound effect on Lucayan sites. Erosion can also be classified as either continuous or discrete. Since relative changes in sea level are insignificant for the period 700 to 1500 A.D. (i.e., prehistoric living surfaces have not been inundated), the major cause of continuous erosion is a shift in marine circulation. For example, in 1960, Hurricane Donna closed the channel between Pine Cay and Water Cay, Caicos Islands; the flow of water off of the banks was redirected into the channel between Pine Cay and Dellis Cay (Figure 5). One result has been the continuous erosion of Pine Cay due to the increased flow of water along its eastern side. Persons familiar with the cay over the past 30 years estimate that up to 2.5 m of coastline have eroded.

Erosion associated with storms has the most dramatic effect. Fifteen Lucayan sites on Delectable Bay, Acklins Island, which extend over 6 km of the shore, were the focus of excavations conducted in 1987 (Keegan, 1988). Excavations quickly revealed that only about 10 m of the original site widths were preserved in the coastal dunes. Since Lucayan sites are typically 30 m or more wide, it is apparent that significant storm damage has occurred. Local informants described three recent hurricanes in which the coastal zone was completely inundated. Geological studies report that waves in excess of 4 m can be

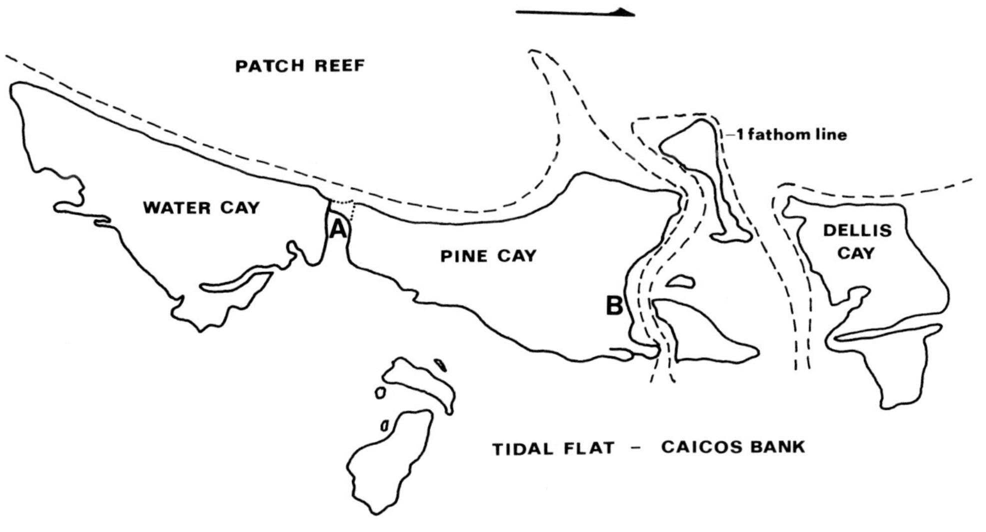

Figure 5. Marine circulation behind Pine Cay. Closure of the channel between Pine Cay and Water Cay (A) redirected the flow off the banks between Pine Cay and Dellis Cay, causing the erosion of the Lucayan site (B).

generated on the Bight of Acklins, an elevation that is equal to the maximum for the adjacent coast. In addition, there are two fossil channels to the south of Delectable Bay that were created by hurricanes, but have since filled with sediments (Doran, 1955). It is apparent that even ''protected'' lee shores are susceptible to drastic modifications.

The Delectable Bay settlements are an archaeological nightmare. Most of the sites are deflated, leaving only artifacts and midden debris scattered on a hard-sand surface; other sites retain less than a 10-m-wide strip of relatively intact deposits, and new sites have been created through the redeposition of site materials along the beach (Keegan, 1988). My lament results from the importance of these sites as the Bahamas only historically recorded village of a chief (Keegan, 1984; Dunn and Kelley, 1989).

CONCLUSION

The present study has been largely descriptive, my purpose being to outline the importance of recent changes in coastal geomorphology in the study of Lucayan settlement patterns. Such considerations have led to the rejection of Sears and Sullivan's (1978) inland/coastal dichotomy, have demonstrated that windward coasts with significant past or present tidal creeks were settled by the Lucayans, have resulted in a modification of survey procedures in areas of actively building dunes, and have provided the basis for inferences about settlement patterns for areas that have been geologically, but not archaeologically, surveyed.

We are continually developing a better appreciation for the fludity of sand coasts. Although it is not yet possible to specify *a priori* what processes will have affected a site, informed speculation can now precede excavations and the effects of geological processes can be recognized in excavation strata. Where continuous deposition or erosion is observed, it is often possible to retrodict the aboriginal shoreline. Finally, we have improved our ability to interpret how recent geological changes influence our interpretations of culture.

REFERENCES

Adams, R.W. (1981) General guide to the geological features of San Salvador. In: *Field Guide to the Geology of San Salvador,* pp. 1–66. Gerace, D.T., Ed. CCFL Bahamian Field Station, San Salvador, Bahamas.

Dietz, R.S., Holden, J.C., and Sproll, W.P. (1970) Geotectonic evolution and subsidence of Bahama platform. *Geol. Soc. Am. Bull.* 81: 1915.

Doran, E. (1955) *Landforms of the Southern Bahamas.* University of Texas Press, Austin.

Dunn, O. and Kelley, J.E., Jr. (1989) *The Diario of Christopher Columbus's First Voyage to America 1492–1493.* University of Oklahoma Press, Norman.

Gillis, W.T. (1977) Biogeography and vegetation. In: *Land Resources of the Bahamas: A Summary,* pp. 13–22. Little, B.G. et al., Eds. Land Resources Study 27, Ministry of Overseas Development, Surrey, England.

Judge, J. (1986) The island of landfall. *Natl. Geogr.* 170(5): 566–599.

Keegan, W.F. (1883a) An Archaeological Reconnaissance of Crooked Island and Acklins Island, Bahamas. Report to the government, Commonwealth of the Bahamas, Nassau.

Keegan, W.F. (1883b) Archaeological Investigations on Mayaguana, Bahamas: A Preliminary Report. Report to the government, Commonwealth of the Bahamas, Nassau.

Keegan, W.F. (1984) Columbus and the City of Gold. *J. Bahamas Hist. Soc.* 6: 34–39.

Keegan, W.F. (1985) *Dynamic Horticulturalists: Population Expansion in the Prehistoric Bahamas.* Ph.D. thesis, University of California, Los Angeles; University Microfilms, Ann Arbor, MI.

Keegan, W.F. (1988) *Archaeological Investigations on Crooked and Acklins Islands, Bahamas: A Preliminary Report of the 1987 Field Season.* Misc. Proj. Rep. 36, Florida Museum of Natural History, Gainesville.

Keegan, W.F. and Mitchell, S.W. (1984a) *The Archaeological Survey of Long Island, Bahamas: Final Report.* PRIDE Publ. Carib. Sci. 1.

Keegan, W.F. and Mitchell, S.W. (1984b) *Archaeological Investigations on Great and Little Exuma, Bahamas.* PRIDE Publ. Carib. Sci. 2.

Keegan, W.F. and Mitchell, S.W. (1986) Possible allochthonous Lucayan Arawak artifact distributions, Bahama Islands. *J. Field Arch.* 13: 255–258.

Keegan, W.F. and Sealey, N.E. (1988) *A Preliminary Archaeological Survey of the Fehling Site (AN-1), Andros, Bahamas.* Misc. Proj. Rep. 40, Florida Museum of Natural History, Gainesville.

Keegan, W.F., Williams, M.W., and Seim, G. (1990) *Archaeological Survey of Grand Turk, B.W.I.* Misc. Proj. Rep. 43, Florida Museum of Natural History, Gainesville.

Little, B.G., Buckley, D.K., Cant, R., Henry, P.W.T., Jefferiss, A., Mather, J.D., Stark, J., and Young, R. (1977) *Land Resources of the Bahamas: A Summary.* Land Resource Study 27, Ministry of Overseas Development, Surrey, England.

Mitchell, S.W. (1986a) Surficial geology of Rum Cay, Bahamas Islands. In: *Proc. 3rd Symp. Geology of the Bahamas,* pp. 231–241. Curran, H.A., Ed. CCFL Bahamian Field Station, Fort Lauderdale, FL.

Mitchell, S.W. (1986b) Sedimentology of Pigeon Creek, San Salvador Island, Bahamas. In *Proc. 3rd Symp. Geology of the Bahamas,* pp. 215–230. Curran, H.A., Ed. CCFL Bahamian Field Station, Fort Lauderdale, FL.

Mitchell, S.W. and Keegan, W.F. (1987) Reconstruction of the coastlines of the Bahama Islands in 1492. *Am. Arch.* 6: 102–108.

Richards, D.G. (1989) Archaeological anomalies in the Bahamas. *J. Sci. Explor.* 2: 181–201.

Sealey, N.E. (1985) *Bahamian Landscapes.* Collins Caribbean, London.

Sealey, N.E. (1987) The Significance of the Geography of the Bahamas in Reconstructing Columbus's Route in 1492. Ms. in possession of author.

Sears, W.H. and Sullivan, S.D. (1978) Bahamas prehistory. *Am. Antiq.* 43: 3–25.

Sullivan, S.D. (1980) An overview of the 1976 to 1978 archaeological investigations in the Caicos Islands. *Fla. Anthropol.* 33: 120–142.

Sullivan S.D. (1981) Prehistoric Patterns of Exploitation and Colonization in the Turks and Caicos Islands. Ph.D. thesis, University of Illinois, Urbana; University Microfilms, Ann Arbor, MI.

Chapter

2

Paleoshorelines and the Prehistory of Barbuda, West Indies

David R. Watters
Division of Anthropology
Carnegie Museum of Natural History
Pittsburgh, Pennsylvania

Jack Donahue
Department of Geology and Planetary Science
and Department of Anthropology
University of Pittsburgh
Pittsburgh, Pennsylvania

Robert Stuckenrath
University of Pittsburgh Radiocarbon Laboratory
Pittsburgh, Pennsylvania

INTRODUCTION

Barbuda, a limestone island of low relief, is located between latitude 17° 31′–17° 45′ N and longitude 61° 44′–61° 53′ W in the northern portion of the Lesser Antilles island chain. The northern Lesser Antilles (also known as the Leeward Islands) are located about 120 km east of the Virgin Islands, the easternmost islands of the Greater Antilles chain. Barbuda is situated in the Outer or Limestone arc (Limestone Caribbees) of the northern Lesser Antilles, some 40 km north of its nearest neighboring island, Antigua (Figure 1). West of Barbuda lie the volcanic islands of the Inner arc in the northern Lesser Antilles. Many aspects of the geology of the Lesser Antilles are still subject to much discussion and additional research is imperative (Donnelly, 1989).

The recognition of the position of ancient shorelines can be a critically important portion of the environmental interpretation of archaeological sites. Of equal importance is the determination of the cause as well as rate of change for a shoreline position. It is essential to know whether either the sea level or a land area adjacent to the ocean has been undergoing progressive change (and how rapidly) during the course of prehistoric occupation in a region. This can vary from a situation as in the Shumagin Islands (see Part II, Chapter 7), where tectonic uplift is extreme, to the case of Barbuda, where both sea level change and land movement are relatively small.

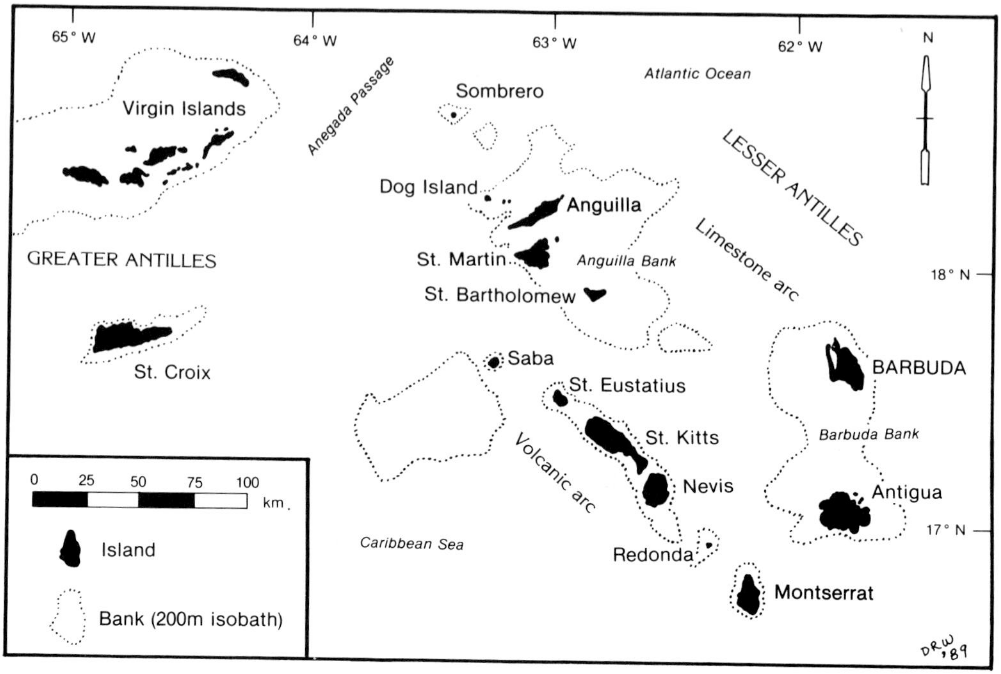

Figure 1. The location of Barbuda with respect to the islands and banks of the northern Lesser Antilles and eastern Greater Antilles.

PHYSICAL SETTING

Barbuda and Antigua are the only emergent features on the Barbuda Bank (Figure 1), one of the larger and shallower banks in the Lesser Antilles. East and north of Barbuda, the bank drops precipitously to the seafloor of the western Atlantic Ocean; west and south of the island, shallow bottoms predominate. Estimates of the Barbuda Bank's area, roughly within the 200-m isobath, range from 3600 km^2 (Martin-Kaye, 1969) to 4000 km^2 (Wigley, 1977); some 80% of the bank is less than 40 m below sea level (Detjen, 1978). The Barbuda Bank is about 90 km long and varies from 24 to 51 km wide. Passages as deep as 1000 m isolate this bank from other banks in the northern Lesser Antilles.

Four marine environments—reef, backreef, sandy bank, and lagoon—occur around Barbuda (Wigley, 1977). Reefs and backreefs are most noteworthy along the east and north coasts, but also occur south of the island. Dominant corals include species of *Acropora, Montastrea, Diploria, Porites, Siderastrea,* and *Millepora,* and coralline algae, notably *Lithothamnium,* is very abundant (Russell and McIntire, 1966; Wigley, 1977). Sea-grass beds, especially *Thalassia testudinum,* dominate the calmer waters west and south of Barbuda, while *Thalassia, Halimeda* (a calcareous alga), and colonies of *Porites* cover many shallow banks of stable substrates within Codrington Lagoon.

Barbuda is fringed with almost continuous beaches of white sand along its south, west, and north coasts. The exposed east shore is quite different, for here extensive stretches of rugged limestone coast, broken only by pocket beaches formed between rocky headlands, face the turbulent Atlantic. Large dune ridges occur inland from the east coast at a number of locations. In western Barbuda, stands of red (*Rhizophora*) and black (*Avicennia*) mangrove and algal flats border the north sector of Codrington Lagoon and Goat Island Flash.

Barbuda contains two major topographic units (Figure 2). In the east-central area of the island is the "Highlands", a plateau-like feature occupying about 25% of the landmass and exhibiting a karstic terrain. Barbuda's highest elevation, near the "Highland House" historic structures (Watters and Nicholson, 1982) in the northern Highlands, has not been determined accurately, but is approximately 42 m. The edge of the Highlands is demarcated by escarpments ranging from 15 to 30 m high, except in the southern sector where the gradient is more gradual. Numerous caves penetrating the escarpment at various heights are the remnants of sea caves. Sinkholes, solution fissures, and caves are common features in the interior of the Highlands.

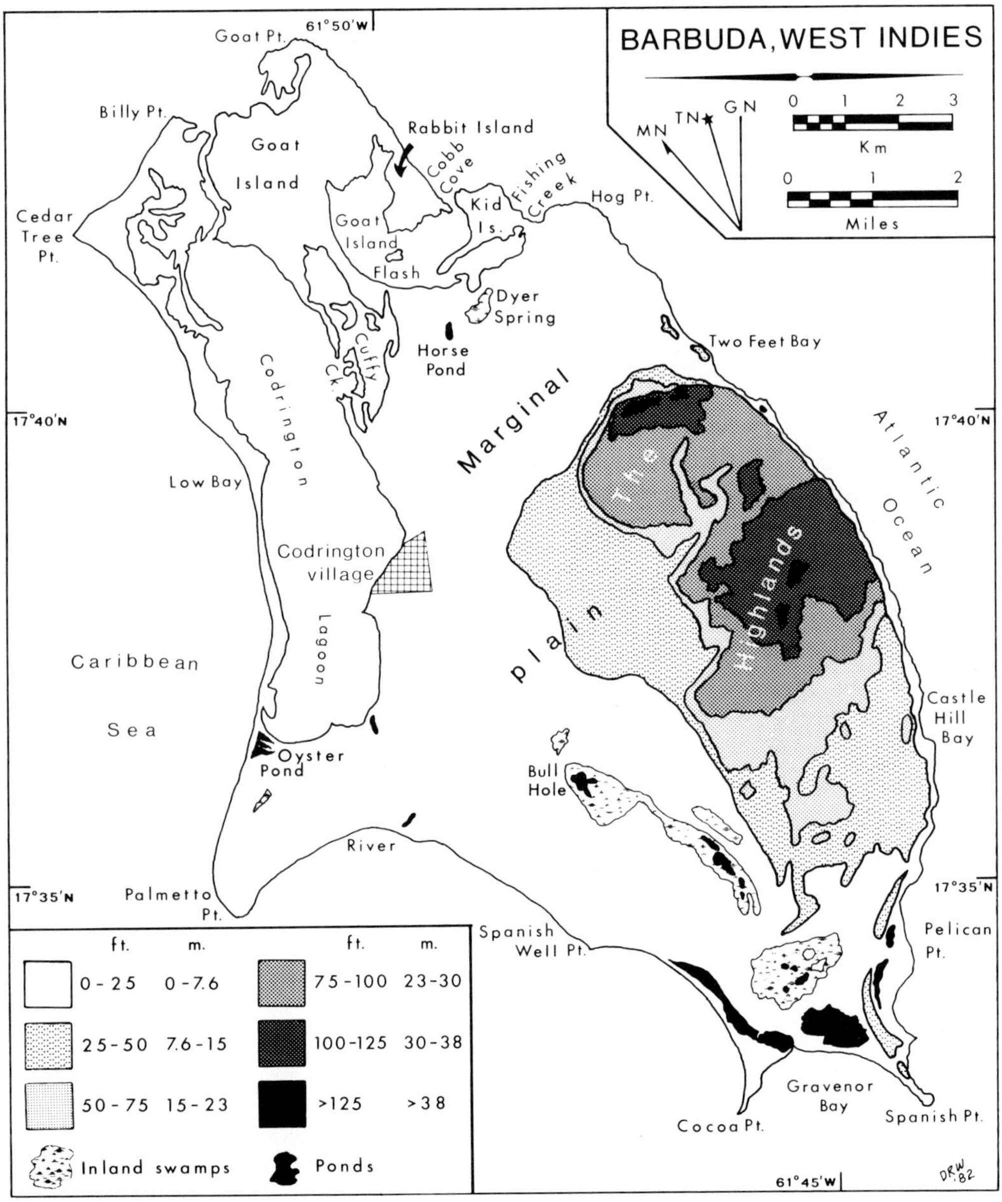

Figure 2. Barbuda's topographic units and elevations.

The other topographic unit, the Marginal plain (Harris, 1965), extends north, west, and, to a lesser extent, south of the Highlands as a relatively level feature of very low relief. Apart from a small area immediately west of the Highlands, almost the entire Marginal plain lies at an elevation of less than 8 m (25 ft). Most of the coast as well as the land bordering Goat Island Flash and Codrington Lagoon is only 1 to 2 m above sea level. A narrow and shallow passage west of Goat Island provides the only permanent entryway into Codrington Lagoon. The narrow sand spit bordering the southwestern part of the lagoon has been temporarily breached by hurricanes, but has always reformed. Ponds and swamps have formed at inland depressions on the Marginal plain and salt ponds occur near the coast.

Soils on Barbuda are for the most part shallow, with deeper soils found only in very restricted regions. Bare limestone protrudes extensively in many areas. Vernon and Lang (1966) define the Barbuda Clay Loam and Highland Ridge Clay Loam series, both of which are restricted to the Highlands, the Blackmere Clay Loam and Codrington Clay series on the Marginal plain, and three "land types", beach sands, salina, and mangrove swamp soil (Figure 3). Rainfall records exist only for the village of Codrington, situated midway on the east shore of the lagoon, and they are not complete. Mather (1971: Appendix 4) shows that the mean of annual rainfalls from 1908 to 1970 (ignoring certain hiatuses) was 35.50 in. (97.79 cm), with a range from 21.96 in. (55.78 cm) in 1930 to 62.19 in. (157.96 cm) in 1942. Precipitation is greatest from August through December and in May. Periodic droughts are not uncommon. Barbuda is without surface streams, and thus most rainfall disperses through infiltration into solution fissures. On the Marginal plain, standing water is common after storms because drainage is so slow.

Barbuda's vegetation is characteristic of an island with low relief, limited rainfall, and shallow soils. Beard (1949) places Barbuda's vegetation within his Evergreen Bushland formation. Harris (1965), in his excellent ecological study, classified the dominant biotic communities as Evergreen Woodland, with the Highlands having a less degraded version than the Marginal plain, and noted the existence of swamp, grassland, and various coastal vegetational communities. In general, Barbuda's vegetation today is xerophytic, diminutive, and dominated by spiny succulents, such as cacti and agave, and thorny trees and shrubs that form impenetrable thickets in many sectors. Large trees are rare on the island and occur mainly in moisture-retaining sinkholes or similar features where roots can tap a reliable water source.

In many ways, Barbuda's terrestrial environments contrast markedly with those seen on the volcanic islands in the Inner arc of the Lesser Antilles, where lush vegetation, productive soils, and greater precipitation are characteristic. In the ocean realm, however, Barbuda possesses an extraordinary diversity of marine and estuarine environments, a diversity that is greater than that observed around the volcanic islands. To a large degree, the islands's geologic history has influenced the particular abiotic habitats and biotic communities that are so characteristic of Barbuda.

GEOLOGIC HISTORY

The geologic formations exposed at the surface on Barbuda consist entirely of moderate to shallow-water limestone sequences ranging in age from Pliocene to Holocene. Nowhere on the island is there an exposure of volcanic rock. The geologic formations consist of the Pliocene Highlands Formation, early-Pleistocene Beazer Formation, late-Pleistocene Codrington Formation, and Holocene Palmetto Point Formation (Figure 4) (Brasier and

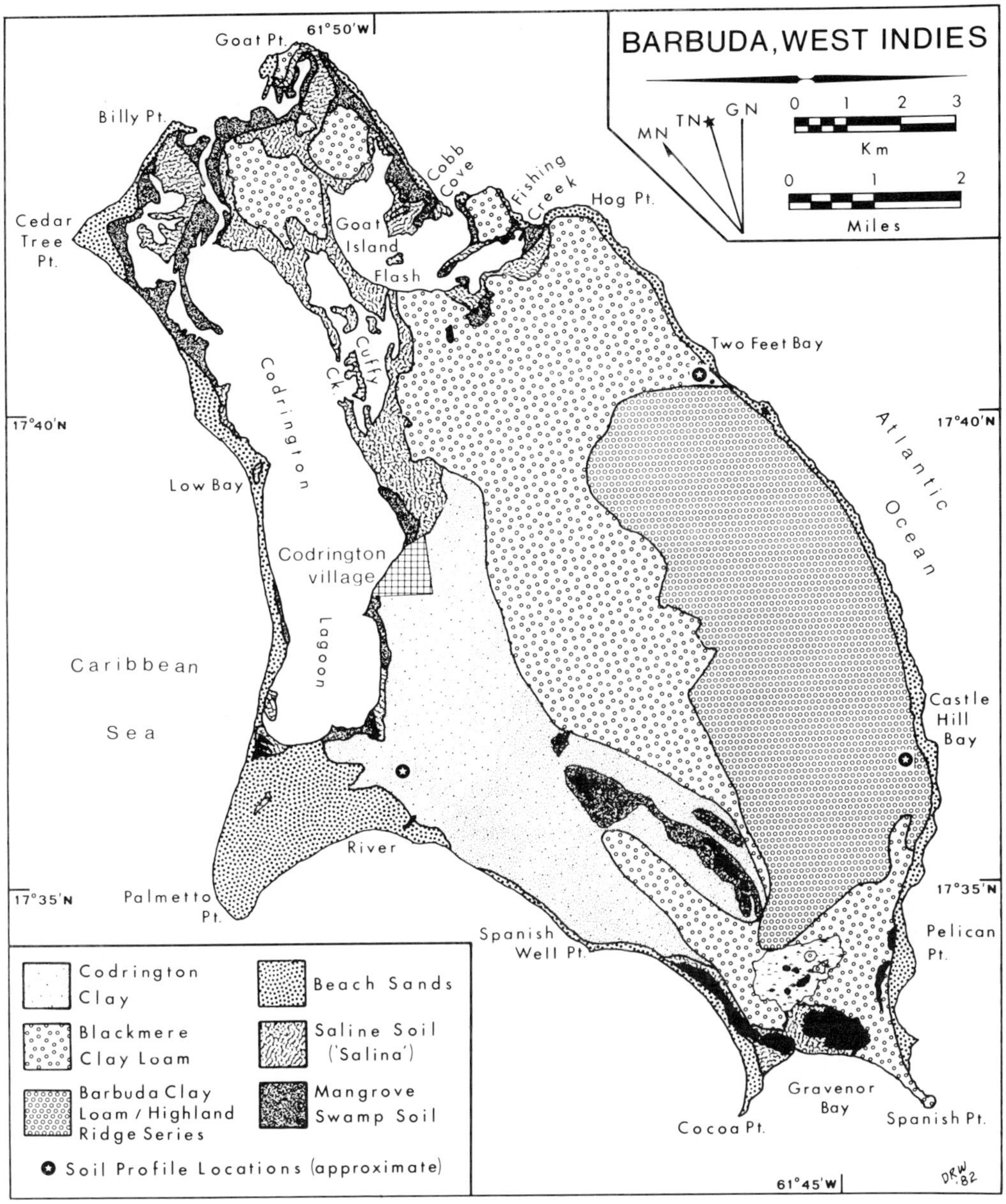

Figure 3. Barbuda's soil units and land types. (After Vernon and Lang, 1966).

Mather, 1975; Brasier and Donahue, 1985). A comparison of Figures 2, 3, and 4 shows that both topography and soil type are well correlated with distribution of the four limestone units. The Highlands Formation is restricted to the high-standing plateau on the east-central part of the island. The Pleistocene Beazer and Codrington Formations comprise the Marginal plain which is especially well developed north, west, and south of the Highlands. Finally, the Palmetto Point Formation makes up Codrington Lagoon, Palmetto Point, and the mangrove and beach-ridge areas northwest of Codrington Lagoon (Figure 4). The distribution of younger limestone units around an older, Pliocene core is a classic offlapping sequence. The asymmetric pattern developed here is caused by Barbuda's position immediately adjacent to the Atlantic Ocean basin. The high-energy eastern shoreline promotes development of a narrow coral reef tract, while the shallow bank north, west, and south of Barbuda's Pliocene core results in prograding, shallow-water marine carbonates.

Barbuda is an excellent tape recorder of eustatic sea level change. This is in distinct contrast to many regions such as the Shumagin Islands (Winslow and Johnson, 1989) where tectonic movements complicate the record of sea level change. Barbuda has remained relatively stable since Pliocene time and the deposition of the Highlands Formation (Brasier and Donahue, 1985). The four periods when marine carbonate deposition occurred are a record of high sea level (Figure 5) with warm climatic conditions. The intervening periods, with a low-sea-level stand, represent periods of cooler temperatures and resultant lower sea level position. Because Barbuda contains an offlapping sequence of carbonate deposits, it is possible to document the change in shoreline position through time. Although most of Barbuda's history occurred before human occupation, it is useful to sketch the geological

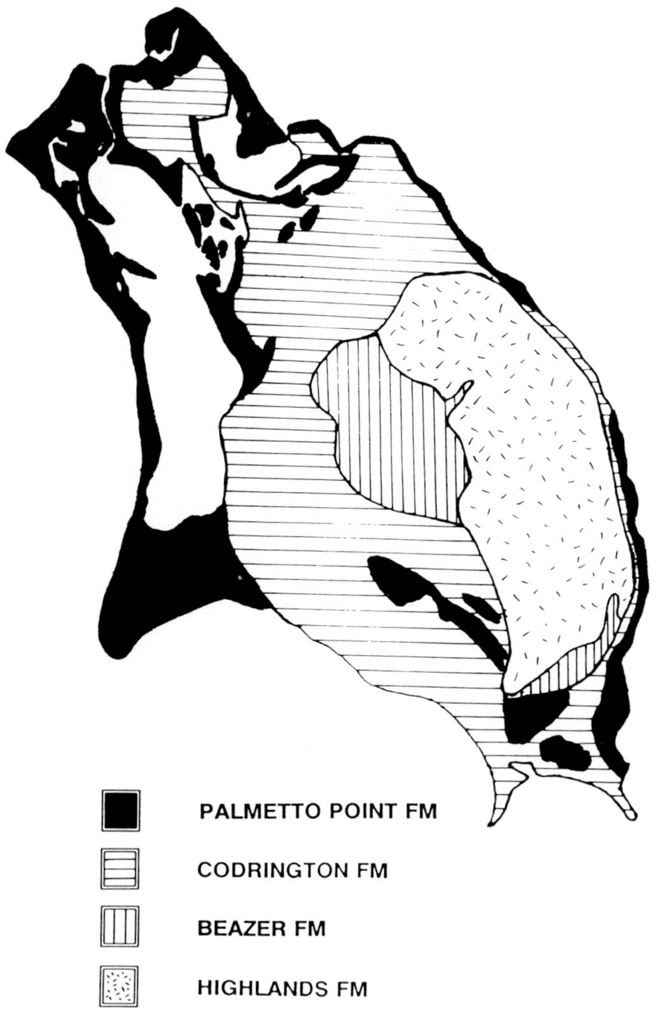

Figure 4. Geologic map of Barbuda. (Adapted from Braiser and Donahue, 1985: Figure 2.)

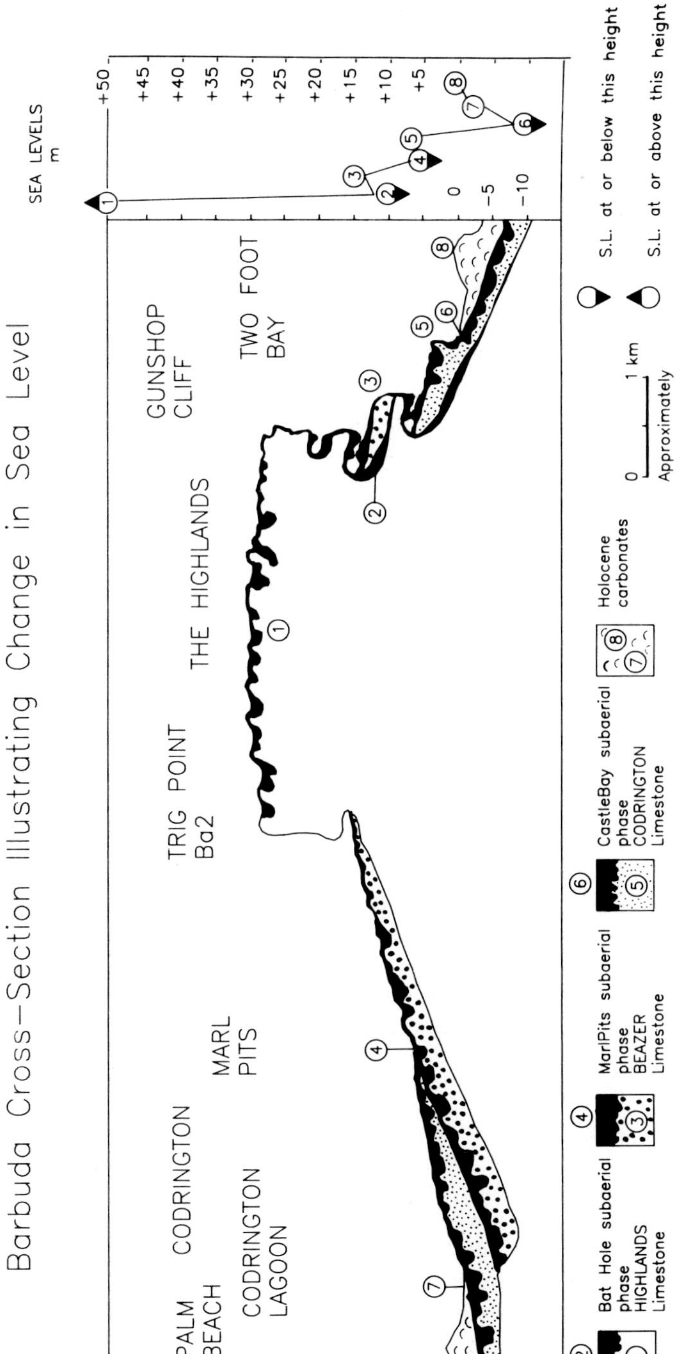

Figure 5. Diagramatic cross-section of Barbuda. (Adapted from Brasier and Donahue, 1985: Figure 3.)

history of the island for a number of reasons. (1) Alternate high- and low-sea-level stands adjacent to the Atlantic Ocean basin were responsible for shaping the present physical configuration of the island. This, in turn, played an important role in determining the distribution of prehistoric sites. (2) Identification of former shoreline positions on a carbonate island such as Barbuda is often a difficult task. A number of criteria have been established on Barbuda to allow recognition of former shoreline position.

A summary of the geologic history, and especially changes in shoreline position, follows and is illustrated in Figures 5 and 6.

Pliocene Highlands Formation (High-Sea-Level Stand)

The geologic age of the Highlands Formation is primarily based on paleontologic criteria, specifically, the presence of benthonic foraminifera such as *Amphistegina* sp. (Brasier and Donahue, 1985). Based on the carbonate facies present within the Highlands Formation and estimated Pliocene sea level positions (Vail et al., 1977; Blackwelder, 1981; Nummedal et al., 1987), the sea level was probably 30 to 50 m higher than at present. Limited field examination of the Highlands Limestone did not reveal the presence of *in situ* coral reef tract or other unquestionable shoreline features.

Late Pliocene-Early Pleistocene Bat Hole Subaerial Phase (Low-Sea-Level Stand)

Global cooling and resultant sea level lowering during middle- to late-Pliocene time (Blackwelder, 1981) resulted in the emergence of the Highlands Formation. An actual regressive facies pattern has not been documented within the Highlands Limestone. Emergence of the limestone is documented by development of karst features and a thin veneer of residual or *terra rossa* soil. During this period of emergence, the 30-m cliff that rings the Highlands developed through the process of wave erosion. Although we do not have subsurface data, field examination in the vicinity of the Highlands cliff suggests that a relatively smooth, level wave-cut platform is present beneath the younger, offlapping Pleistocene limestone on Barbuda. Field examination also revealed that the upper surface of the Highlands Formation slopes very gently southward. Brasier and Donahue (1985) suggested that mild tectonic tilting occurred during the Bat Hole Subaerial Phase. The actual amount of sea level drop cannot be determined from the data presently available to us, but it undoubtedly fell below its present position.

Early Pleistocene Beazer Formation (High-Sea-Level Stand)

The best documentation of shoreline position during Beazer time is found along the east coast of Barbuda. During at least one and perhaps two early Pleistocene high-sea-level stands, sea caves were eroded into the eastern cliff of the Highlands. Caves developed at two distinct elevations, 15 and 10 m above present sea level (Brasier and Donahue, 1985). Field examination documented the presence of Beazer Age beach sediments within and immediately adjacent to the caves. On the northwest side of the Highlands, a low shoreline notch, less than one m in height, has been eroded in the Highlands Formation, at an elevation between 15 and 10 m above present sea level. A thin veneer of Beazer Age beach sediments is found west of this notch. The absolute age for development of the erosional feature and associated Beazer Age beach sediments cannot be determined at present. Deposition of these sediments most probably is the result of one, and perhaps two, Pleistocene interglacial high-sea-level stands.

Middle- to Late-Pleistocene Marl Pits Subaerial Phase (Low-Sea-Level Stand)

One good exposure of the contact between the Beazer and Codrington Formations exists just east of the village of Codrington at an historic excavation known as the marl pits. The development of solution features, a thin residual soil, and a change from aragonite to calcite mineralogy within corals and mollusks in the upper 2 to 3 m of the Beazer Formation can be used to document subaerial exposure on the Marginal plain after Beazer deposition. In other areas of the Marginal plain, the Beazer Formation is covered or overlapped by the Codrington Formation. The same situation holds along the narrow east coast where Codrington Limestone covers probable deposits of Beazer age. How much of the Beazer Formation was removed by subaerial erosion prior to deposition of the Codrington is unknown. The absolute age and the amount of sea level lowering that occurred during the Marl Pits Subaerial Phase is not known. We can only state that it occurred during at least one glacial or stadial period before late Pleistocene time.

Late Pleistocene Codrington Formation (High-Sea-Level Stand)

Since the Codrington Formation is the youngest unit on the eastern, narrow coastal strip and the Marginal plain, it allows the best documentation of former shoreline position. A sea level position about 4 to 5 m above present initiated deposition of Codrington carbonate sediments. Well-exposed facies document the presence of a coral reef tract, lagoon, and beach environments east of and parallel to the Highlands escarpment. North, west, and south of the Highlands, faunal and floral carbonates demonstrate the presence of a shallow-water marine environment. A slow drop in sea level through Codrington time resulted in a gradually expanding shoreline position, especially to the north, west, and south of the Highlands. The last phase of Codrington deposition resulted in parallel beach ridges, the outermost of which contains the *Strombus* line discussed later in this chapter. Although the data are somewhat tenuous, there is a suggestion that the beach-ridge sequence extends to the west, below present sea level (Brasier and Donahue, 1985). If this is correct, then Codrington deposition continued beyond the present land area of Barbuda in response to a falling sea level. Again, the absolute age of the Codrington Formation has not been determined. The most likely period of time for this high stand is the last interglacial of about 125,000 years ago (Hays et al., 1976; Mercer, 1978) which resulted in a sea level rise of about 6 m.

Late Pleistocene Castle Bay Subaerial Phase (Low-Sea-Level Stand)

The last subaerial exposure of Barbuda probably occurred during the Wisconsinan glacial stage after 125,000 years ago. The best documentation for this sequence occurs along the southeast coast in the vicinity of Castle Hill Bay. Solution features and thin paleosol development can be seen on the surface of the Codrington reef facies. More impressive is the development of eolian dunes extending down to the vicinity of Spanish Point (Figure 6). These carbonate sand dunes have been lithified, in contrast to Holocene eolian dunes located inland of the present shore both north and south of the Highlands. Further documentation of the late Pleistocene age for the lithified dunes exists in the form of calcified root casts within the sediment, development of a paleosol on the dune surface, and the presence of pulmonate snails associated with the paleosol.

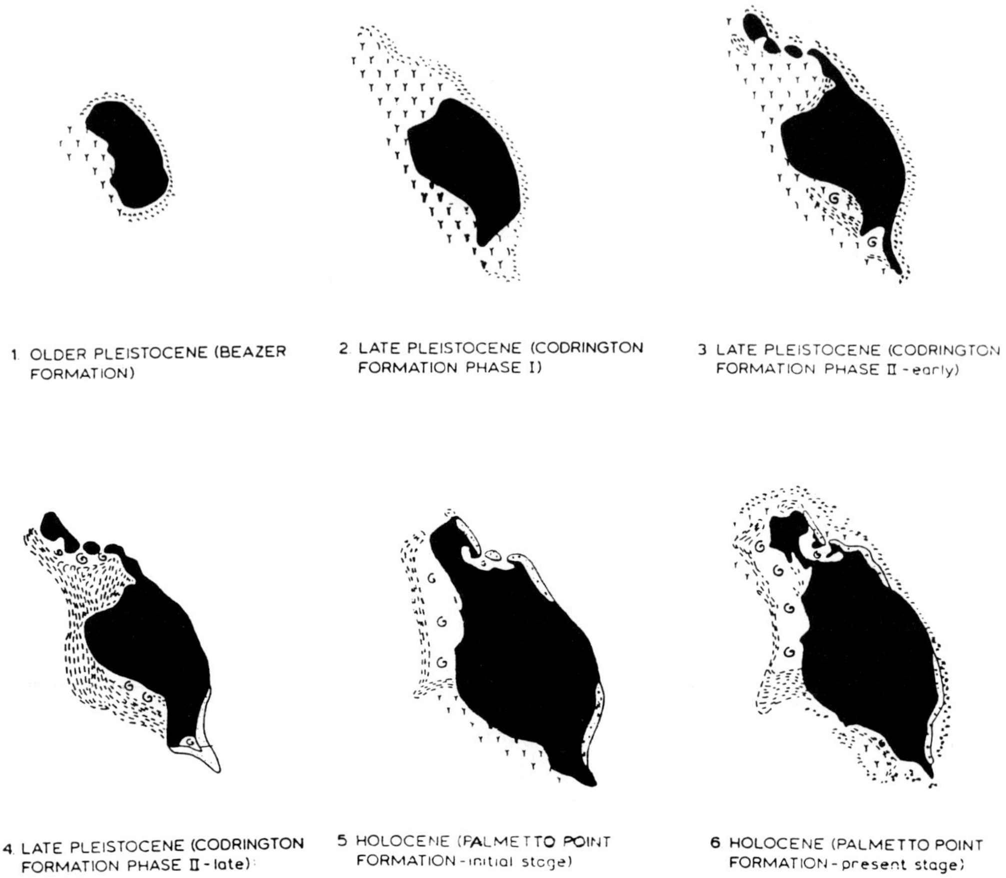

Figure 6. Paleogeographic maps reconstructing the evolution of Barbuda's paleoshorelines. (Adapted from Brasier and Donahue, 1985: Figure 9.)

Holocene Palmetto Point Formation (Present-Sea-Level Stand)

Both Codrington Lagoon and Palmetto Point developed in mid- to late-Holocene time by the formation of a southward-extending sand spit along the western edge of Codrington Lagoon and the southward progradation of progressive beach ridges to form Palmetto Point. Beach ridges and/or cheniers generally form adjacent to a river mouth where abundant terrigenous sediments are transported along shore from a delta (Augustinus, 1989). The development of parallel sets of beach ridges or cheniers is generally attributed to longshore accumulation of sediments and their subsequent reworking and buildup as ridges by high spring tides and/or storm surges. Beach-ridge plains such as that observed on Palmetto Point are not well documented in the literature. Perhaps the best-described sequence is that documented by Stapor (1975) along the Florida panhandle. In the case of Barbuda, longshore drift along the south coast, from east to west, has provided a continual sediment supply to the area of Palmetto Point. Reworking of the sediment, probably by infrequent storm surges, has generated a series of parallel beach ridges. At least four different parallel sets of ridges have been generated (Russell and McIntire, 1966). A second longshore current, flowing from north to south along the west coast of Barbuda, has truncated the ridges on their western end. This same current is responsible for the southern hook at the western end of Palmetto Point. The development of Palmetto Point is undoubtedly related to the middle- to late-Holocene rise in sea level. Thus, the initial development of Palmetto Point would have begun some time after about 5000 years ago. Beach-ridge deposition (albeit minor) also took place at Cedar Tree Point on Barbuda's northwest coast (Figure 3).

PREHISTORIC ARCHAEOLOGY

Evidence of occupation by "Archaic Age" peoples, dating at least as far back as 3000 B.C., is found on a number of islands in the Lesser Antilles, as well as some Greater Antilles islands where an even earlier (to 5000 B.C.), but poorly understood "Lithic Age" occupation also is reported (e.g., Rouse, 1986; Rouse and Allaire, 1978). Archaic sites in the Caribbean are relatively scarce, few have been studied in any detail, and their artifact assemblages are by no means uniform (Veloz Maggiolo and Vega, 1982). Archaic sites, which lack ceramic materials, sometimes are termed "preceramic" sites, implying they were deposited prior to the appearance of ceramic technology in the Caribbean region. Other terms appearing in the literature, such as nonceramic or aceramic, sometimes are used in the same sense as preceramic, but at other times merely signify a site lacking ceramics, without regard to its time period or cultural affiliation.

Beyond the problems raised by terminological inconsistency are two more fundamental issues related to Caribbean aceramic sites. First is the possibility that some sites, deemed to be later Archaic sites, may contain limited quantities of ceramics, resulting from the exposure of Archaic peoples to the ceramic technology being introduced into the region by later colonizers. Such sites would not fit the model of all Archaic sites being aceramic, nonceramic, or even preceramic. The second issue is the possibility that some sites without ceramics may date to a time period when people in the Caribbean had ceramic technology. For example, more recent aceramic sites might be specialized activity areas such as shellfish collecting and processing locales, where ceramics may have played no role in the activity. These sites could be correctly labeled aceramic or nonceramic, but not preceramic (cf., Lundberg, 1985).

The Ceramic Age occupation of the Caribbean, which followed some 3000 years of Archaic occupation, is better known. Initial colonization of the Lesser Antilles by ceramic-bearing peoples moving northward from South America took place during the second or third century B.C. (Rouse, 1989). This Saladoid colonization is distinguished by two distinctive forms of pottery decoration, white-on-red (WOR) painting and zoned-incised-crosshatching (ZIC), that are regarded as diagnostic markers of the Saladoid ceramic series (Rouse and Allaire, 1978) and, more recently, the Cedrosan Saladoid ceramic subseries (Rouse, 1986). Without going into detailed discussion, it should be noted that some Caribbean archaeologists (Chanlatte Baik, 1981; Rodríguez, 1989) now question the association of both forms of pottery decoration with the Saladoid colonization and, more generally, the spatial, temporal, and cultural issues related to "early ceramic populations" in the region (see various articles in Siegel, 1989). Despite the debate over particulars, the important point is that the Ceramic Age in the Caribbean region began at least by 300 B.C.

Our current understanding of the cultural sequence in the Lesser Antilles supposes initial colonizations, probably by more than one nonceramic Archaic group and possibly from different origin areas, by 3000 B.C. and perhaps earlier, followed by the first colonization of ceramic-bearing Saladoid peoples by 300 B.C., if not somewhat earlier. By about 600 to 700 A.D., distinctive styles attributed to the Saladoid ceramic series begin to disappear from the archaeological record in the northern Lesser Antilles. In their place are found somewhat nondescript "post-Saladoid" ceramics, with plain (undecorated) pottery becoming by far the dominant form, although some sherds exhibit red "washes" and incised elements. In general, post-Saladoid ceramics have yet to be well delineated or even seriously studied in the northern Lesser Antilles.

Archaeological research on Barbuda confirms the presence of Archaic and Ceramic Age (Saladoid and post-Saladoid) sites on the island. Based on our current knowledge of the distribution of sites and the range of artifact assemblages, it seems the first *significant* occupation of Barbuda occurred in the second half of the first millennium A.D., during the "post-Saladoid" period.

Site Distribution and Characteristics

Prehistoric sites have been identified on Barbuda through the combined efforts of Barbudans, members of the Antigua Archaeological Society (AAS), and professional archaeologists. Before 1978–1979, when a one-year archaeological survey and testing project took place on Barbuda and Montserrat (Watters, 1980a), there was almost no published information available about the prehistory of Barbuda, although data on some site locations and characteristics had been recorded by the AAS. By 1979, eight prehistoric sites were known; more recent research has increased the inventory to 14 sites (Figure 7). Table I provides basic information about these 14 sites, one of which is a questionable prehistoric site.

The questionable site, Spanish Point (BA2), which originally was recorded by the AAS, consists of accumulations of *Strombus gigas* (Queen conch) shells observed along the west side of the Spanish Point peninsula on the east edge of Gravenor Bay. After observing the location in 1979, Watters (1980a; Watters et al., 1984) argued that these shell accumulations should not be designated as a prehistoric site because there is no confirmed evidence that the shells were culturally modified, nor were any artifacts found in association with the shells. This was the first time the issue of designating shell accumulations as prehistoric sites arose, but it was by no means the last case. The status of the Spanish Point shell accumulations needs to be reconsidered in light of some of the research findings reported in the present article.

Figure 7 shows that known prehistoric sites on Barbuda are clustered in the east part of the island and generally near the Highlands, while the rest of Barbuda is largely without sites. The absence of prehistoric sites in the west sector of Barbuda's Marginal plain, including areas beside Codrington Lagoon and Goat Island Flash, on Goat Island and Palmetto Point, and in general along the north, west, and south coasts, is not the result of survey bias. During the 1978–1979 project, eight east-west transects were walked from coast to coast (barring impenetrable areas); transects crossed terrain (Figure 7, inset map) from Goat Island in the north to Gravenor Bay in the south (Watters, 1980a; Watters and Scaglion, 1980). No prehistoric sites were observed in those portions of the western Marginal plain that were traversed by the eight transects. Western Marginal plain areas between the transects were not observed in 1978–1979, and thus the absence of sites in those areas could not be verified. However, during further research on Barbuda in 1983 and 1984, when previously unobserved areas of the western Marginal plain were examined, no prehistoric sites were located. Therefore, it can be said with considerable confidence that prehistoric sites on Barbuda tend to cluster in the eastern part of the island and, in most instances, adjacent to or in the Highlands.

Table I indicates that there is considerable variation among site characteristics. Three sites, BA1 (Indiantown Trail), BA3 (Sufferers), and BA10 (Pelican Bay), are large open sites and, based on the quantity and variety of materials observed on the surface, each site is regarded as a principal habitation site, most likely representing a major village. In two test pits at the two excavated sites (BA1 and BA3), cultural deposits reached almost 70 cm in depth. In each instance, a sterile subsoil marked the base of the cultural deposits. The only other excavated open site (BA5, Highland Road), which had widely distributed surface materials, was found to have a shallow (about 20 cm) cultural deposit directly over limestone bedrock, making it a less likely candidate for a village site.

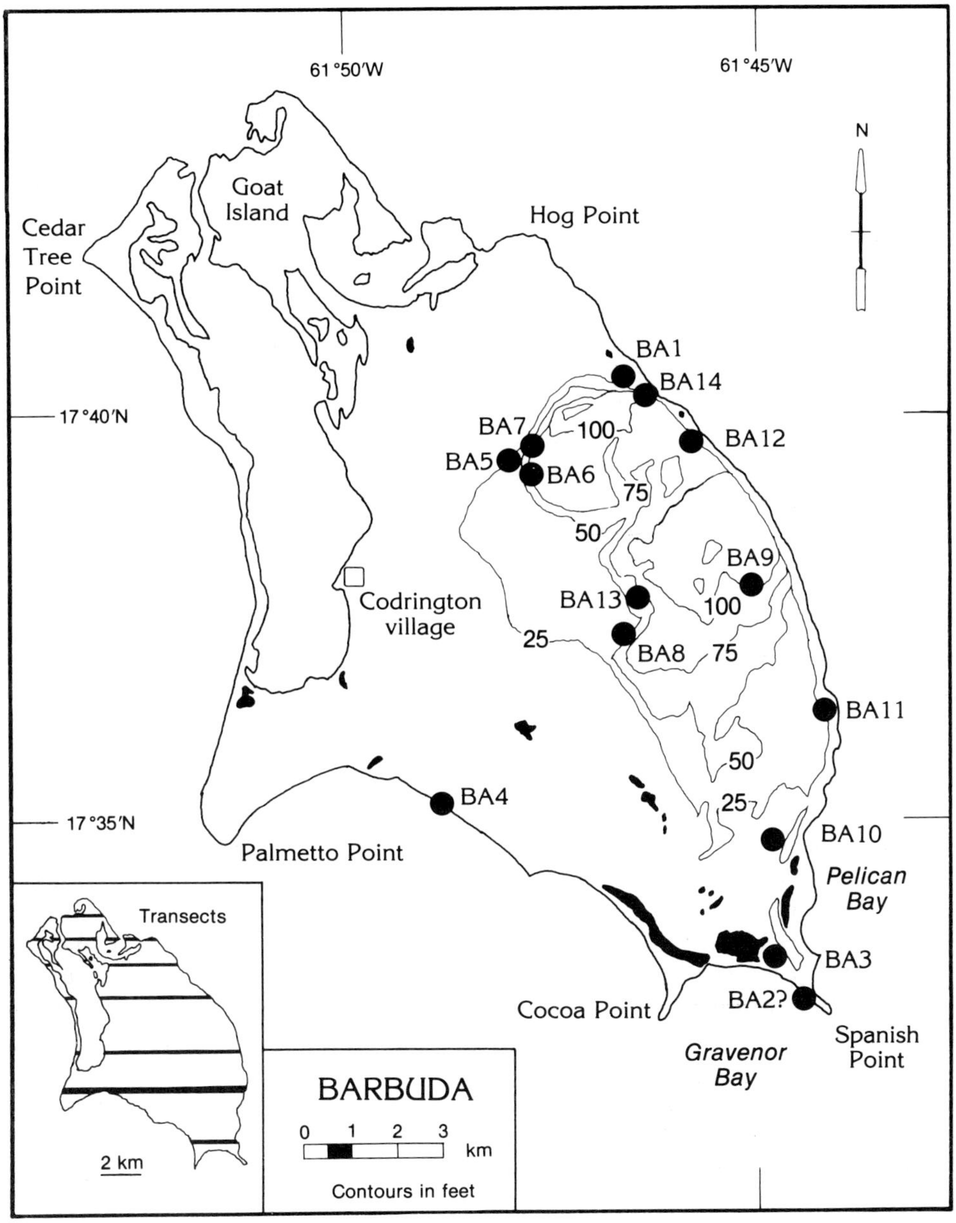

Figure 7. Prehistoric site locations on Barbuda (with inset map of cross-island transects).

TABLE I. Barbuda's Prehistoric Sites

Site Number and Name	Terrain	Cultural Attribution[1]	Research	Remarks
BA1 Indiantown Trail	Open	Post-Saladoid	Tested	Village
BA2 Spanish Point	Open	Unknown	Surveyed	Questionable site
BA3 Sufferers	Open	Saladoid and post-Saladoid	Tested	Village
BA4 River	Open	Archaic	Surveyed	No ceramics
BA5 Highland Road	Open	Post-Saladoid	Tested	?Village
BA6 Ravine Rockshelter	Closed	?Post-Saladoid	Surveyed	
BA7 Overview Cave	Closed	Post-Saladoid	Tested	
BA8 Eureka	Open	?Post-Saladoid	Surveyed	
BA9 Dark Cave	Open[2]	Unknown	Not observed	Has ceramics
BA10 Pelican Bay	Open	Post-Saladoid	Surveyed	Village
BA11 Double Cave	Closed	?Post-Saladoid	Surveyed	
BA12 Windy Root Cave	Closed	?Post-Saladoid	Tested	
BA13 Hilltop	Open	?Post-Saladoid	Surveyed	
BA14 Northeast Rockshelter	Closed	?Post-Saladoid	Surveyed	

[1] ?Post-Saladoid = probable Post-Saldoid
[2] The Dark Cave site actually is an open site located near that cave according to the AAS.

There are five other open sites. As noted, BA2 (Spanish Point) is a questionable site. BA8 (Eureka) and BA13 (Hilltop) cover small areas and, based on nearby protruding bedrock, most likely contain cultural deposits of limited depth. Artifacts were very limited in quantity and diversity (mainly scatters of sherds, lithics, and shells), suggesting that BA8 and BA13 represent transitory sites, possibly temporary camps or work sites. BA9, Dark Cave, which actually is an open site near the cave, was discovered and recorded by the AAS. Since BA9 has not been observed by the authors, the site attributes are uncertain and its location in Figure 7 is only approximate. It is reputed to be ''moderately sized''. The final open site, the River site (BA4), being anomalous in a number of ways, is discussed later.

Five closed sites, rockshelters or caves, are listed in Table I. Two sites (BA6, Ravine Rockshelter and BA7, Overview Cave) are located in the west escarpment of the Highlands in the vicinity of the Highlands Road (BA5) site. Two other sites occur in the Highlands' east escarpment, one to the north (BA12, Windy Root) and the other south (BA11, Double Cave). The final site (BA14, Northeast Rockshelter) is in the north escarpment near the major habitation site of Indiantown Trail (BA1). The five closed sites contain minimal cultural deposits and also are classed as transitory sites. These sites occupy sea caves formed in the geologic past at different heights along the escarpments. It is likely that other sea cave rockshelter sites eventually will be discovered elsewhere in the escarpments fringing the Highlands.

Of the 13 confirmed sites, seven (BA6, BA7, BA9, BA11, BA12, BA13, and BA14) are located atop the Highlands or in its escarpments and all are regarded as transitory (excluding BA9, which was not observed and cannot be evaluated). Three (BA1, BA5, and BA8) are located on the Marginal plain in very close proximity to the base of the Highlands escarpments. Another site (BA10) is somewhat more distant, but is still fairly near the low-lying southern sector of the Highlands. Two sites, BA3 (Sufferers) and especially BA4 (River), are located further from the Highlands.

Of the three major habitation sites, BA1 is adjacent to, BA10 is relatively close to, and BA3 is furthest away from the Highlands (Figure 7). These three sites have a number of attributes in common, and it is the combination of these characteristics that seems to be the major factor in controlling settlement patterns. First, each is an open site relatively close to the coast, within about 700 m of the present shoreline. Second, marine and coastal environments near each site are diverse but consistent, ranging from reef to sea-grass beds to sandy beaches to rocky headlands. Analysis of faunal remains from the Indiantown Trail site (BA1) shows heavy use of marine vertebrates, especially fish (Watters et al., 1984), as well as shellfish. Third, each site is in the lee of a landmass that deflects the prevailing winds and also accentuates water runoff. BA1 is protected by the mass of the Highlands, while BA3 and BA10 are situated in back of topographic high points (8-m-high limestone ridges) on what otherwise is a relatively flat Marginal plain. Fourth, the soil at each site is deeper than in the surrounding areas; these three sites are located on Blackmere Clay Loam.

River Site (BA4)

The most anomalous site on Barbuda is the River site (BA4) on the south coast. It is anomalous in terms of its physical characteristics and its cultural attributes. It is the only site known from the western Marginal plain, being situated some 5 km south-southwest of the nearest sector of the Highlands (Figure 7). With the possible exception of the questionable BA2 site on Spanish Point, BA4 has the lowest elevation above sea level (<2m) and closest proximity to the shoreline, being immediately adjacent. It is the only site located on Codrington Clay soil. BA4 is the only site located in the vicinity of the *Strombus* line, a point of particular interest that is discussed below. River is the only known prehistoric site on Barbuda that lacks ceramic artifacts. Based on two radiocarbon dates (presented for the first time in this report), BA4 is now interpreted as an Archaic Age site.

The River site is crossed by the Cocoa Point Road, the main route paralleling the south coast from the Cocoa Point peninsula westward to the east edge of Palmetto Point, where it joins a road running north to the village of Codrington. BA4 is located just where the Cocoa Point Road diverges into two branches, and the beach fork actually traverses the site (Figure 8). The site itself and the surrounding limestone terrain is generally level. The western part of the site is covered by dense scrub, while the eastern sector is open scrub and short grass, possibly caused by road construction and kept that way by vehicular traffic. Rocky coast abuts most of the site, although sandy beaches are nearby. BA4 consists mostly of beach sand, but gleyed soil and limestone bedrock occur between the parallel roads. Based on observed surface artifacts, the River site is roughly oval in shape, about 65 m wide by 80 m long, and encompasses an area roughly estimated at 3800 m^2 (Watters, 1980a: Table 7). The extent of BA4, notably to its southeast, may have been artificially expanded by road construction.

The fact that River yielded no ceramics, despite surface surveys conducted at three different times, is of particular interest since every other Barbuda site (except questionable BA2) contained pottery remains. Surface observations reveal that the western part of BA4 has deeper sediments and denser vegetation, especially when compared with the shallow soil and protruding limestone in the eastern sector. To judge from one exposure facing the coast, the maximum depth of sediment, mainly beach sand, over the Codrington Formation limestone bedrock is only about 30 cm. Because River has not been test excavated, the depth of the cultural deposit is unknown. BA4 certainly is the prime candidate among Barbuda's known sites for being a true nonceramic site. Groundstone tools also were not observed at River.

The two artifact categories recovered from the surface at River are chipped stone materials and shell tools, notably the conch celt made from the outer whorl of *Strombus gigas*. Other mollusk remains, especially the West Indian Top-shell *Cittarium pica,* were observed within the site, while on the shoreline just to its south (Figure 8) were several

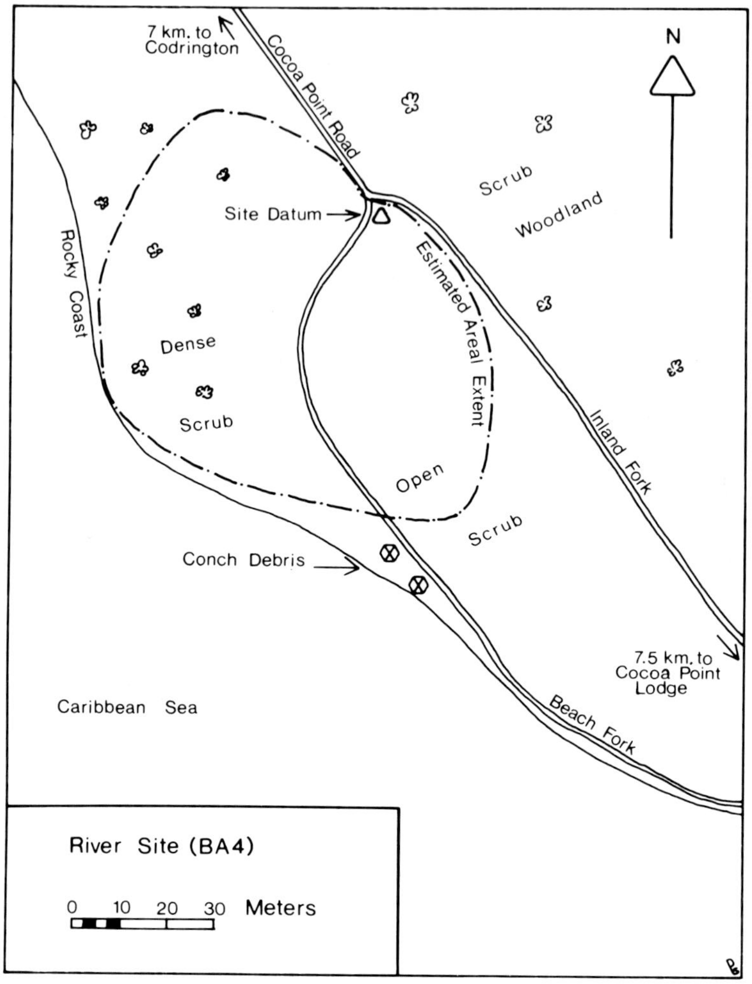

Figure 8. Map of the River (BA4) prehistoric site.

piles of *Strombus* shell debris. None of the mollusk remains were collected except for the shell tools and preforms.

The River site, an anomalously located nonceramic site, can be interpreted in two ways. River could be a ''limited activity area'' used by a Ceramic Age population to collect and process shellfish resources, which are tasks that might not have involved the use of pottery. In this case, River would be an ''aceramic'' site from a later cultural period. After the 1978–1979 research, Watters (1980a) advanced this interpretation, arguing that no Archaic Age sites were known on Barbuda, all other sites (n = 6) then known (ignoring questionable BA2) contained ceramics, and that few Archaic Age sites were known in the northern Lesser Antilles. The reported Archaic Age sites were from Antigua (Davis, 1974) and St. Kitts (Goodwin, 1978). Alternatively, BA4 could be a ''preceramic'' site deposited prior to the time ceramic technology was brought to Barbuda.

To resolve the issue of the aceramic or preceramic status of the River site and to determine its age, two surface-collected *S. gigas* celts (Figure 9), definitely manufactured by humans, were submitted for radiocarbon dating. These shell tools yielded radiocarbon ages of 3650 ± 35 years: 1700 B.C. (PITT 0717) and 3825 ± 25: 1875 B.C. (PITT 0731) (Table II). The dates place the celts within the Archaic Age in the Caribbean rather than the Ceramic Age. The two radiocarbon dates and the lack of pottery argue that River is a true preceramic site, thus providing the first evidence for an Archaic Age occupation of Barbuda.

The River site is the only prehistoric site in the vicinity of the *Strombus* line, a lithified beach ridge in southwest Barbuda on which shell scatters were observed. BA4 is about 1 km east of the southern terminus of the lithified ridge. Thus, the two radiocarbon dates

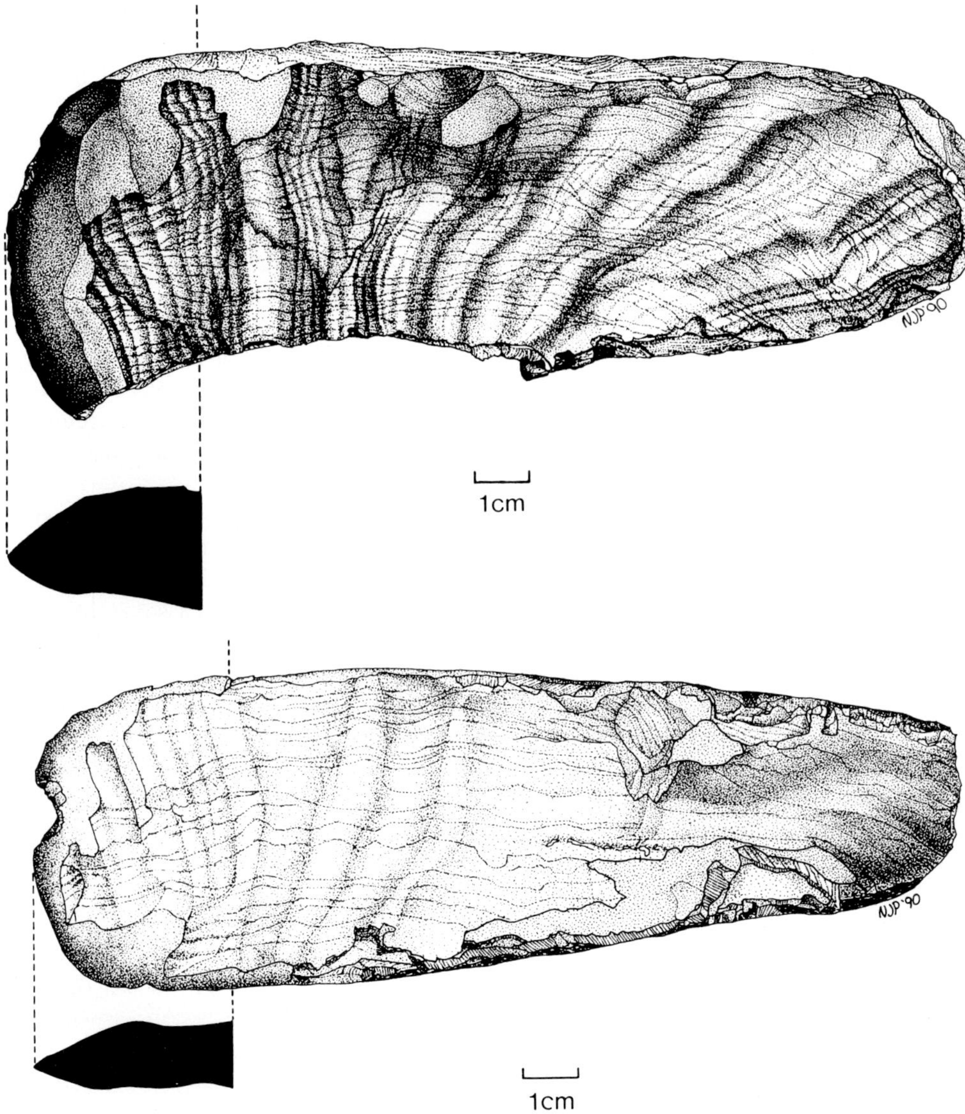

Figure 9. Two River site (BA4) *Strombus gigas* shell celts sent for radiocarbon dating. The top celt (BA4-RC1) dates to 3650 ± 35 B.P. (PITT 0717); the bottom celt (BA4-RC2) dates to 3825 ± 25 B.P. (PITT 0731). The black silhouettes indicate differences between ground edge angles.

from BA4 may bear on the question of deposition of *Strombus* line shells. The large set of issues relating to the *Strombus* line are dealt with at length below.

Research since 1980 on nearby Antigua has increased the inventory of known Archaic Age sites to over 40 and it has provided more information about artifact assemblages and locational data (Nodine, 1989). Those findings, when published, should provide useful comparative data for further research on the River site as well as for surveys to locate other possible Archaic sites on Barbuda.

Island Chronology

To date, only one site (BA4) on Barbuda, dating between 1700 and 1875 B.C., can be attributed to the Archaic, only one site (BA3, Sufferers) has yielded Saladoid ceramics in any numbers, and all remaining sites are classed as either post-Saladoid or probable post-Saladoid sites (Table I).

The sole confirmed Saladoid site on Barbuda is the village site at Sufferers (BA3), located on the south coast near the head of Gravenor Bay (Figure 7), the area of Barbuda that is closest to Antigua. The location of this site may well relate to its proximity to Antigua where an early Saladoid occupation, which persisted, is confirmed, and from which initial colonization of Barbuda by ceramic-bearing populations would have been feasible. Interestingly, however, the ceramics found at the Sufferers site reflect a late version (Mill Reef style) of the Saladoid ceramic series rather than earlier Saladoid forms which are known from Antigua. No radicarbon dates exist for Sufferers, but the Mill Reef style of ceramics suggests that the Ceramic Age occupation of Barbuda took place no earlier than the middle part of the first millennium A.D. This suggests that the Saladoid occupation of Barbuda occurred later than on many other islands, especially volcanic islands, and involved only the Sufferers site.

Plain (undecorated) pottery, the dominant variety at the remainder of Barbuda's sites, tends to be thicker, more friable, and generally less well made than most Saladoid sherds. Based on the abundance of plain pottery and the absence of Saladoid sherds, nine sites currently are attributed to the post-Saladoid period on Barbuda, although some sites with limited ceramic samples are so classed provisionally (Table I). Indiantown Trail (BA1) and Pelican Bay (BA10) are the major post-Saladoid habitation sites. It should be noted that BA3 (Sufferers) was interpreted previously as having a post-Saladoid component due to the existence of many undecorated sherds. However, more recent analysis of thin sections from six BA3 sherds, which documented a marked similarity between the Saladoid and purported post-Saladoid ceramic tempers at BA3 (Donahue et al., 1990), raises questions about the validity of the post-Saladoid component. All other Barbuda sites generally lack

TABLE II. Radiocarbon Dates (Prehistoric Sites)

Sample Number	Lab Number	^{14}C Years B.P.	Material
River (BA4)			
BA4-RC1	PITT 0717	3650 ± 35	*Strombus* celt
BA4-RC2	PITT 0731	3825 ± 25	*Strombus* celt
Indiantown Trail (BA1)			
BA1-RC1	Beta 18492	910 ± 220	Charcoal (0.1 g)
BA1-RC2	PITT 0594	445 ± 30	Whole *Cittarium pica*
BA1-RC3	PITT 0595	1065 ± 45	Whole *Strombus gigas*

Figure 10. An example of a "punched hole" *Strombus gigas* shell from the Indiantown Trail (BA1) site. The lip (outer whorl) is broken off.

sherds with typical Saladoid decoration, such as WOR painting and ZIC incising, although the very shallow deposit at Highland Road (BA5) did yield three tiny ZIC sherds.

Apart from the two shell-tool radiocarbon dates from River (BA4), the other dates from Barbuda are from the Indiantown Trail site, where three samples gave radiocarbon ages of 910 ± 220 years: 1040 A.D. (Beta 18492) on charcoal (0.1 g), 445 ± 30 years: 1505 A.D. (PITT 0594) on *Cittarium pica* shell, and 1065 ± 45 years: 885 A.D. (PITT 0595) on a "punched-hole" *S. gigas* shell (Table II). The punched-hole technique, which is well documented in shell samples from numerous Caribbean prehistoric sites, left a very distinctive hole in the apex of the *S. gigas* shell as clear evidence of human exploitation (Figure 10). The hole facilitated extraction of the animal from its shell.

Significant occupation of Barbuda took place during the post-Saladoid period in the last half of the first millennium A.D., judging from the greater number of sites, the abundance of undecorated pottery (and lack of sherds exhibiting Saladoid WOR or ZIC decoration), and the few radiocarbon dates. Anguilla, a low-lying, dry limestone island very similar physiographically to Barbuda, also has numerous post-Saladoid sites. However, greatly increased numbers of post-Saladoid sites apparently is a pattern not restricted to limestone islands because Wilson (1989) has documented a similar situation on Nevis, one of the islands in the Volcanic arc of the northern Lesser Antilles (Figure 1).

STROMBUS LINE

Characteristics

One feature of the landscape of southwest Barbuda is the *Strombus* line, a slightly elevated limestone ridge trending south-southeast from near the southeast corner of Codrington Lagoon to the south coast of Barbuda (Figure 11). The linear feature borders the east side of the Palmetto Point peninsula between UTM coordinates PQ237476 and

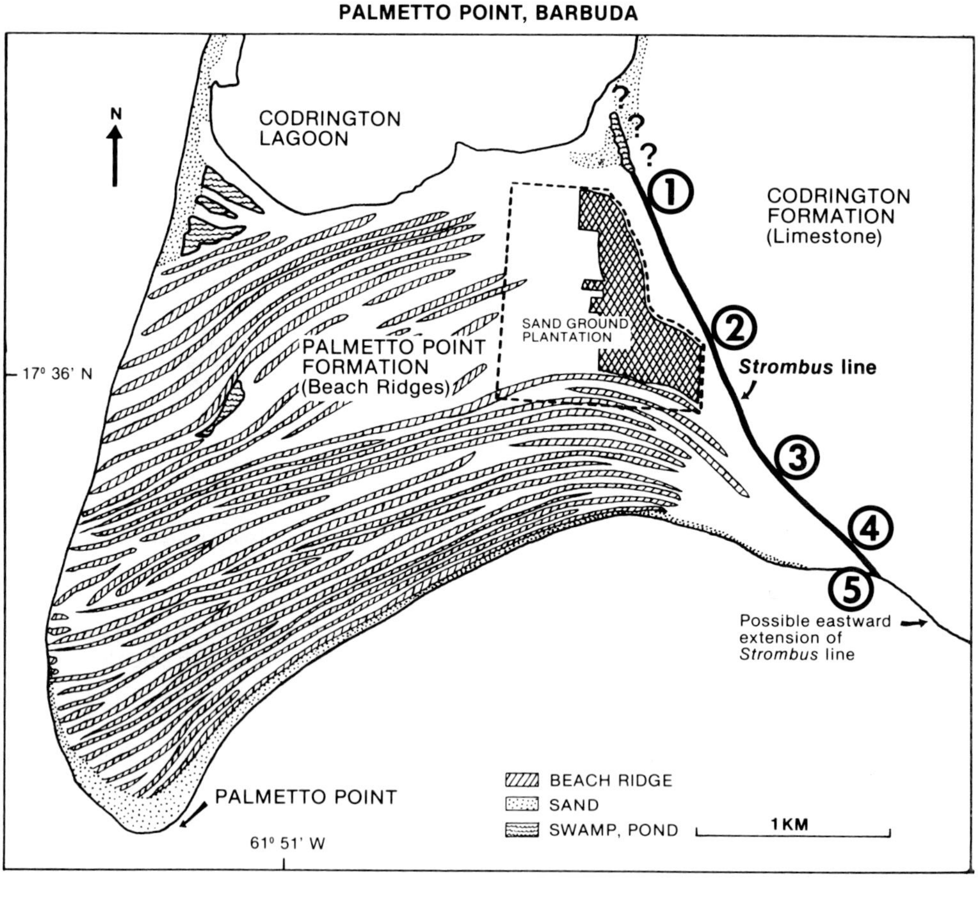

Figure 11. Palmetto Point peninsula and the adjacent *Strombus* line, with locations of shells collected for radiocarbon dating: 1 is North Sand Ground Plantation (NSGP), 2 is Sand Ground Plantation (SGP), 3 is Singer Cave Road (SCR), 4 is Sandman (SM), and 5 is Jetty Area (JA) on the south coast.

PQ250453. The feature is a lithified beach ridge extending for about 3 km and it can be readily observed on the ground, even though its height above surrounding terrain is not uniform along its entire length, being up to about 1 m high in some sectors but only slightly elevated in other areas. Along the northern portion of the *Strombus* line where it approaches Codrington Lagoon, the ridge gradually diminishes in height, becomes indistinct, and no longer forms a discernible surface feature. Construction work related to the building of a jetty on the south coast has damaged and largely obscured the southern terminus of the ridge.

The lithified ridge is an undisputed line of demarcation for geologic formations and soils, separating Codrington Formation limestone to the east from unconsolidated sands of the Palmetto Point Formation to the west (Brasier and Mather, 1975; Brasier and Donahue, 1985). It also separates the Codrington Clay Series soil to the east from the "beach sands land type" defined by Vernon and Lang (1966). These distinctions are very evident on the ground, where to the east one observes the black, almost waxy Codrington Clay soil interspersed with protruding limestone, while to the west are seen only the grayish sandy deposits marking the east edge of Palmetto Point peninsula. Vegetation on each side of the ridge is equally distinctive, with degraded Evergreen Woodland to the east and coastal vegetation to the west. Immediately west of the northern half of the *Strombus* line lies Sand Ground Plantation, which is heavily planted with coconut palms. Codrington Clay soil occurs west of the limestone ridge (see Figure 3) only at the northern part of the *Strombus* line, near Sand Ground Plantation in the most northeasterly sector of Palmetto Point peninsula.

From a geologic perspective, the lithified beach ridge is of particular interest because it formed part of the ancient western shoreline of Barbuda, and it had to exist before Palmetto Point peninsula prograded. The ridge was formed during the last phases of deposition for the Codrington Formation. The beach ridge parallels the coast southeast of Palmetto Point so that a distinct ridge exists along the present shoreline. Proceeding north from the jetty along the junction between the Codrington Formation and Palmetto Point (Figure 4), the limestone ridge is less distinct. This is because the Codrington beach ridges are not parallel to the line of junction. Erosion of the beach-ridge sequence occurred after the deposition of this area. The entire *Strombus* line, except for its extremities, is now located well inland from any coastline.

Natural vs. Cultural Deposition

Archaeological interest in the *Strombus* line is sparked by the shells found on the limestone ridge, not by the ridge itself. Distributed along the length of the lithified ridge are shell scatters containing many broken and some whole specimens. In certain areas, especially toward the south, shells are somewhat densely concentrated to the point that they almost cover the ridge. Elsewhere along the feature, shell accumulations are quite sparse. The *Strombus* line is not a solid line of shells, but instead consists of areas of denser accumulations which are interspersed among sectors having few shells. The discontinuous nature of this distribution becomes especially evident toward the northern terminus where the shell scatter virtually disappears. Shells generally occur on the ridge itself, but not on the flat terrain on either side of the ridge. Almost all of the observed shells, both whole and fragmentary, are of the genus *Strombus,* and whole shells that could be identified confidently to species level are the Queen conch, *S. gigas*.

The occurrence of shells on the lithified beach ridge has raised a number of issues relating to their deposition and age. The central issue involves competing hypotheses about the process by which the shells were deposited on the ridge, and the authors have disagreed on this basic point since their joint research in 1984. The fundamental point of the disagreement is summarized in one phrase: "shell deposition by cultural or natural process". In what may seem to be a reversal of roles, the archaeologist argues a natural process deposited the shells on the ridge, whereas the geologist contends the shells were culturally deposited, and each has some evidence to support his hypothesis.

Before further discussing the two hypotheses, it is worth noting some points on which the authors concur and some on which they disagree. They agree the lithified beach ridge is a past shoreline of western Barbuda, which existed before Palmetto Point peninsula was created, and that the ridge does demarcate geologic formations, soil units, and vegetational communities. Although the geologist and archaeologist agree on the existence of the ridge, they do not agree on its extent. The archaeologist has the ridge terminate on the present south coast near the jetty construction area. The geologist has it continue further to the southeast toward Cocoa Point (see eastward extension noted in Figure 11), essentially paralleling the south coast as do the beach ridges within the Codrington Formation. They also do not agree about whether shell scatters occur along this extension. Their differences of opinion about the southeasterly extension mainly result from the cursory nature of observations made in this area. They are in agreement about the existence of the ridge and shell scatters between Codrington Lagoon and the vicinity of the jetty area on the south coast (Figure 11), the 3-km area where they concentrated their study in 1984.

Three points of contention underlie the disagreement between the authors: (1) the implications of the taxonomic uniformity of the shells, (2) the conditions of fragmented and whole shells, and (3) the absence of prehistoric artifacts on the *Strombus* line. The basis of the first point of contention is the fact that virtually all of the shells observed on the limestone ridge are from one taxon, the genus *Strombus,* and this taxonomic uniformity can be interpreted in two ways. The archaeologist contends that this taxonomic uniformity provides a supporting argument for a long-term natural deposition process, during which shells from dead animals were gradually swept to shore and deposited along the paleoshoreline. Also, short-term events, such as hurricane surges, may have rapidly transported additional specimens to the shoreline. As Palmetto Point later gradually built outward, the shell accumulations would have been stranded at their inland locations on the paleoshoreline.

From the geologist's perspective, this same taxonomic uniformity of the shell accumulations argues against natural deposition. This view contends that single-genus accumulations do not reflect natural depositional processes. Natural deposits washed ashore would consist of a mixture of remains from various taxa, such as other mollusks or corals, instead of containing specimens from only one genus. In other words, the *undiversified* character of the shells on the *Strombus* line is what is the most "unnatural" from a geologist's standpoint. Taxonomic uniformity is regarded as evidence for human selectivity which subsequently resulted in cultural deposition of the shells. According to this interpretation, the single-genus accumulations of *Strombus* shells observed along the paleoshoreline are argued to be identical to recent accumulations of *Strombus* shells along the present shoreline. Such accumulations are caused by a modern harvesting technique, where empty conch shells are discarded slightly inland from the shoreline. Today, meat is removed from the conch by a blow from a machete, resulting in an elongated gash in the same area of the spire where the hole was punched in the prehistoric extraction technique.

Research on other Codrington Formation beach ridges, located further inland, to see if shell accumulations are present and if so, their constituents, which could bear on the taxonomic uniformity issue, was not conducted during the 1984 work.

The second point of contention concerns the condition of the shells, both fragmented and whole, observed on the *Strombus* line. From the archaeologist's standpoint, if these mollusks were exploited by humans, the shells should reveal some kind of evidence of cultural modification. However, evidence for such cultural modification is lacking, both on shell samples collected and those observed on the *Strombus* line. In particular, the characteristic punched hole, resulting from the prehistoric extraction technique, was not present in a single shell observed or collected from the *Strombus* line. The absence of any punched-hole shells is especially noteworthy because they are regularly found at prehistoric sites in the Caribbean. In addition, shell tools (see Dacal Moure, 1978; Keegan, 1984) were not observed among the *Strombus* line specimens. From the archaeologist's perspective, the absence of culturally modified specimens among *Strombus* line shells, especially in view of the abundance of shells, supports the argument for a natural deposition process. If the accumulations were culturally deposited, it is logical to expect that at least some shells would show evidence of human modification.

From the geologist's point of view, the lack of punched-hole shells is seen as negative evidence and, therefore, cannot be used to support the natural deposition argument. It also has been suggested (Brasier and Donahue, 1985) that some of the fragmented shells, for example, those missing apices, may be evidence for an earlier butchering technique that predates the punched-hole method of extraction.

The final point of contention concerns artifacts. To the archaeologist, another argument against the *Strombus* line shells being culturally deposited is the fact that no prehistoric artifacts were recovered anywhere on the lithified ridge itself. If humans had exploited the conch resources, it seems logical to propose that some kind of durable artifact, such as lithic tools, should remain as evidence of human activity. None were observed on the *Strombus* line. The geologist dismisses the lack of artifacts on the *Strombus* line by contending that more intensive surveys eventually will locate such artifacts and, furthermore, points to the River site, which is relatively close to the *Strombus* line, arguing that it could bear on the issue of natural vs. cultural deposition of the shells, a point that is discussed below.

Radiocarbon Dates

With the competing hypotheses from the 1984 research in mind, it was decided to determine the ages of *Strombus* line shell specimens by radiocarbon dating. Of particular interest were determinations of (1) when the specimens died and (2) whether the range of death ages was broad or restricted. Such dates could relate to several issues, ranging from establishing whether the shells were contemporaneous with human occupation of Barbuda to the geologic issue of when Palmetto Point began to prograde.

Twelve shell samples collected by the authors in 1984 from locations 1 through 4 along the *Strombus* line (Figure 11) were submitted for radiocarbon dating. Four samples, one from each location, were sent to the Smithsonian Institution Radiocarbon Laboratory in 1984 and 1985. In 1989, four more samples from the same four locations were sent to the University of Pittsburgh Radiocarbon Laboratory (formerly the Smithsonian Laboratory). In 1990, four more samples from the four locations were submitted to the Pitt Radiocarbon Laboratory. The four samples sent to the Smithsonian and first four samples sent to Pitt were fragmented parts of *Strombus* sp. shells; the last four samples sent to Pitt were whole or largely intact *S. gigas* shells (Figure 12). Thus, there are three series of samples from four locations on the *Strombus* line. Locations were purposefully selected to ensure that coverage, north to south, was as complete as possible on the lithified ridge. The northernmost part, where the ridge and shell scatter disappear from view, is the most evident gap in the coverage (Figure 11).

Figure 12. Example of a largely intact *Strombus gigas* shell from the Singer Cave Road location on the *Strombus* line lithified ridge.

In addition, a single *S. gigas* specimen from location 5, on the present south coast of Barbuda, was sent to the University of Pittsburgh Radiocarbon Laboratory in 1989. This whole shell, which had a punched hole in its apex, was the only specimen that definitely was culturally modified. Location 5 (Figure 11) is slightly west of the presumed southern terminus of the *Strombus* line, near the area where the limestone ridge flattens into the surrounding terrain and where bulldozing, associated with building the jetty and other structures, scoured the land from the shore to about 300 m inland. Bulldozed piles of conch just north and west of the construction area attest to the extent of this damage. As a result, an association between the south shore shell specimen and the *Strombus* line cannot be verified.

Results of the three series of radiocarbon assays are shown in Table III, where three dates from each of four locations on the *Strombus* line are presented in a north-to-south orientation (North Sand Ground Plantation or NSGP, Sand Ground Plantation or SGP, Singer Cave Road or SCR, and Sandman or SM: Figure 11) and before the single date from the south coast Jetty Area (JA). The first observation for the 12 *Strombus* line dates, considered together, is that they are widely variable, with ages ranging from the oldest (SGP1) of 5480 ± 100 (SI 6879) to the youngest (SGP3) of 1755 ± 75 (PITT 0719) radio-

TABLE III. Radiocarbon Dates (*Strombus* Line and South Coast)

Sample Number	Lab Number	^{14}C Years B.P.	Material
A. *Strombus* Line			
North Sand Ground Plantation (Figure 11, location 1)			
NSGP1	SI 6695	3340 ± 70	*Strombus* fragment
NSGP2	PITT 0590	3555 ± 45	*Strombus* fragment
NSGP3	PITT 0718	2100 ± 35	Whole *Strombus*
Sand Ground Plantation (location 2)			
SGP1	SI 6879	5480 ± 100	*Strombus* fragment
SGP2	PITT 0592	2900 ± 50	*Strombus* fragment
SGP3	PITT 0719	1755 ± 75	Whole *Strombus*
Singer Cave Road (location 3)			
SCR1	SI 6696	4085 ± 85	*Strombus* fragment
SCR2	PITT 0591	2825 ± 75	*Strombus* fragment
SCR3	PITT 0720	1930 ± 65	Whole *Strombus*
Sandman (location 4)			
SM1	SI 6880	3150 ± 55	*Strombus* fragment
SM2	PITT 0593	2645 ± 50	*Strombus* fragment
SM3	PITT 0721	3350 ± 55	Whole *Strombus*
B. South Coast			
Jetty Area (location 5)			
JA1	PITT 0589	1075 ± 60	Whole *Strombus* with punched hole

carbon years, a disparity of some 3725 radiocarbon years between two samples from the same location. The range of death dates for all 12 mollusks on the *Strombus* line is very broad.

When the oldest and youngest ages at each location are examined, radiocarbon year disparities range from 1455 years at North Sand Ground Plantation (NSGP2 and NSGP3), to 3725 years at Sand Ground Planatation (SGP1 and SGP3), to 2155 years at Singer Cave Road (SCR1 and SCR3), to 705 years at Sandman (SM2 and SM3). Table III documents a lack of congruence among radiocarbon ages derived from samples from the same location. Table III also shows a lack of congruence among samples from all four locations. Thus, the range of mollusk death dates from across the *Strombus* line is very broad, although by location the breadth varies from 3725 to 705 radiocarbon years.

Moreover, the expected pattern of a progressively younger sequence of dates trending north to south, to match the southward progradation of beach ridges on Palmetto Point, simply is not evident among the radiocarbon dates from samples from the four locations. Although the second series of dates (NSGP2, SGP2, SCR2, and SM2) follows that pattern, the first series does not, and the third series, which it was hoped would duplicate the second, proved to be even more out of sequence than was the first series (Table III). Largely intact shells (third series) do not match the anticipated pattern; one shell-fragment series (second) does match but another fragment series (first) does not.

The punched-hole specimen (a whole shell) from location 5, Jetty Area (Figure 11), being younger than all *Strombus* line shells, confounds the issue even more. With a radiocarbon age of 1075 ± 60 years (PITT 0589), it is 680 years younger than the youngest *Strombus* line sample (SGP3), but it is 4405 years younger than the oldest *Strombus* line sample (SGP1) (Table III).

Before discussing the radiocarbon dates from the 13 *Strombus* specimens with respect to the River site and formation of Palmetto Point, several comments on procedure are warranted. Each shell sample was cleaned with HCl, removing 25 to 30% of the outermost shell to avoid post-mortem surface exchange with surrounding carbonates, before carbon dioxide evolution with H_3PO_4. The carbon dioxide was then converted to methane in a static reactor at 450°C in the presence of a ruthenium catalyst to provide the medium for gas proportional counters. All samples were counted at least twice for periods of not less than 2000 min. Ages were calculated using the Libby half-life (5568 years).

No "tree-ring" or "calendric" corrections have been applied to the ages obtained from these samples. While a number of laboratories are now engaged in precise measurements of variations of radiocarbon ages from true age, there is only the most general agreement among those laboratories; until there is agreement upon discrete corrections, it would be presumptuous of the user to apply one correction scheme above all others as the most valid. For example, the application of corrections provided by Stuiver and Becker (1986) simply increases all ages greater than 1000 years by amounts of from 60 to 600 years and does little to solve the problem at hand.

Equal, if more subtle, vexation results from inherent metabolic isotopic fractionation within the shellfish dated; shells will provide apparent radiocarbon ages some 400 years younger than those of contemporaneous trees on nearby land. At the same time, marine organisms growing in well-mixed ocean layers will be exposed to dissolved carbon dioxide somewhat depleted with respect to atmospheric radiocarbon levels, and will yield apparent ages 300 to 400 years older than those of nearby trees. The two effects offset each other, and the radiocarbon ages for such shells may be taken at face value. The same mechanism, however, may not be true for shells of intertidal habitat (such as *Strombus*) since the mixed-layer buffer may not be as strongly present, and the apparent shell ages may well be 150 to 200 years too young (Stuiver and Borns, 1975).

Archaeological Implications

Eleven of the 12 samples from the four *Strombus* line locations yielded dates that fall within the generally accepted time span of Archaic Age occupation in the West Indies (i.e., more recent than about 3000 B.C.), and the twelfth date (SI 6879 from Sand Ground Plantation), with an age of 5480 ± 100 radiocarbon years, is only slightly earlier. Thus, the mollusks died at times coeval with the Archaic Age occupation of the Caribbean region. Moreover, ten of the 12 *Strombus* line dates are younger than the two shell-tool dates of 3650 ± 35 and 3825 ± 25 radiocarbon years from the River site, thereby confirming that those 10 mollusks died while Archaic Age peoples were inhabiting Barbuda. The contemporaneity of the radiocarbon dates for the River site shell tools and some *Strombus* line mollusks, the proximity (about 1 km apart) of the site and lithified ridge, and the nonceramic nature of the BA4 site are factors that can be argued to support the geologist's hypothesis that the *Strombus* line shells were culturally deposited. Indeed, those are telling points, yet they assume rather than demonstrate a relationship between these two locations. In addition, from the archaeologist's perspective, two points of contention, the absences of confirmed culturally modified shells and of artifacts on the *Strombus* line, are still unresolved. They probably cannot be resolved without further fieldwork to intensively survey the *Strombus* line. Certainly the likelihood that at least some of the *Strombus* line shells were deposited as the result of cultural processes is enhanced by the coincident radiocarbon dates at BA4 and the *Strombus* line.

On the other hand, the sole example of an undisputed culturally modified shell, the Jetty Area punched-hole shell, clearly is more recent than the *Strombus* line shells. It falls within the time span of the Ceramic Age occupation of Barbuda in the first millennium A.D., and is coincident with the radiocarbon dates from the ceramic site of Indiantown Trail (BA1) (Table II), the only other suite of radiocarbon dates from Barbuda's sites. As previously noted, any relationship between the Jetty Area shell and the shells on the lithified ridge is tenuous because construction damage has obscured the southern terminus of the *Strombus* line. The Jetty Area shell may have no relationship to the *Strombus* line and it also may have no relationship to the Archaic site at River. The Jetty Area shell is 2550 to 2775 radiocarbon years younger than the River site shell-tool dates.

Three *Strombus* line shell dates are of interest because of their relatively young ages. Interestingly, all three dates were derived from the third series of submitted specimens, the largely intact shells. These samples (PITT 0718 at 2100 ± 35, PITT 0719 at 1755 ± 75, and PITT 0720 at 1930 ± 65 radiocarbon years) all are more recent than the beginning of the Ceramic Age colonization of the Lesser Antilles in about 300 B.C. Also of interest, the samples come from the three northernmost locations on the *Strombus* line (Table III and Figure 11). Although these three dates fall within the Ceramic Age colonization of the Lesser Antilles as a whole, they still predate the first confirmed occupation of Barbuda by ceramic-bearing populations, as reflected in the Mill Reef style of Saladoid pottery at Sufferers (BA3), in the latter half of the first millennium A.D. These three shells could be of particular interest because *if* they were deposited on the *Strombus* line by humans, then they may indicate that an Archaic Age preceramic population persisted on Barbuda into the first centuries of the first millennium A.D. Alternatively, and once again assuming the deposition was by humans, these shells could also be interpreted as representing the remains of an occupation of the island by ceramic-bearing peoples at a time even earlier than currently posited.

In summarizing these archaeological implications, ten of the *Strombus* line shell dates are contemporaneous with or younger than the two shell-tool radiocarbon dates from the River site, which indicates that Archaic peoples were inhabiting Barbuda when those mollusks died. Whether the Archaic Age peoples at BA4 actually exploited and deposited the *Strombus* line shells remains a point of contention, not because of the radiocarbon dates, but because cultural modification of the shells is unconfirmed. The radiocarbon dates certainly lend support to the geologist's cultural deposition hypothesis for *Strombus* line shells, but they cannot resolve the separate issue of cultural modification.

Geological Implications

Radiocarbon dating of the *Strombus* line shell specimens also has implications for Barbuda's geologic history, most notably in terms of providing data about (1) the youngest possible age of the lithified beach ridge on which the shells rested and (2) the initial progradation time of Palmetto Point peninsula. Assuming that the shells were deposited on the ridge when it formed part of Barbuda's western paleoshoreline, the shells provide a temporal datum point which is more recent than the time of deposition of the ridge, although exactly how much it predated them could not be determined. As long as it is assumed that the shells deposited on the ridge came from adjacent nearshore waters, the process (cultural or natural) by which those shells came to rest on the ridge is irrelevant. The oldest-dated shell provides a datum point of 5480 ± 100 (SI 6879), which means the lithified beach ridge must predate that datum. The broad range of shell dates from each location suggests that shells were being deposited on the ridge for a long time. Unfortunately, the oldest date is considerably older than the other *Strombus* line shell dates, being some 1395 radiocarbon years older than the next-oldest shell (Table III). If the date is erroneous, the extrapolated "older than" age for the ridge also would be incorrect. If the ridge was already located inland before the shells were deposited, necessitating human or natural transport (e.g., hurricane surges), the shells could be much younger than the ridge. In this case, an extrapolated "older than" date for the ridge could be much too young.

The *Strombus* line shell dates have different implications for Palmetto Point peninsula (Figure 11). Here, the dates could provide a baseline after which Palmetto Point began to prograde. In this case, it is the most recent shell date that is relevant since Palmetto Point, in theory, would have started prograding only after deposition of the youngest shell on the lithified ridge. The youngest shell date, from the Sand Ground Plantation locale, is 1755 ± 75 radiocarbon years (PITT 0719) or roughly 195 A.D.

If this date is even approximately correct, is has several profound implications. It means that Palmetto Point peninsula (1) prograded some 2 km southward in about 1750 radiocarbon years, (2) did not exist when the Archaic peoples occupied the River site at about 1800 B.C., and (3) was forming, but certainly had not reached, its current configuration during the Ceramic Age occupation of Barbuda in the latter first millennium A.D. If this scenario holds, the landscape and seascape of southwest Barbuda today are radically different from what was observed by the Archaic Age and, to a lesser degree, the Ceramic Age peoples who inhabited the island in the past.

There is one serious reservation about the *Strombus* line shell dates and their possible relationship to progradation of Palmetto Point. Figure 11 shows that the orientation of the Palmetto Point peninsula beach ridges generally is along a northeast-to-southwest axis and that they are roughly perpendicular to the *Strombus* line lithified ridge. Progradation of the coastline is southeasterly. In theory, the oldest beach ridges should occur in the north, with progressively younger beach ridges toward the south. The beach ridges, deposited as the coastline prograded south, should have stranded sections of the lithified ridge, beginning in the north and successively moving southward. In theory, therefore, shells deposited on the northern part of the lithified ridge should be older than those in the southern part. Table III shows that the three youngest dates are from the three northern locations on the *Strombus* line, while the fourth (the southernmost) location at Sandman has no generally young dates.

A number of factors may bear on this issue. First, the northern terminus of the lithified ridge is not a discernible surface feature. Beach ridges, which should have been deposited first in this area, are largely indistinct as well. Second, the beach ridges in Sand Ground Plantation were destroyed when the agricultural station was built. Third, beach ridges near the south terminus of the lithified ridge are indistinct and have been damaged by construction activities. Thus, the sequence of beach-ridge development is best observed further west on Palmetto Point (Figure 11) and not immediately adjacent to the lithified ridge. In fact, the beach ridges generally do not directly abut the lithified ridge. That the expected pattern of progressively younger shell dates from north to south is not present among the 12 shell samples remains bothersome. The incongruence of shell radiocarbon dates, at each of the four locations and along the *Strombus* line, could relate to the reworking of beach-ridge materials as Palmetto Point prograded and to the redeposition of shells.

Palmetto Point Beach-Ridge Development

The timing for the development of the Palmetto Point beach-ridge plain must be placed within two time constraints. Palmetto Point itself may not have developed before emplacement of the shells found on the *Strombus* line. Excluding the oldest date, 5480 ± 100 from a *Strombus* shell fragment at Sand Ground Plantation (SI 6879), all the dates fall between 4085 and 1755 radiocarbon years ago (Table III). Thus, we may be dealing with about a

2300-year period when *Strombus* shells were being accumulated either naturally or through prehistoric harvesting on the southwestern shoreline of Barbuda. However, Russell and McIntire (1966) suggest that Palmetto Point beach ridges may have begun developing in response to a Holocene sea level rise. Thus, the northern portion of Palmetto Point, their accretionary sets A and B (Russell and McIntire, 1966: Figure 3) were present before 3000 years ago. If Russell and McIntire (1966) are correct in their interpretation, the *Strombus* line does not mark the position of a pre-2000-year-old paleoshoreline. The shells located on the line were transported either culturally or naturally a distance of between 0.5 and 1.5 miles (0.8 to 2.5 km) from the shoreline of the time.

A second important point made by Russell and McIntire (1966), and confirmed by our field observations, is that the Palmetto Point beach-ridge plain extended further west during its initial development and, in fact, through a large part of its history. As illustrated in Figure 13, Palmetto Point is presently prograding southward, with its shape controlled by two longshore current systems. A current along the south coast moves from east to west and provides carbonate sediment to the Palmetto Point area. Reworking of this sediment by storm surges results in the generation of progressively younger beach ridges. A second current, along the west coast, moves from north to south and is primarily erosive in nature. Thus, the beach ridges are truncated on their western end. Field examination of the western

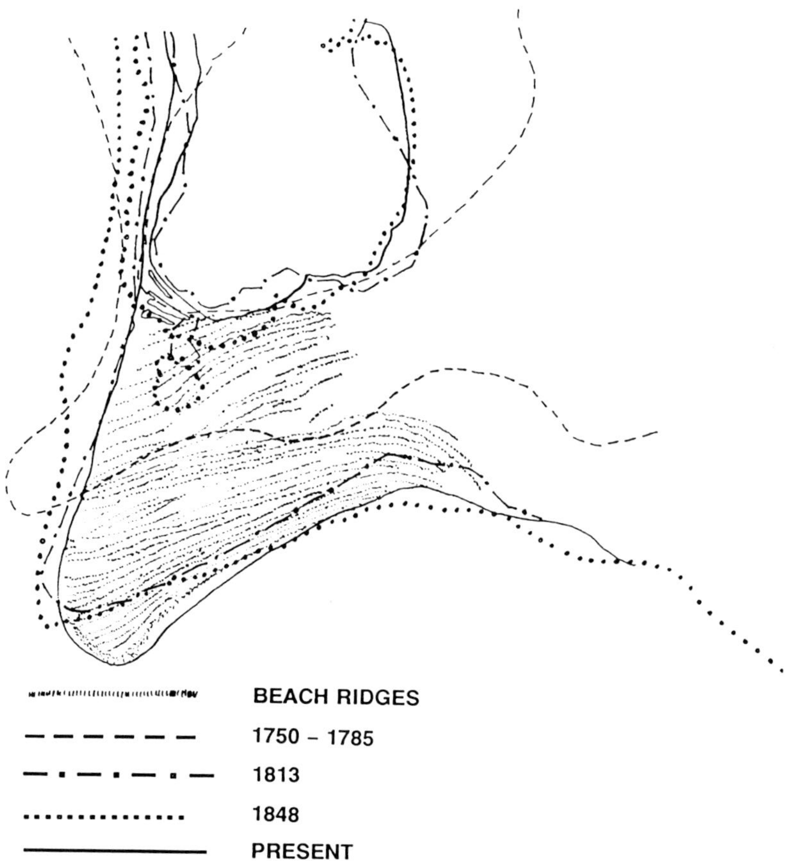

Figure 13. Palmetto Point's shorelines based on historic charts and maps.

shore of Palmetto Point shows the presence of an approximately 5 to 10-m scarp which cuts through each beach ridge, revealing the presence of cross-bedding as the internal sedimentary structure of the ridge. Thus, as Palmetto Point developed, it has prograded southward and, at the same time, been eroded along its western flank.

Examination of available historic maps of Barbuda gives some indication of progradational growth rates in the last 100 to 150 years. Figure 13 shows shoreline positions indicated on charts or maps from 1750 to 1785, 1813, 1848, and 1971 (based on air photos from 1958). The 1750 to 1785 chart is hand drawn and probably quite inaccurate. The 1813 and 1848 maps are British Admiralty charts and should be accurate. Disregarding the 1750 to 1785 chart, the southwestern portion of Palmetto Point has prograded about 500 m in 145 years. If we naively assume that this is the progradational rate for Palmetto Point, then the feature is only 700 to 800 years old.

In summary, the development of Codrington Lagoon and Palmetto Point is most probably related to a Holocene sea level rise and predates 3000 years ago. Accumulation of shells on the *Strombus* line began about 4000 years ago. If Russell and McIntire (1966) are correct, then beach-ridge development with progradation of ridges along the lithified Codrington Formation westernmost ridge was contemporaneous with or predated accumulation of shells on the *Strombus* line. Finally, based on data from historic maps, the progradational rate for Palmetto Point in the last 150 years is significantly higher than it was in the past 3000 to 4000 years.

Similar Features

Three other areas with *Strombus* shell accumulations have been observed on Barbuda. First, the Spanish Point "site" (BA2) has accumulations of shells with no associated cultural material. More research at BA2 may recover such cultural materials, which would confirm its designation as an archaeological site. Given its proximity (Figure 7) to the major habitation site at Sufferers (BA3), it is possible that BA2 may be associated with that Ceramic Age site. However, BA2 also may be an Archaic site. Second, Brasier and Donahue (1985) noted accumulations of *Strombus* along the south coast leading almost to Cocoa Point. As previously noted, they regard this as an eastward extension of the *Strombus* line, a point not agreed to by the archaeologist. Third, an accumulation of shell was observed inland from Pelican Bay, along a cemented eolian dune sequence on the east coast, and fairly close to the post-Saladoid Pelican Bay (BA10) site (Figure 7). As none of these three shell accumulations has been studied in any detail, whether they are cultural or natural deposits on paleoshorelines remains uncertain.

The final point concerns the possibility of intentional postdepositional alteration of *Strombus* accumulations. An historic period structure on the northwest corner of Gravenor Bay has been interpreted as a kiln, possibly used to burn shell to make lime (Watters, 1980b). Historic records reveal that Barbuda exported 556 hogsheads of lime in 1743 (Lowenthal and Clarke, 1977). One possible raw-material source for making reduced lime would have been the accumulations of *S. gigas* shells resulting from conchs exploited by prehistoric peoples. If the quantity shipped in 1743 is in any way typical of production in other years, then one must consider the possibility that major damage occurred to prehistoric conch-shell accumulations in the historic period. If this happened, then the limited piles of conch seen on Barbuda today may be no more than the remnants of much larger piles accumulated in the past. In the West Indies, the practice of burning conch for lime persisted into this century. Verrill (1915) commented on conch piles he observed on St. Croix in the Virgin Islands that were maintained to produce burnt lime.

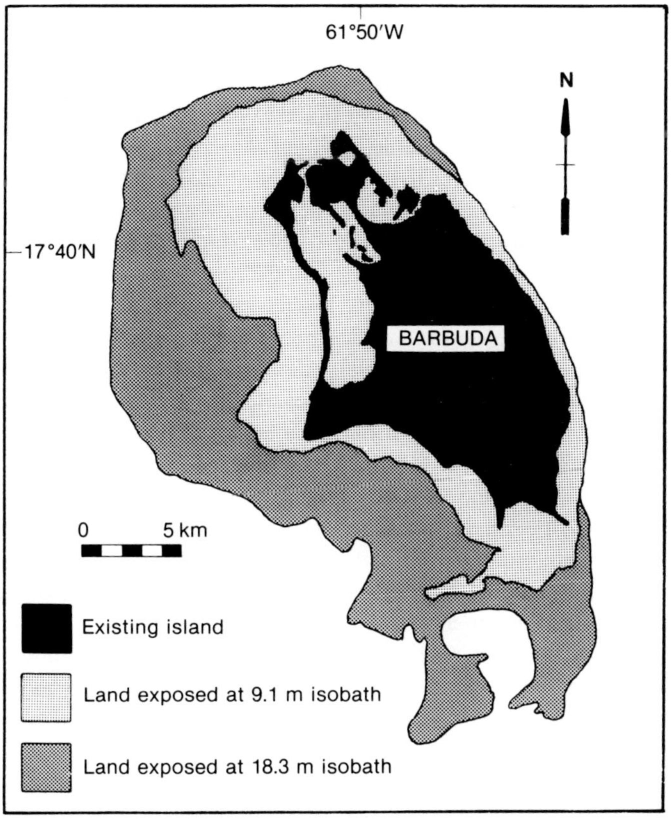

Figure 14. The expanses of presently submerged land around Barbuda that would be exposed if the sea level dropped to the 9.1- and 18.3-m isobaths. (After Defense Mapping Agency Hydrographic Center, 1920; fathoms converted to meters.)

SUBMERGED LANDS

Lower Holocene sea levels may have significance for human colonization of the West Indies, and inundated prehistoric sites may be present on the insular shelves surrounding the islands (Nicholson, 1976; Ruppé, 1980). Figure 1 shows that a drop in sea level to the 200-m isobath in the northern Lesser Antilles and eastern Greater Antilles would expose vast expanses of land presently submerged, connect many currently separated islands, and create such newly emergent islands as the large bank west of Saba; the previous existence of the vastly larger insular landmasses in the Lesser Antilles has been of interest to biogeographers, paleontologists, and other scientists (Morgan and Woods, 1986; Watters, 1989). The outer (Limestone) arc of the Lesser Antilles, in particular the Anguilla and Barbuda Banks, would have much greater amounts of emergent land than the Volcanic inner arc. However, it is not necessary to evoke a drastic 200-m drop in sea level to see significant changes in local landmass.

A much smaller drop in sea level results in an impressive amount of exposed land when examined on an individual island basis. Figure 14 shows the existing island of Barbuda and the expanses of land that would be exposed if the sea level dropped by 9.1 m (5 fathoms; 30 ft) and 18.3 m (10 fathoms; 60 ft). Barbuda sits on the northeastern edge of the Barbuda Bank, and to the east, the shelf edge slopes steeply to ocean depths exceeding 2000 m (Detjen, 1978), a situation that accounts for the small amount of submerged land to the east of the island where the two isobaths coincide (generally corresponding to today's reefs). Shallow bottoms predominate to the island's south, west, and, to a lesser degree, north. Russell and McIntire (1966) estimate the area within the 10-fathom (18.3 m) line as about 2.75 times that of the land and lagoon existing today. Barbuda would connect with Antigua if the sea level dropped by about 35 m, while other islands in the northern Lesser Antilles would be joined by an even smaller drop in sea level, such as Anguilla (Figure 1) connecting to St. Martin at about 20 m (Watters and Rouse, 1989: Figure 2).

The issue of lower Holocene sea levels probably has little direct bearing on initial Ceramic Age occupation of the Lesser Antilles because it occurred so recently (about 300 B.C.), at a time when the sea level was very close to its present position. However, a lower sea level could have major importance for the earlier colonization of the Lesser Antilles by Archaic Age peoples, which took place by 3000 B.C., and for the even earlier Lithic Age colonization (to 5000 B.C., but only documented in the Greater Antilles). Indeed, insular shelves around the islands could contain the earliest evidence of human colonization of the West Indies, *if* this took place when the shelves were exposed by a lower sea level, *if* the inhabitants occupied paleoshorelines that subsequently were submerged, and *if* the sites survived the impact of transgressing seas (Watters and Rouse, 1989). Although most Caribbean archaeologists are aware of the possible existence of inundated sites, virtually no field research has been done to detect, sample, or excavate such sites. One exception is Vega's (1981) report on a submerged site near Puerto Rico.

In the Lesser Antilles, the issue of investigating submerged sites is complicated by the scarcity of relevant paleoecological data. One paper on the paleoecology of the entire Caribbean region readily points out the limitations in the Lesser Antilles. Carbone (1980: Figure 2) uses an impressive array of data sets—pollen profiles, soil studies, deep-sea cores, cave deposits, and coastline changes—in his model of Holocene environments and climates in the Caribbean region, yet he does not incorporate a single data set for the Lesser Antilles islands. This also calls to mind Donnelly's (1989) characterization of the eastern Caribbean as a geologic "problem area". Tectonic activity on the eastern margin of the Caribbean Plate has affected and continues to affect the islands and banks resting on the Lesser Antilles Ridge (Uchupi, 1975). However, the responses by islands and banks do not appear to be uniform along the entire ridge; in particular, there are differences between the Limestone and Volcanic arcs. Moreover, there still is no sea level curve developed for the Lesser Antilles archipelago. All of this conspires to place any discussion of potentially submerged archaeological sites on, at best, a hypothetical level.

There are, however, some interesting Caribbean sea level data that may be relevant, especially in view of Brasier and Donahue's (1985) contention that Barbuda has long been a stable landmass, with changes in its shoreline position primarily being controlled by sea level fluctuations. Fairbanks (1989: Figure 2) recently presented a 17,000-year glacio-eustatic sea level curve, corrected for uplift (34 cm per 1000 years) and apparent age of sea water (400 years subtracted from measured radiocarbon ages), derived from late-Pleistocene and Holocene coral reefs (*Acropora palmata*) drilled off the south coast of Barbados, an island some 500 km south-southeast of Barbuda. The most recent part of this curve, which incorporates *A. palmata* age-depth data from four other Caribbean islands, is of particular interest to archaeologists because it shows that at 5000 years ago, a time roughly coeval with the postulated 3000-B.C. Archaic occupation of the West Indies, the rise curve flattens out at about −7 m. Prior to 5000 years ago, the sea level was much lower (e.g., −25 m at 8000) and the rate of rise was accelerated.

According to Fairbanks' curve, the sea level would have been about −5 m at 3700 to 3800 years ago, the Archaic Age occupation of the River site (BA4). This would place Barbuda's 3800-B.P. south coast some undetermined distance south of the BA4 site. Whereas today the south coast is immediately adjacent to the site, at 3800 B.P. the coast would have been further south. The River site would have been located some distance inland from the south coast. The relationship of the River site to the west coast of Barbuda at 3800 years ago is problematical because of questions about whether Palmetto Point peninsula existed at 3800 B.P. and, if it did exist, what its position was.

Another way to look at Fairbanks' curve is in relation to the two isobaths shown in Figure 14. The 9.1-m isobath would correspond to about 5500 years ago, a time period that is an acceptable "ballpark" figure for Archaic occupation of the West Indies. On Fairbanks' curve, the 18.3-m isobath corresponds to 7500 years ago, a time generally regarded as predating human colonization of these islands. Nevertheless, from a hypothetical viewpoint, we suggest that the large embayment on the 18.3-m isobath, due south of the present island of Barbuda (Figure 14), would be a logical place to begin looking for inundated sites. The aforementioned accelerated rate of sea level rise (in this case, some 9 m in about 2000 years) is evident between the two dates postulated for these two isobaths.

SUMMARY

The geoarchaeological research discussed in this article focuses on past and present coastal and nearshore areas of Barbuda. Depositional processes that began around the Pliocene core of the island, creating the classic offlapping sequence, continue to the present. Change in sea level position rather than land movement was the primary factor controlling deposition. Holocene accretionary deposits in the form of beach ridges are evident along the southwest coast at Palmetto Point and, to a lesser degree, on the west coast near Cedar Tree Point.

Although significant occupation of Barbuda by prehistoric peoples occurred in the latter half of the first millennium A.D., during the post-Saladoid period, a somewhat earlier Saladoid colonization is revealed at one site, Sufferers (BA3), the only site that to date has yielded Saladoid ceramic remains in any number. A coastal orientation, encompassing a variety of marine, littoral, and terrestrial environments, is evident at the three major habitation sites: Sufferers (Saladoid, possibly with a post-Saladoid component), Indiantown Trail (post-Saladoid), and Pelican Bay (post-Saladoid). Analysis of fauna from Indiantown Trail (BA1) shows extensive use of marine vertebrates, notably fish, but including sea mammals such as the manatee (Watters et al., 1984), as well as numerous shellfish, chitons, and corals. The abundance of marine vertebrates at Indiantown Trail stands in sharp contrast to their relative scarcity at Trant's, a major habitation site on the volcanic island of Montserrat, where a terrestrial orientation seems to predominate (Steadman et al., 1984). Variation in the distribution of marine and estuarine habitats, reflected in such factors as coral reef distribution between "reef-rich" Barbuda and "reef-poor" Montserrat (Watters and Rouse, 1989: Figure 1), clearly relates to the diversity and quantity of marine organisms recovered from prehistoric sites.

The earlier occupation of Barbuda by Archaic Age peoples, at least by 1800 B.C., has significant implications for a number of issues involving sea level fluctuation and changing shorelines. If Fairbanks' sea level record from Barbados can be applied, even approximately, to Barbuda, it suggests that currently submerged nearshore bottoms, to a water depth of as much as 5 m, were exposed as dry land at 1800 B.C., when Archaic peoples inhabited the River site. It also suggests that the River site was not immediately adjacent to the shoreline, as it is now, but instead was positioned somewhat inland from the 1800-B.C. south coast paleoshoreline. It further suggests that even older Archaic sites, if they exist at all, would most likely be found on extensive shallow bottoms to Barbuda's south, west, and, to a lesser degree, north, if the sites survived later transgressing seas.

Equally significant for issues of sea level fluctuation and change in paleoshorelines are the radiocarbon dates derived from shell samples from the *Strombus* line. The relationship between the shell dates and progradation of the Palmetto Point peninsula is complicated by probable reworking of sediment deposits and possible redeposition of shells. However, at 3800 B.P., when Archaic Age peoples occupied the River site, Palmetto Point peninsula would have been an attenuated version of what is seen today, and it *may* not have existed at all. In either case, the onset of progradation of Palmetto Point is significant for aspects of the geologic history of Barbuda, such as the time of enclosure of the Codrington Lagoon by the sand spit to its west (Figure 3). Progradation of Palmetto Point also is significant for the island's prehistory in terms of the paleoshorelines seen by Archaic Age occupants and, to a lesser degree, by the later Ceramic Age inhabitants.

Radiocarbon dates derived from shell and, in one instance, from charcoal (at BA1) play a key role in the geoarchaeological research discussed in this paper. Figure 15 summarizes the dates for the *Strombus* line, south coast, and two prehistoric sites and distinguishes the materials dated by different laboratories. It is clear that the majority of the shell

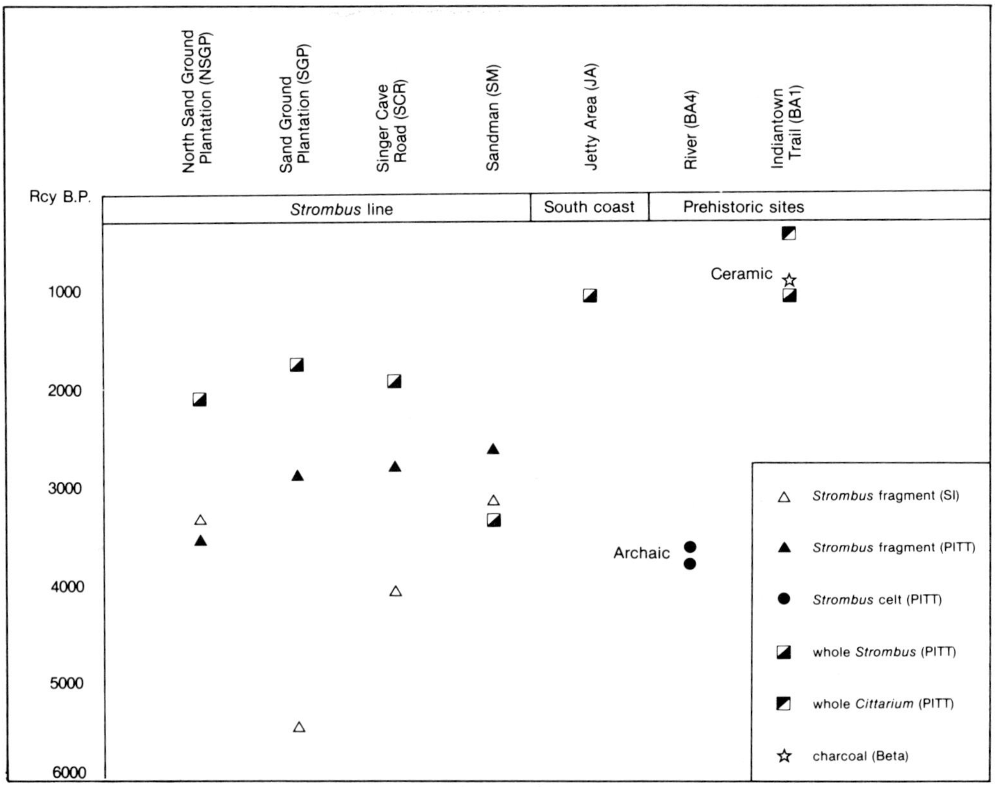

Figure 15. Summary chart of radiocarbon-dated specimens from the *Strombus* line, south coast, and prehistoric sites of Barbuda.

samples from the *Strombus* line are coeval or younger than two shell celts from the Archaic site at River (BA4). Figure 15 shows that fragmented shells from the *Strombus* line tend to have dates older than those from whole shells (with the Sandman locale as the exception). The punched-hole shell from the Jetty Area on the south coast dates to about the same time as the Ceramic site at Indiantown Trail (BA1).

The *Strombus* line lithified beach ridge with its accumulated shells has generated a disagreement among the authors, the basis of which is the process (cultural or natural) by which the shells were deposited on the ridge.

Radiocarbon dates on 12 shells from four locations along the *Strombus* line confirm that, in most instances, the mollusks died at times coeval with or more recent than the Archaic Age occupation of the River site, a point that can be justifiably construed as supporting the geologist's contention of cultural deposition of the shells. However, the radiocarbon dates cannot address the issues raised by the archaeologist concerning the lack of definite cultural modification of any *Strombus* line shells and the absence of artifacts found in association with those shells.

Although the disagreement is not resolved at this time and neither hypotheses is refuted, this article has alluded to some parameters for future research that could help resolve the issues: (1) examination of older Codrington Formation beach ridges for evidence of shell accumulations, (2) verification of the single-genus taxonomic uniformity of *Strombus* line shells, (3) determination of cultural or natural breakage patterns for the shells, (4) resolution of the issue of whether the punched-hole extraction technique is present in both Archaic Age and Ceramic Age sites, (5) intensive survey of the lithified ridge for associated artifacts or sites, (6) additional radiocarbon dates from fragmented and whole shells of the *Strombus* line, Jetty Area, and River site, (7) test excavation of the preceramic River site to characterize and document Archaic Age artifact assemblages on Barbuda, (8) study of lithified ridges having similar shell accumulations found elsewhere on Barbuda, (9) survey of Palmetto Point peninsula for prehistoric sites, and (10) testing of nearshore bottoms south of Barbuda for evidence of submerged Archaic Age sites.

NOTE ADDED IN PROOF

Since this manuscript was submitted, the AAS has recorded a second Archaic site on Barbuda. The aceramic site is located south of the village of Codrington on the east side of Codrington Lagoon. If this site really is Archaic Age, it means both Archaic sites are located on the western Marginal plain, far removed from the Ceramic Age sites near the Highlands.

ACKNOWLEDGMENTS

This paper results from various phases of research conducted during the past decade and supported by funds from a number of sources, including a Fulbright-Hays Doctoral Dissertation Research Abroad Fellowship (DHEW: OE) and an Andrew Mellon Predoctoral Fellowship at the University of Pittsburgh, a Marine Policy and Ocean Management Postdoctoral Fellowship at the Woods Hole Oceanographic Institution, and the Anne L. and George H. Clapp Charitable and Education Trust, the M. Graham Netting Research Fund, and the Edward O'Neil Research Fund of the Carnegie Museum of Natural History. Geologist Martin D. Brasier of the University of Hull accompanied Watters and Donahue on the

1984 research trip to Barbuda and provided valuable insight and spirited discussion about the *Strombus* line issues. James B. Richardson, III, brought the Fairbanks article to our attention. James B. Petersen provided insightful comments. Nancy Perkins, Gail Richards, and Melinda Carmichael of the Exhibits Division of The Carnegie Museum of Natural History produced many of the graphics for this paper. The authors wish to thank the people of Barbuda for their assistance during various phases of the project, and especially acknowledge the contribution of Morris (Tomac) Nedd, a truly superb field guide, and Desmond Nicholson of the AAS. The authors express their gratitude to Lucille Lewis Johnson, volume editor, for her willingness to grant them extended time to conduct additional radiocarbon dating and revise the manuscript, and to the reviewers.

REFERENCES

Augustinus, P.G.E.F., Ed. (1989) Cheniers and Chenier Plains: a general introduction. *Mar. Geol.* 90: 219–229.

Beard, J.S. (1949) The natural vegetation of the Windward & Leeward Islands. In *Oxford Forestry Memoirs* 21. Clarendon Press, Oxford.

Blackwelder, B.W. (1981) Late Cenozoic marine deposition in the United States Atlantic Coastal Plain related to tectonism and global climate. *Paleogrogr. Paleoclimatol. Paleoecol.* 34: 87–114.

Brasier, M. and Donahue, J. (1985) Barbuda—an emerging reef and lagoon complex on the edge of the Lesser Antilles island arc. *J. Geol. Soc. London.* 142: 1101–1117.

Brasier, M.D. and Mather, J.D. (1975) The stratigraphy of Barbuda, West Indies. *Geol. Mag.* 112: 271–282.

Carbone, V.A. (1980) The paleoecology of the Caribbean area. *Fla. Anthropol.* 33(3): 99–119.

Chanlatte Baik, L.A. (1981) *La Hueca y Sorcé (Vieques, Puerto Rico): Primera Migraciones Agroalfareras Antillanas.* Privately printed, Santo Domingo, Dominican Republic.

Dacal Moure, R. (1978) *Artefactos de Concha en las Comunidades Aborígenes Cubanas.* Museo antropológico Montané publicación 5.

Davis, D.D. (1974) Some notes concerning the Archaic occupation of Antigua. In *Proceedings of the Fifth International Congress for the Study of Pre-Columbian Cultures of the Lesser Antilles,* pp. 65–71.

Defense Mapping Agency Hydrographic Center. (1920) Barbuda. Chart 25605, 12th Ed. (corrected to 1974), scale 1:73, 745 (based on British Admiralty Chart 1997, surveyed in 1848).

Detjen, P. (1978) The Antigua-Barbuda insular platform: introduction. In *W-42 Cruise Report,* p. 54. Gaines, A.J., Ed. Sea Education Association, Woods Hole, MA.

Donahue, J., Watters, D.R., and Millspaugh, S. (1990) Thin section petrography of northern Lesser Antilles ceramics. *Geoarchaeology.* 5: 229–254.

Donnelly, T.W. (1989) Geologic history of the Caribbean and Central America. In *Geology of North America—An Overview,* pp. 299–321. Bally, A.W. and Palmer, A.R., Eds. *The Geology of North America,* Vol. A. Geological Society of America, Boulder, CO.

Fairbanks, R.G. (1989) A 17,000-year glacio-eustatic sea level record: influence of glacial melting rates on the Younger Dryas event and deep-ocean circulation. *Nature.* 342: 637–642.

Goodwin, R.C. (1978) The Lesser Antillean Archaic: new data from St. Kitts. *J. Virgin Islands Archaeol. Soc.* 5: 6–16.

Harris, D.R. (1965) *Plants, Animals and Man in the Outer Leeward Islands, West Indies: An ecological Study of Antigua, Barbuda and Anguilla.* Univ. Calif. Publ. in Geogr. 18.

Hays, J.D., Imbrie, J., and Shackleton, N.J. (1976) Variation of the earth's orbit: pacemaker of the Ice Ages. *Science.* 194: 1121–1132.

Keegan, W.F. (1984) Pattern and process in *Strombus gigas* tool replication. *J. New World Archaeol.* 4(2): 15–24.

Lowenthal, D. and Clarke, C.G. (1977) Slave-breeding in Barbuda: the past of a Negro myth. *Ann. N.Y. Acad. Sci.* 292: 510–525.

Lundberg, E.R. (1985) Interpreting the cultural associations of aceramic deposits in the Virgin Islands. *J. Field Archaeol.* 12: 201–212.

Martin-Kaye, P.H.A. (1969) A Summary of the geology of the Lesser Antilles. *Overseas Geol. Min. Resour.* 10(2): 172–206.

Mather, J.D. (1971) *A Survey of the Groundwater Resources of Barbuda.* Institute of Geological Sciences, Hydrogeological Department, London.

Mercer, J.H. (1978) Glacial development and temperature trends in the Antarctic and South America. In *Antarctic Glacial History and World Paleoenvironments,* pp. 77–93. VanZindereen Bakker, L.M., Ed. A.A. Balkema, Cape Town, South Africa.

Morgan, G.S. and Woods, C.A. (1986) Extinction and the zoogeography of West Indian land mammals. *Biol. J. Linnean Soc.* 28: 167–203.

Nicholson, D.V. (1976) The importance of sea levels—Caribbean archaeology. *J. Virgin Islands Archaeol. Soc.* 3: 19–23.

Nodine B. (1989) Preceramic site survey finishes. *Newsl. Hist. Archaeol. Soc. Antigua Barbuda* 24: 2.

Nummedal D., Pilkey, O.H., and Howard, J.D., Eds. (1987) *Sea-Level Fluctuation and Coastal Evolution.* Spec. Publ. 41, Society of Economic Paleontologists and Mineralogists, Tulsa, OK.

Rodríguez, M. (1989) The zoned incised crosshatch (ZIC) ware of early Precolumbian Ceramic Age sites in Puerto Rico and Vieques Island. In *Early Ceramic Population Lifeways and Adaptive Strategies in the Caribbean,* pp. 249–266. Siegel, P.E., Ed. BAR Int. Ser. 506.

Rouse, I. (1986) *Migrations in Prehistory: Inferring Population Movements from Cultural Remains.* Yale University Press, New Haven.

Rouse, I. (1989) Peoples and cultures of the Saladoid Frontier in the Greater Antilles. In *Early Ceramic Population Lifeways and Adaptive Strategies in the Carbibean,* pp. 383–403. Siegel, P.E., Ed. BAR Int. Ser. 506.

Rouse, I. and Allaire, L. (1978) Caribbean. In *Chronologies in New World Archaeology,* pp. 431–481. Taylor, R.W. and Meighan, C.M., Eds. Academic Press, New York.

Ruppé, R.J. (1980) Sea level rise and Caribbean prehistory. In *Proc. 8th Int. Congr. Study of Pre-Columbian Cultures of the Lesser Antilles,* pp. 331–337. Lewenstein, S.M., Ed. Ariz. St. Univ. Anthropol. Res. Pap. 22, Arizona State University, Tempe.

Russell, R.J. and McIntire, W.G. (1966) *Barbuda Reconnaissance.* La. St. Univ. Coast. Stud. Ser. 16, Louisiana University Press, Baton Rouge.

Siegel, P.E., Ed. (1989) *Early Ceramic Population Lifeways and Adaptive Strategies in the Caribbean.* BAR Int. Ser. 506.

Stapor, F.W. (1975) Holocene beach ridge plain development, northwest Florida. *Z. Geomorphol.* 22: 116–144.

Steadman, D.W., Watters, D., Reitz, E., and Pregill, G. (1984) Vertebrates from archaeological sites on Montserrat, West Indies. *Ann. Carnegie Mus.* 53: 1–29.

Stuiver, M. and Becker, B. (1986) High-precision decadal calibration of the radiocarbon time scale, AD 1950–2500 BC. *Radiocarbon.* 28: 863–910.

Stuiver, M. and Borns, H.W., Jr. (1975) Late Quaternary marine invasion in Maine: its chronology and associated crustal movement. *Geol. Soc. Am. Bull.* 86: 99–104.

Uchupi, E. (1975) Physiography of the Gulf of Mexico and Caribbean Sea. In *The Ocean Basins and Margins,* Vol. 3, *The Gulf of Mexico and the Caribbean,* pp. 1–64. Nairn, A.E.M. and Stehli, F.G., Eds. Plenum Press, New York.

Vail, P.R., Mitchum, R.M., Jr., and Thompson S. (1977) *Seismic Stratigraphy and Global Changes of Sea Level, Part 4.* Am. Assoc. Pet. Geol. Mem. 26, American Association of Petroleum Geologists, Tulsa, OK.

Vega, J.F. (1981) Excavation of a Submerged sub-Taino Site in Puerto Rico. M.A. thesis, Florida Atlantic University, Boca Raton.

Veloz Maggiolo, M. and Vega, B. (1982) The Antillean Preceramic: a new approximation. *J. New World Archaeol.* 5(2): 33–34.

Vernon, K.C. and Lang, D.M. (1966) *Soil and Land-Use Survey No. 19B: Barbuda.* Regional Research Centre, University of the West Indies, Trinidad.

Verrill, A.H. (1915) *Isles of Spice & Palm*. D. Appleton and Company, New York.

Watters, D.R. (1980a) Transect Surveying and Prehistoric Site Locations on Barbuda and Montserrat, Leeward Islands, West Indies. Ph.D. thesis, University of Pittsburgh, Pittsburgh.

Watters, D.R. (1980b) Observations on the historic sites and archaeology of Barbuda. *J. Archaeol. Anthropol.* 3: 125–154.

Watters, D.R. (1989) Archaeological implications for Lesser Antilles biogeography: the small island perspective. In *Biogeography of the West Indies: Past, Present, and Future,* pp. 153–166. Woods, C.A., Ed. Sandhill Crane Press, Gainesville, FL.

Watters, D.R. and Nicholson, D.V. (1982) Highland House, Barbuda: an 18th century retreat. *Fla. Anthropol.* 35: 223–242.

Watters, D.R., Reitz, E., Steadman, D., and Pregill, G. (1984) Vertebrates from archaeological sites on Barbuda, West Indies. *Ann. Carnegie Mus.* 53: 383–412.

Watters, D.R. and Rouse, I. (1989) Environmental diversity and maritime adaptations in the Caribbean area. In *Early Ceramic Population Lifeways and Adaptive Strategies in the Caribbean,* pp. 129–144. Siegel, P.E., Ed. BAR Int. Ser. 506.

Watters, D.R. and Scaglion, R. (1980) Utility of a transect survey technique in Caribbean prehistoric studies: applications on Barbuda and Montserrat. In *Proc. 8th Int. Congr. Study Pre-Columbian Cultures of the Lesser Antilles,* pp. 338–347. Lewenstein, S.M., Ed. Ariz. St. Univ. Anthropol. Res. Pap. 22, Arizona State University, Tempe.

Wigley, P. (1977). Facies analysis of Holocene carbonate sediments and Tertiary(?)-Pleistocene limestones on and around Barbuda, West Indies. In *Reefs and related carbonates—Ecology and Sedimentology,* pp. 299–312. Frost, S.H., Weiss, M.P., and Saunders, J.B., Eds. Stud. Geol. 4. American Association of Petroleum Geologists, Tulsa, OK.

Wilson, S.M. (1989) The prehistoric settlement pattern of Nevis, West Indies. *J. Field Archaeol.* 16: 427–450.

Winslow, M.A. and Johnson, L.L. (1989) Prehistoric human settlement patterns in a tectonically unstable environment: outer Shumagin Islands, southwestern Alaska. *Geoarchaeology.* 4: 297–318.

Mainland

Lucille Lewis Johnson
Department of Anthropology
Vassar College
Poughkeepsie, New York

INTRODUCTION

Both of the chapters in this section are primarily geological. Owen Mason reverses the usual relationship between geology and archaeology by using the dates and location of archaeological sites to date geological beach ridges, while Suguio, Martin and Flexor use the location of sites and C-13 content of shells in middens to confirm and extend geologial information about sea level change in Brazil.

On the other hand, these two papers differ in that the sites occupied by the ancient Alaskans and Brazilians had very different relationships to the shore even though both peoples were marine oriented. In Alaska, the ancient people lived on the closest beach ridge to shore which could lift their sites above the reach of the waves and the permafrost perched water table. Thus, this is an ideal situation for geologists to use archaeological evidence for dating geological features. Owen Mason points out that, in areas where people camped inland from the coast, on lagoon or stream shores, the relationship of the sites to shoreline features is more indirect and difficult to understand.

Suguio, Martin and Flexor accept the challenge, looking at lagoon margin settlements to see what can be accomplished. Sambaquis are prehistoric shell middens which are situated around bays, paleobays, and estuaries on the south coast of Brazil. The shell middens are dated and the sea level during their period of occupation is estimated by looking at the distance of the sambaqui from present sea or lagoon margin, by evaluating the nature of the substrate of the site, by measuring the elevation of the substrate above present sea level, by examining the species of mollusk present in the midden and its environmental constraints, and by establishing the dc-13(PDB) content of the shells. This final measure is particularly interesting as it establishes the salinity of the water from which the shells came and, inferentially, the salinity of the water in the vicinity of the site when it was occupied. As these sambaquis represented tons of shells, it is assumed that they are located close to the beds from which the shellfish were gathered.

Mason and Suguio et al. derive quite different sea level models as can be seen in the following comparative chart:

MASON	SUGUIO ET AL.
	7000–5100 BP submergence
	5100–3800 BP emergence
4000–3300 progradation	3800–3600 BP submergence
3300–2000 transgressive dunes	3600–2700 BP emergence
	2700–2500 BP submergence
2000–1200 progradation	2500–present emergence
1200–600 transgressive dunes	

Mason states that, in Alaska, prograding coastlines correlate with peat growth and tree line expansion, suggesting rising sea levels. Thus, his periods of progradation should indicate submergence in Suguio et al.'s scheme. Though this is partially true, Suguio et al. see more detail and variability in early post-glacial sea level change while Mason monitors more variation later. These differences may be due both to the longer settlement history in Brazil and to the sensitivity of Alaska beach ridge chronology. These models and their implications for understanding the local conditions to which people must adjust are paradigmatic of the need to derive sea level curves for each area independently.

Chapter
3

A Geoarchaeological Methodology for Studying Prograding Coastal Sequences: Beach-Ridge Geomorphology in Kotzebue Sound, Alaska

Owen K. Mason
Alaska Quaternary Center
University of Alaska, Fairbanks
Fairbanks, Alaska

ABSTRACT

Beach-ridge systems develop on tectonically stable shorelines supplied with a surplus of sediment in excess of tidal and eustatic fluctuations. Despite numerous studies of modern facies, the Holocene chronostratigraphy of such prograding deposits is rarely documented. However, depositional units within beach-ridge systems may be readily defined and cross-correlated using an interdisciplinary approach combining aerial photo interpretation with evidence from archaeology, sedimentology, granulometry, pedology, and landform evolution. At two Kotzebue Sound localities, Cape Espenberg and Choris Peninsula, archaeological and geological C-14 assays (n = 41) provide upper-age estimates on ridge formation and allow the delineation of chronostratigraphic units. Four distinct depositional units are defined at Cape Espenberg, reflecting an alteration between (1) progradation marked by low beach ridges separated by wide swales and (2) vertical dune growth atop beach ridges. Rapid progradation occurred from 4000 to 3300 B.P. and from 2000 to 1200 B.P., while transgressive dunes built 3300 to 2000 and from 1200 to the present. A similar pattern is found on the gravel ridges of Choris Peninsula. Differences in storm frequency and wind intensity explain variations in the sedimentation regime.

INTRODUCTION

Clastic shorelines respond rapidly and often catastrophically to shifts in wave energy and wind direction produced by atmospheric disturbances (Carter, 1988). Progradation, the net addition of shoreline, results if a surplus of sediment is coupled with a negligible rate of relative sea level rise (Curray, 1964). As a consequence, shore parallel lines of coarser deposits or beach ridges may accumulate horizontally, forming wide coastal plains over the course of millennia. Beach-ridge plains may provide proxy records of paleowind direction (Moore and Giddings, 1961; Curray et al., 1969), the intensity of the El Niño phenomenon (Sandweiss, 1986), alluvial history (DePratter and Howard, 1977), and sea level trends (Searle and Woods, 1986). Inferences derived from beach-ridge plains are applicable to paleoclimatic and paleoshoreline reconstructions on all continents, from Mozambique to Norway and Brazil to Australia (Bird and Schwartz, 1985).

Despite the volume of studies on the sedimentology and facies relations of spits and beach ridges (Reinson, 1984), comparatively few studies concern their Holocene history or cross-correlations. When geologists study beach sedimentology, they are usually concerned with ''hard-rock'' geological models. In the last 20 years, the emerging interdisciplinary synthesis of quaternary studies has focused more interest on recent, unlithified sediments to derive paleoclimatic and paleoenvironmental reconstructions. In this sudy, I present a methodology for delineating and interpreting chronostratigraphic units at two beach-ridge complexes in Kotzebue Sound. My methods should assist other researchers in describing similar phenomena from barrier island and beach-ridge plains.

The geoarchaeological approach is interdisciplinary (combining archaeology, botany, geology, and geomorphology), but primarily uses dated archaeological remains atop prograding coastal sediments to provide a chronology of coastal and cultural evolution. The approach is limited to the examination of surficial deposits and contrasts with the more standard geological practice of deep coring (Curray et al., 1969; Kraft et al., 1985). Despite their chronological limitation to geomorphic settings which postdate the stabilization of sea level, surficial investigations are expedient under conditions subject to logistical constraints, as in Alaska and other cold regions, or in sparsely inhabited, roadless locations dependent on costly air-based logistics.

Use of the geoarchaeological method can overcome one of the major difficulties in historical studies of coastal deposits: the scarcity of datable materials. For example, within the numerous beach ridges on the Outer Banks of North Carolina, datable carbonaceous material is often rare, especially in older ridges where most shell may have dissolved (Fisher, 1967). The use of archaeological radiocarbon assays allows the dating of individual depositional units and, ultimately, the cross-correlation of several complexes (Mason, 1990; Mason and Ludwig, 1990). I provide an example of this methodology by cross-correlating beach-ridge histories at Cape Espenberg and Choris Peninsula. Similar geomorphic settings are widespread along ''trailing edge'' continental margins on the Gulf and Atlantic coasts of the U.S. and Mexico (Curray et al., 1969), and even on the more active margins of the Pacific Northwest (Bird and Schwartz, 1985).

RESEARCH STRATEGIES, APPLICABILITY, AND PROBLEM ORIENTATION

Prograding beach-ridge plains form along micro- and mesotidal coasts subject to stable or falling sea levels which are supplied with abundant sediment (Curray, 1964; Kraft and

Chrzastowski, 1985). A beach ridge is a clastic storm-related facies, deposited beyond the influence of subsequent lower sea levels prevailing during fairweather conditions (Carter, 1988). Sequences or sets of beach ridges provide proxy climatic records in several ways: (1) variations in the spacing of beach ridges can provide an estimate of storm recurrence intervals (Tanner, 1988), (2) the altitude of the beach facies in dunes or that of gravel ridges may be used to estimate past storm or eustatic sea levels (Roep, 1986; Sandweiss, 1986), and (3) shifts in orientation of ridge sets may be linked to changes in wind direction (Moore and Giddings, 1961; Curray et al., 1969).

Beach ridges, sometimes covered by coastal dunes, are widespread in northern Alaska along the tectonically stable Chukchi Sea coast and contain numerous records of human occupation (Giddings and Anderson, 1986). In the years before radiocarbon dating, Arctic archaeologists discovered that beach-ridge position was useful as a survey stratagem to find contemporary-aged settlements and as an estimation of relative age (Collins, 1937; Giddings, 1967). However, the paleoclimatic or paleosea-level implications of beach ridges were not fully realized by these archaeological pioneers. Although Giddings and Moore (1961) did hypothesize that shifts in the position of the Polar Front controlled the sedimentation regime at Cape Krusenstern, they did not cross-correlate deposits from more than one locality. By cross-correlating deposits from more than one complex, it is possible to establish the regional nature of paleoclimatic changes. The chronological delineation of depositional units and their cross-correlation is tied to the archaeological record; therefore, a consideration of the context of the archaeological database is a necessary preface to beach-ridge studies.

THE BASIC POSTULATE OF BEACH-RIDGE ARCHAEOLOGY: ETHNOGRAPHIC ANALOGY

The basic postulate of the geoarchaeological beach-ridge method is that maritime hunters located their settlements on beaches adjacent to the open coast, preferentially on prominent headlands or barrier islands. The nineteenth century ethnographer Edward W. Nelson (1891) first linked abandoned Inupiat and Yup'ik (''Eskimo'') settlements to former shoreline positions. In the 1930s, Henry B. Collins (1937) used settlement location in relation to shoreline position to construct a relative cultural chronology on St. Lawrence Island. In the 1950s, the beach-ridge paradigm guided the survey efforts of J. Louis Giddings and allowed him to reconstruct nearly 4000 years of cultural history in northwest Alaska (Giddings, 1966).

However, the situation on other prograding coastlines may not be similar because resources and, hence, settlement preferences may differ. For example, these differences have hindered the progress of beach-ridge archaeology on the Atlantic seaboard. Along the coastlines of northwest Florida and the Carolinas, the resource base of sheltered lagoons is greater than that of the exposed beach. As a consequence, aboriginal groups settled on the lagoons to gather supplies of shellfish and shrimp (Larson, 1980). This settlement pattern does not facilitate a beach-ridge geoarchaeological approach because the lagoonal shoreline remains comparatively static as the beach-ridge plain builds seaward, away from the lagoon. In situations where successive human cultures settled on the static lagoonal anchor of the beach-ridge plain, archaeological C-14 dates may yield an inaccurate reflection of coastal evoluation. This was not considered by Florida geologists who used archaeological C-14 dates on the lagoon shore to establish the onset of a coastal depositional sequences (Stapor, 1975). If the beach-ridge method is to be used in applying archaeological data to geologic problems, then resource availability and its relationship to coastal processes must be examined (Meehan, 1982).

STUDIES OF ALASKA BEACH RIDGES

In the region from Point Hope to Cape Nome, the breadth and duration of research varies widely from one beach-ridge complex to another (Figure 1). While Giddings spent four seasons (1958 to 1961) at Cape Krusenstern undertaking intensive surveys and excavations and collecting numerous C-14 samples, he visited Choris, Cape Espenberg, Wales, and Sisualik only briefly and obtained only a few C-14 dates from these latter areas. Even so, the radiocarbon chronology at Krusenstern (n = 33) is heavily biased toward enigmatic archaeological materials, leaving much of the depositional history of the complex radiometrically undated (Mason and Ludwig, 1990). Although Pt. Hope witnessed three different intensive archaeological excavations, in 1939–1941 (Larsen and Rainey, 1948), the late 1960s (Hosley, 1972), and in the 1970s (Shinkwin, 1977), few C-14 dates pertain to the age of ridge formation. Bockstoce (1979) excavated a substantial number of houses at Safety Sound, east of Cape Nome, but his limited date list (n = 9) provides only an outline of its depositional history (cf., Mason, 1990).

Archaeological inventorying in the later 1980s by the National Park Service (NPS) resulted in a resurgence of beach-ridge archaeology in Kotzebue Sound (Schaaf, 1988; Harritt, 1989, 1990). My own research at Espenberg and Choris was conducted in concert with the surveys and excavations of the NPS.

GEOLOGIC SETTING OF BEACH RIDGES IN NORTHWEST ALASKA

In northwest Alaska, beach-ridge plains occur at seven realignments in coastal orientation between the Bering Strait and Point Hope (Figure 1). The geomorphic configuration of a particular beach-ridge complex is determined by the grain size and lithology available offshore and updrift. The southeast Chukchi Sea coast is divided into two lithic regions: one of sand, the other of gravel. Sandy beaches are limited to three stretches: (1) along the north Seward Peninsula coast, (2) at Sisualik near the mouth of the Noatak River, and (3) along the eastern shore of Kotzebue Sound below the eroding Baldwin Peninsula. Gravel beaches predominate along the remainder of the rocky coast from southern Kotzebue Sound to Point Hope. I studied one beach-ridge complex from each region: Cape Espenberg, a sand spit, and Choris Peninsula, a series of gravel ridges (Mason, 1990).

The construction of a beach-ridge plain requires several preconditions: a tectonically stable shoreline, quasi-stable sea levels, low tidal fluctuations, and/or a local surplus of sediment (Curray, 1964; Hayes, 1979; Kraft and Chrzastowski, 1985). In Kotzebue Sound, data from seismic studies offshore and drill holes on land indicate no appreciable tectonic uplift during the Holocene (Eittreim et al., 1977; Hopkins, 1988). The region was not glaciated during the late Pleistocene (Péwé, 1975), so isostatic factors have not affected the formation of beach-ridge systems either. In terms of eustatic sea levels, the Chukchi Sea shoreline has been near its modern position (± 1.5 m) since about 5000 to 4000 B.P., based on oceanographic cores (McManus et al., 1983) and inferred indirectly from the contemporaneous onset of deposition at nearly all the Chukchi Sea beach-ridge plains (Hopkins, 1967). Short-term sea level fluctuations in the Chukchi Sea (e.g., tides and

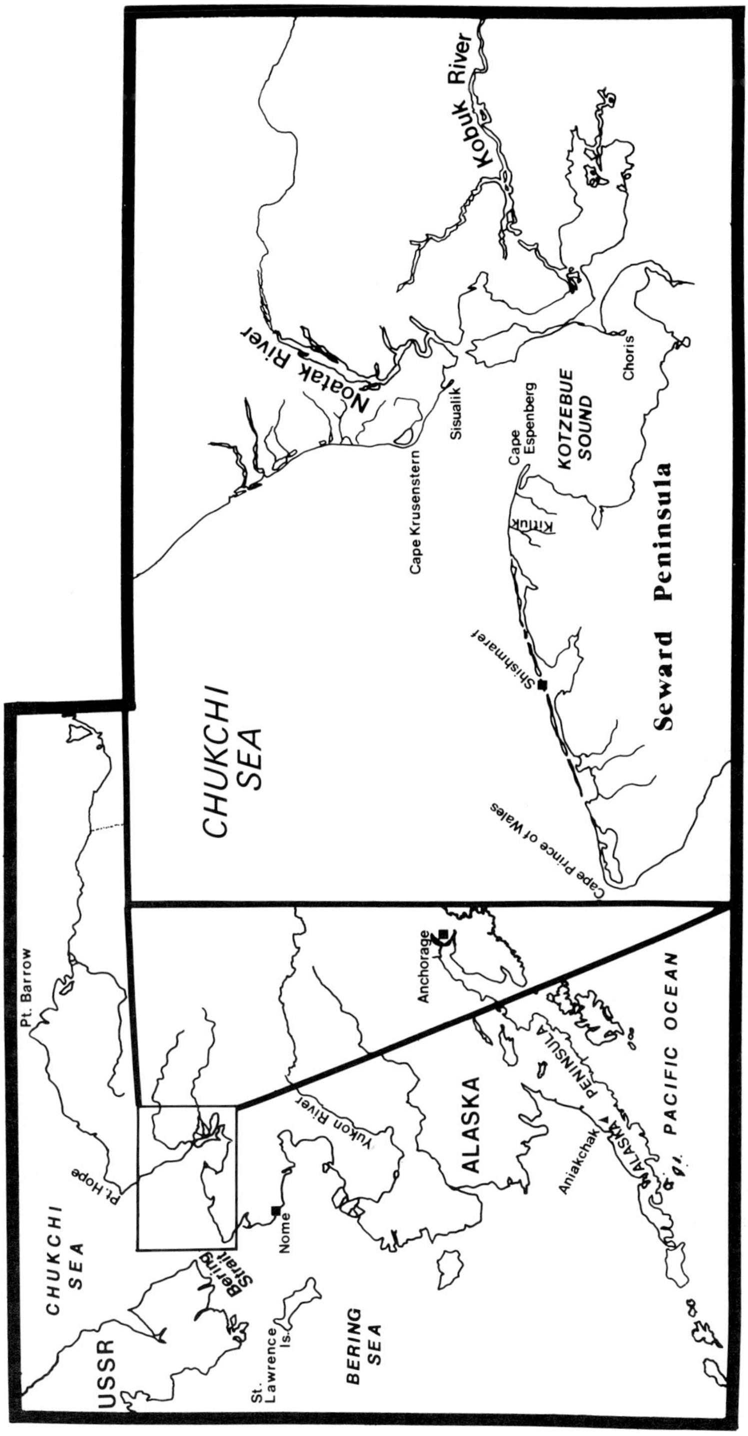

Figure 1. Location of study area. Beach-ridge plains in western Alaska.

meteorologically induced pressure changes) range less than 1 m annually. However, an exceptional storm surge may produce waves of up to 4 to 5 m above sea level (Peratrovich & Nottingham, Inc., 1982; Wise et al., 1981). The Chukchi Sea is ice covered for nearly 8 months and subject to open water processes only in the summer and autumn (July to October).

The shallow (less than 80 m below sea level) Chukchi Sea continental shelf, offshore from the northern Seward Peninsula, has an abundant source of sand for marine currents to rework landward into barrier islands and spits (Creager and McManus, 1966). Mainland bluffs of tephraeous or glaciomarine sand and silt provide an equally important source of sediment for building beach ridges, especially bluffs updrift from Cape Espenberg, and the Baldwin Peninsula on the east shore of Kotzbue Sound, updrift from Choris Peninsula (Hopkins, 1988).

On the sandy coasts of Seward Peninsula, beach ridges are often ornamented with dunes. Therefore, investigations into progradational history must distinguish the modes of ridge formation (marine and/or eolian) and interpret the internal stratigraphy of the dunes, describing primary structures, paleosols, and cryogenic alterations.

BEACH-RIDGE FORMATION AND CLAST SIZE: SAND VS. GRAVEL

In cross-correlating sedimentary deposits from the sand ridges of Cape Espenberg and gravel ridges of Choris, it is necessary to consider that sand and gravel respond differently to storm-wave climates (Shepard, 1973; Wright et al., 1979; Carter and Orford, 1981; Orford, 1987). Gravel forms higher, steeper (up to 24° in the case of cobbles), energy-reflective beaches, allowing high-amplitude waves to break close to the shore. Steep gravel beaches form a ramp and clasts are pushed up to the crest. Only storms with the highest energies can drive clasts up the beachface, so that the gravel system is able to absorb numerous smaller storm events without producing a distinctly new ridge. Hence, the process of constructing a gravel beach ridge involves the cumulative effects of many high-magnitude storms. The amount of time involved in this process is documented at Safety Sound, east of Cape Nome, where the same ridge contains vertically stratified archaeological materials about 3500 years old and about 2400 to 2000 years old (Bockstoce, 1979). Because gravel beach ridges form under wave action alone, their position more closely reflects paleoshoreline position than sandy ridges.

In fine or very fine sand, beach profiles are often nearly planar, with slopes of only 1 to 3° (Shephard, 1973) due to the energy-dissipative capabilities of sandy sediments (Wright et al., 1979). Flat beach faces damp the energy of storm waves and allow greater landward penetration of storm waves, favoring the development of washover deposits and transgressive dunes (Ritchie and Penland, 1988). In this way, the eolian transport of sand onshore may distort the simple correspondence between linear ridges and former shore position.

STUDY AREAS: THE ESPENBERG SAND SPIT AND THE CHORIS GRAVEL RIDGE PLAIN

Cape Espenberg

The Cape Espenberg beach-ridge complex is a mainland-attached spit issuing from the northern limit of the Seward Peninsula, on the Arctic Circle (66° 33′ N. lat.) in northwest

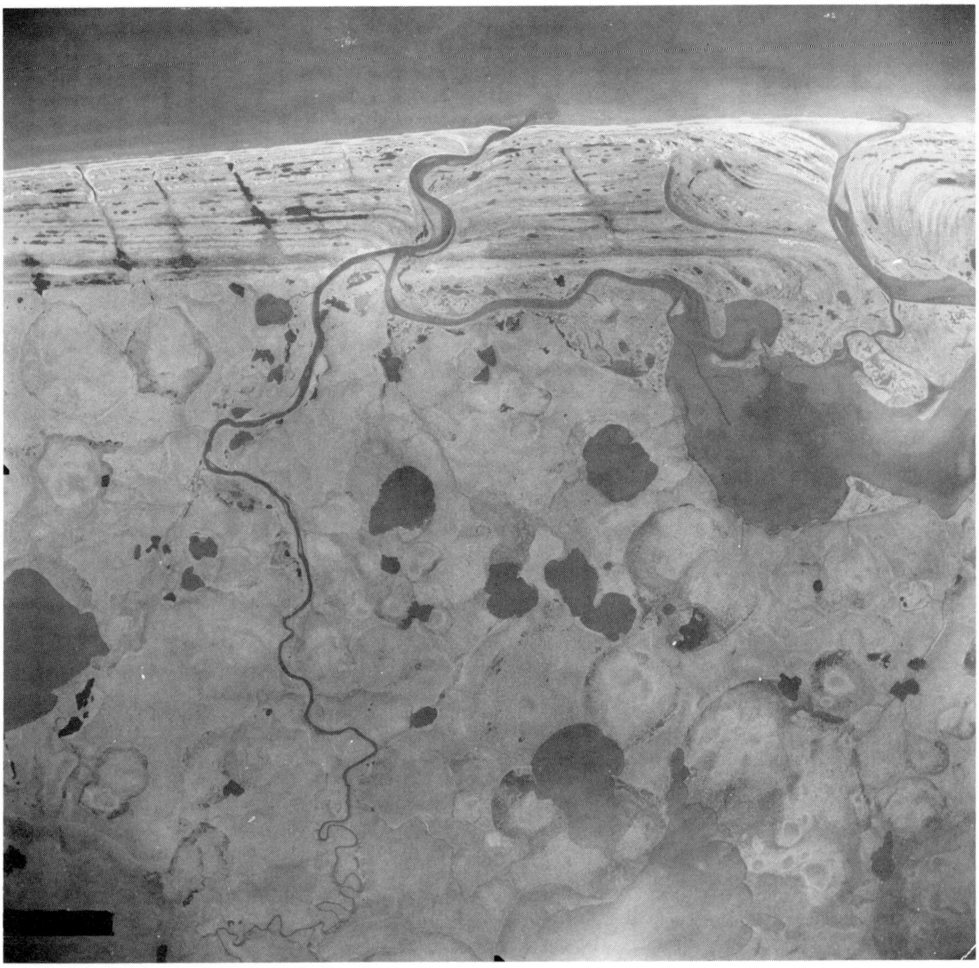

Figure 2. Aerial photograph of Espenberg complex. Pleistocene surface lies to the south, notable by several cycles of relict and active circular thaw lakes. By contrast, the Holocene-aged Espenberg beach and dune ridges are revealed by high-reflectance linear ridges separated by darker, wetter swales. A large tidal delta formed between two subcomplexes. A series of offshore bars may be observed in the nearshore zone.

Alaska (Figure 1). The Cape Espenberg spit complex extends 29 km west to east (at 164° 30′ to 165° 15′ W.) and varies in width from 1 to 2 km. Cape Espenberg combines aspects of a beach-ridge plain and a recurved baymouth spit as a result of its location at the terminus of a northeastward longshore transport cell and at the entrance to Kotzebue Sound. At its western attachment to the mainland, the Holocene-aged Espenberg ridges are seaward of a slight 10 to 15 m scarp, the disconformable contact between the ridge complex and the older, Pleistocene silts and sands of the mainland. The mainland contains numerous thaw lakes and dry lake basins (Figure 2). By contast, the Espenberg ridge complex alternates between shore-parallel ridges and swales of variable width, and records a progradational sequence since the comparative Holocene stabilization of sea level after 5000 to 4000 B.P. (McManus et al., 1983; Hopkins, 1967).

Four major drainages, rivers or tidal channels, cross-cut the Espenberg ridge complex, the north-flowing Espenberg River being the largest (Figure 3). West of the Espenberg

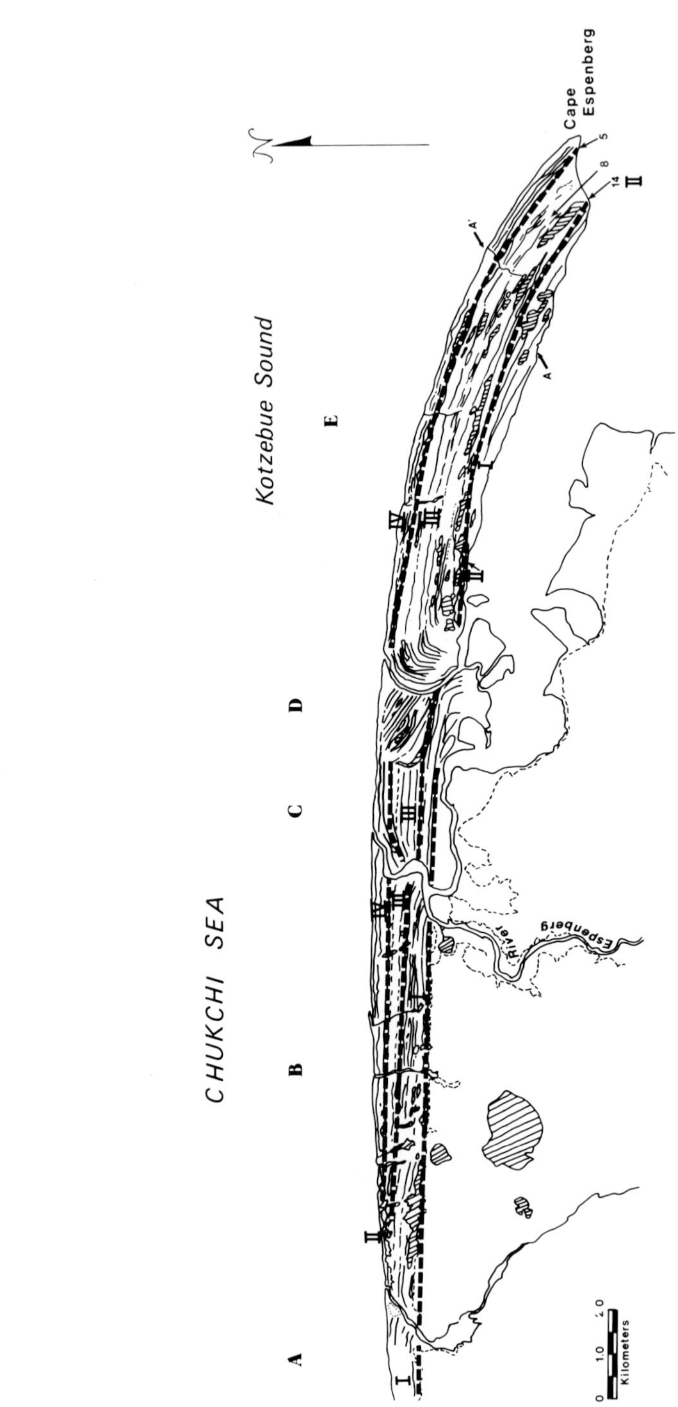

Figure 3. Depositional units at Cape Espenberg. Letters (A,B,C,etc.) are used to subdivide the complex. Arabic numerals (1, 2, 3, etc.) increase with ridge distance from the sea. Roman numerals indicate depositional units, from oldest (I) to youngest (IV).

River, the complex is attached to the mainland, while east of the river, the complex has formed as a series of islands defined by three tidal channels. Flow through the cross-cutting inlets or channels affected the seaward addition of ridges and, as a result, the pattern, number, and spacings of ridges varies between the channel-subdivided sedimentary packages, termed subcomplexes: A, B, C, D, and E. Subcomplexes A and B are at the western limit of Espenberg spit and form a wedge of sediment opening to the northeast from the eroding mainland Kitluk River coast. In its midsection, at the C subcomplex, the Espenberg coast undergoes a gradual realignment, curving southward. By the most easterly E subcomplex, the coast is oriented to the southeast. Nearly half of the longitudinal extent of the Espenberg beach-ridge plain is within the insular E subcomplex. The entire Espenberg ridge complex forms a single coastal cell of sediment transfer, with erosion prevailing in the west and deposition in the eastern part of the complex (Mason, 1990).

Choris Peninsula

Choris Peninsula lies 85 km southeast of Cape Espenberg, across Kotzebue Sound (Figure 1). Choris Peninsula is a knob of Paleozoic schist (Patton and Miller, 1968) up to 100 m above sea level, connected by a tombolo to the sinuous Baldwin Peninsula to the north. Choris Peninsula itself encloses the smaller, shallow embayment of Eschscholtz Bay to the east (Figure 1). Within three former embayments of Choris Peninsula, coarse to fine sands, granules, and small cobbles built beach-ridge plains (Figure 4). The three complexes are oriented to differing directions, either to the west (Choris A subcomplex), the southwest (Choris B subcomplex), or east (Choris C subcomplex). The number of ridges varies between the subcomplexes from nine in the west-oriented A subcomplex to 15 in the east-oriented C subcomplex. Only the westernmost Choris A subcomplex will be discussed here (cf., Mason, 1990 for discussion of the others).

Three principal sources contribute to the formation of ridges at Choris Peninsula: local bluff erosion, offshore sediments, and updrift bluff erosion. Large schist clasts (12 to 40 cm in length) in the ridges are eroded from the bedrock bluffs near the ridge complexes. Bottom sediments offshore are silts and fine sand (Creager and McManus, 1966); the latter forms a significant component in some ridges. Exotic materials, e.g., black and red chert cobbles, derive from the erosion and southward longshore transport of Baldwin Peninsula glacial outwash deposits (McCullough et al., 1965; Hopkins, 1990, personal communication) located 50 km to the north of Choris Peninsula.

THE GEOARCHAEOLOGICAL METHOD OF DATING AND CORRELATING BEACH RIDGES

The stratigraphic description and correlation of beach-ridge deposits requires a special approach. Beach-ridge deposits may be termed a depositional facies of a lithostratigraphic unit (North American Commission, 1983), being stratified sediments, derived from common lithic materials, resulting from common genetic processes, with easily definable contacts, and (unlike many lithostratigraphic units) related to a particular time period. I delineate depositional units at Cape Espenberg and Choris Peninsula using various criteria: vegetational and geomorphic differences evident in aerial photos, topographic and morphologic variations in sand ridges, and granulometric analyses. Chronologic referents are provided by radiocarbon dates; in this study primarily derived from archaeological investigations.

CHRONOLOGIC REFERENTS

Radiocarbon Samples (n = 41)

Thirty-seven radiocarbon assays are available from Cape Espenberg (Figure 5), while only four are available from Choris Peninsula (Figure 6). Most of the samples are from archaeological contexts. At Espenberg, nearly all the C-14 samples were collected from archaeological components. Radiocarbon samples were collected during an archaeological survey from exhumed or eroded paleosols, exposed by blowout erosion or at natural cutbanks (Schaaf, 1988). About half of the Espenberg C-14 assays (n = 18, or 49%) were obtained from archaeological excavations conducted by the NPS in 1989 and 1990 (Harritt, 1989, 1990). The collection biases of NPS archaeologists favored charcoal over wood when a variety of materials were available, although a single date was run on a surface sample of tar-like, charred, sea mammal oil-bound sand that was associated with diagnostic cultural

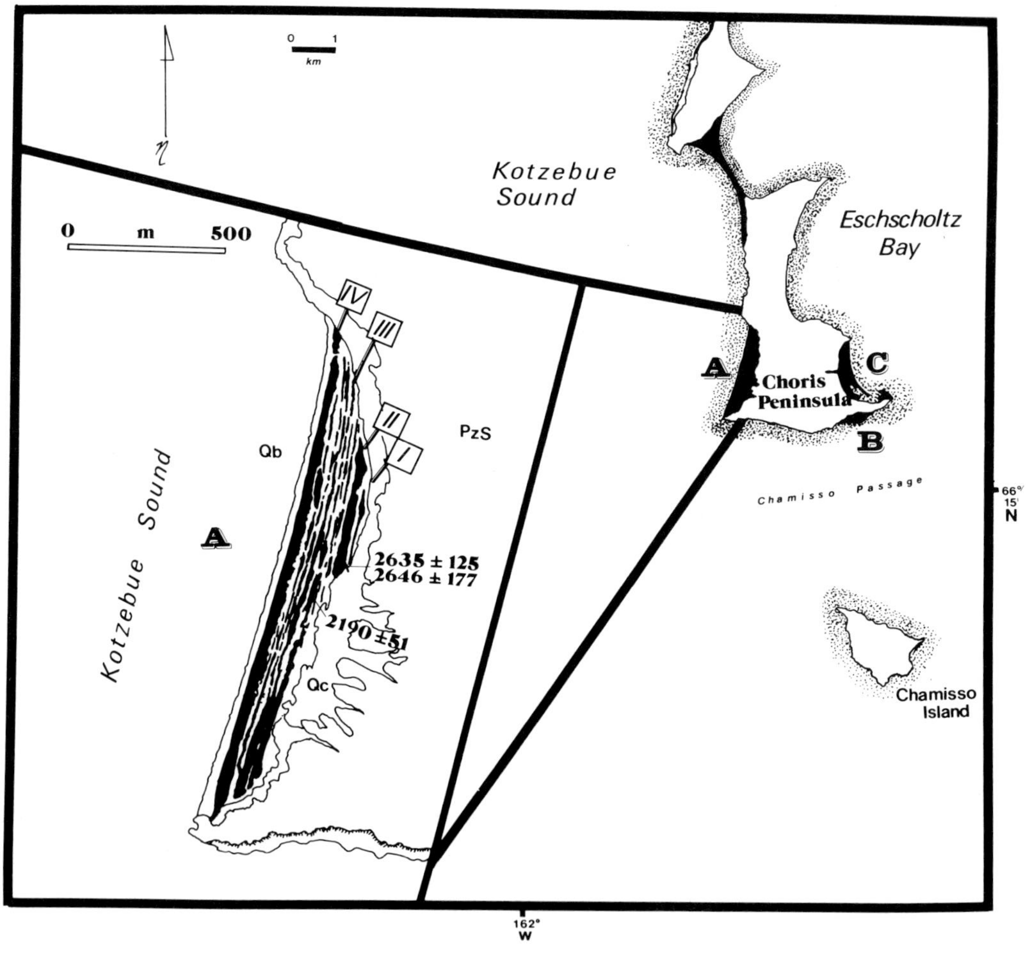

Figure 4. Choris beach-ridge complexes, with an inset of subcomplex A. Roman numerals indicate depositional units, from oldest (I) to youngest (IV). Dates from Giddings and Anderson (1986: 30). Geologic units: PzS refers to Paleozoic limestone, Qb to Quaternary-aged beach ridges, and Qc to Quaternary colluvial deposits.

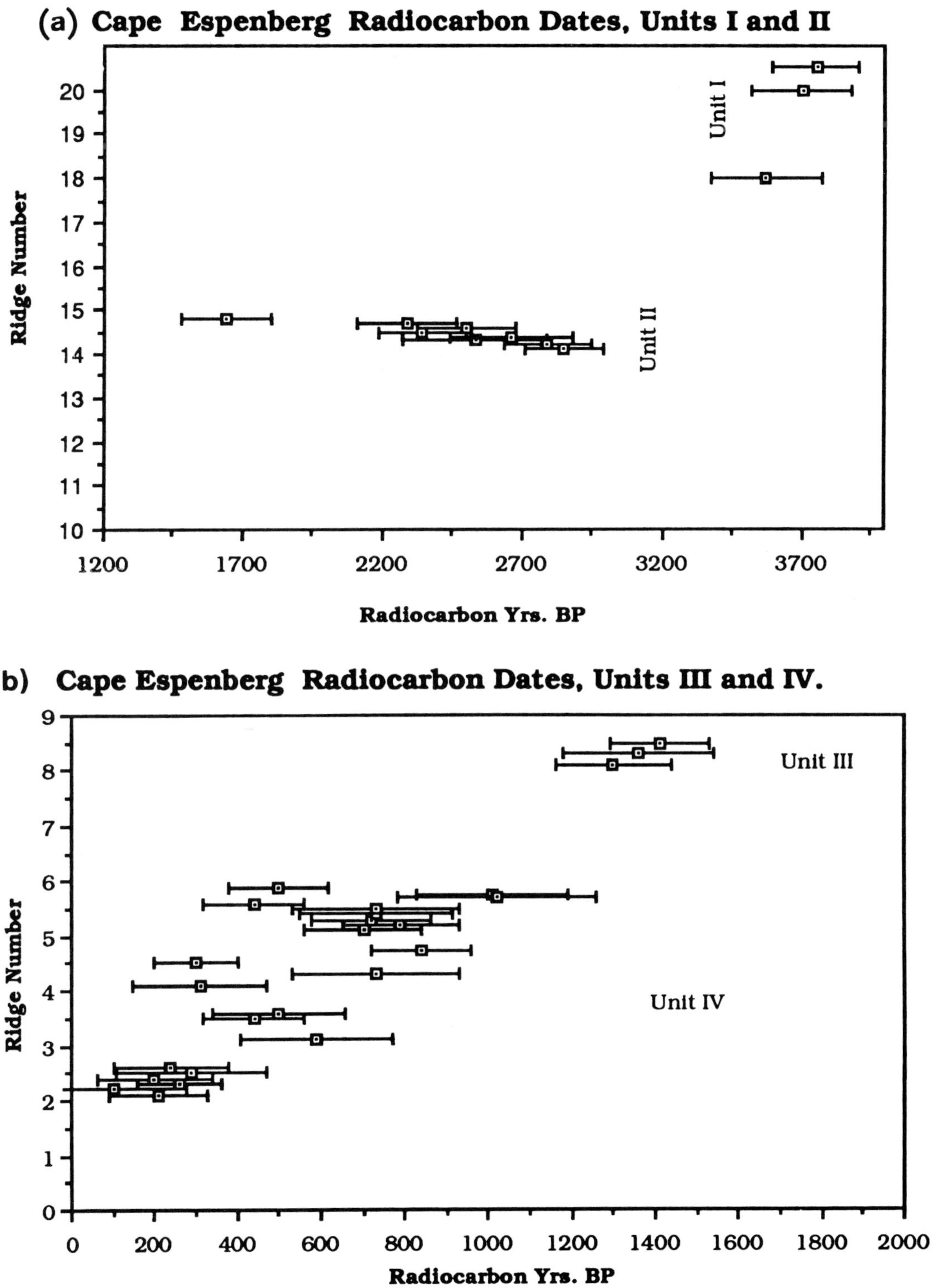

Figure 5. Radiocarbon dates from Cape Espenberg. (a) Units I and II: (b) units III and IV. Dates are graphed with the two-sigma value plotted as an error bar. Note the scale difference between the two graphs. Ridge number is plotted on the y-axis.

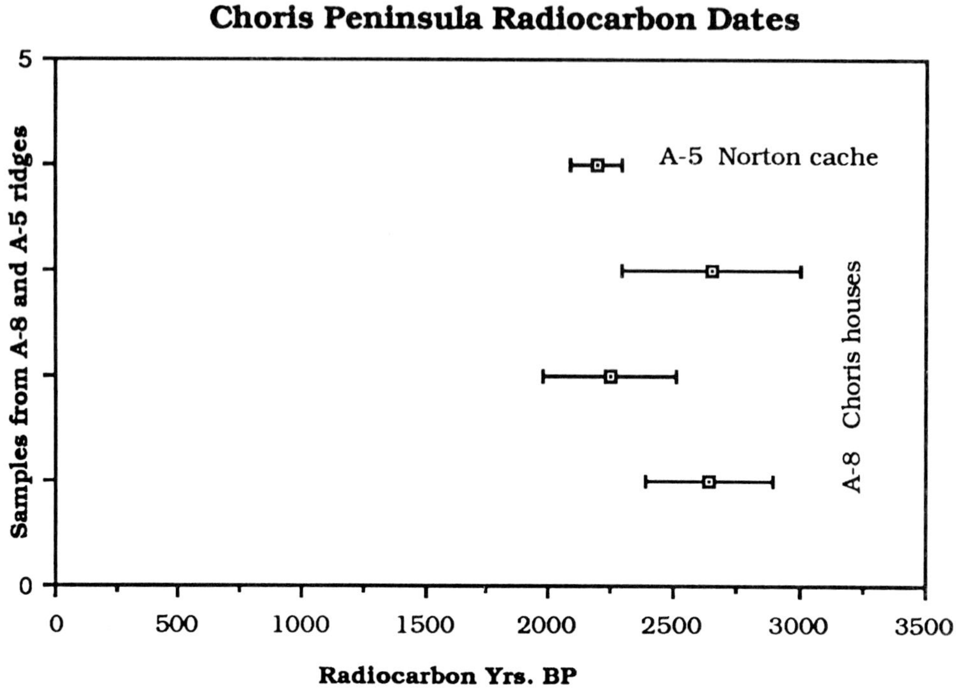

Figure 6. Radiocarbon dates from Choris Peninsula. Dates are graphed with the two-sigma value plotted as an error bar.

materials (Schaaf, 1988). Several geological samples (n = 5) were obtained at Espenberg from either cutbank exposures of grass horizons (n = 3) or the problematic context of marine shells (n = 2) found on the surface of ridges more than 300 m from the modern shore, presumably beyond the carrying distance of marine birds (Mason, 1990). At Choris, the materials submitted by Giddings for C-14 assays included charcoal, antler, and wood, all obtained from archaeological excavations (Giddings and Anderson, 1986: 30).

Tephra

The occurrence of airfall tephra (volcanic ash) within surficial deposits aids in the correlation and dating of sediments (Westgate and Briggs, 1985). Fine-grained volcanic ash is an ejecta, produced during a singular explosive event and subject to redistribution by winds over an extensive region. For western Alaska, a 4000 to 3400-year-old marker bed is provided by the widespread tephra associated with a caldera forming event at the Aniakchak volcano (Miller and Smith, 1987; Riehle et al., 1987).

Archaeological Cultures

The regional cultural chronology plays a substantial role in the dating and correlation of beach ridges in northwest Alaska. The cultural sequence, summarized in Table I, extends over 4000 years and probably represents the occupation of northwest Alaska by the ancestors of Eskimo (Yup'ik and/or Inupiat) people (Giddings and Anderson, 1986; Dumond, 1987). Diagnostic artifacts and house types provide critical time referents and may be used as

TABLE I. Archaeological Cultures of Northwest Alaska, Used in Relative Dating for Beach-Ridge Complexes

Culture	Diagnostic Characteristics	Dates[a] (B.P.)	Distribution
Arctic Small Tool tradition (ASTt) (Denbigh Flint Complex)	"Exquisitely" flaked microlithic tools: burnated flakes, endblades, & microblades; oval to sq. houses with stone-lined hearths, fire-cracked rocks; seasonal camps on coast: seal hunting	4500 (?) to 3000	Alaska Peninsula to Greenland Type site: Cape Denbigh
Old Whaling	Toggling bone harpoon; circular houses with entry; notched bifacial points and endscrapers, whalebone debris, ground basalt oil lamps	3200–3000	Kotzebue Sound, Wrangel Is.
Choris	Cord & Linear-stamp ceramics, Lg. diagonally flaked pts.—"lanceolate", no burins; Lg. oval houses; ground slate, flaked adzes, labrets & incised bonework, barbed harpoons, pecked stone oil lamps; lg. (winter) camps in mts.—stone-lined hearths	3000–2500	NW Alaska, (poss.) Chukotka Type site: Choris Peninsula (SLK-007)
Norton	Linear, check and plain ceramics; great diversity of proj. pts., side and end blades; discoidal scrapers, notched stones (for nets); clay & stone lamps, Ground adz blades; sq. subterranean houses with central hearth & tunnel entry	3000–1500	Alaska Peninsula to Mackenzie Delta Type site: Cape Denbigh Site numerous on coast & lg. in scale
Old Bering Sea/ Punuk	Elaborate, linear incised harpoon heads; round houses: long entry, check-stamp ceramics; shouldered pts. of flaked slate; whalebone burials; use of iron engraved tools, enigmatic incised winged objects	1800–1000	St. Lawrence Is., Chukotka Type sites: Gambell, Punuk Islands (U.S.) Ekven/Uelen (U.S.S.R.)
Ipiutak	Open-work ivory carving, inset eye burials, anthropomorphic design, LACKS pottery, ground slate; Houses sq., lack tunnels; lithics: side & endblades; lg. coastal settlements	1700–1000	NW Alaska, Brooks Range So. limit: Norton Sound Type site: Pt. Hope
Birnirk	Multispurred harpoons, linear designs; clay pots: curvilinear stamp; ground slate weaponheads	1400–1000	NW Alaska, N Chukotka, coastal only Type site: Pt. Barrow
Western Thule (incl. Kotzebue Period)	Multiroomed houses; slate and chipped stone pts., barbed harpoons; simple linear decoration, pottery circle or spiral decor; high percent of organic artif. birch bark, beaver-tooth knives	1300 (?) to 200	Alaska Peninsula to Greenland Type site: Cape Denbigh

[a]Dates used differ from Anderson's usage, for reasons explained in Mason and Ludwig, 1990: Anderson uses Onion Portage dates selectively and posulated the end of ASTt at ca. 3500 B.P., a break used to define the beginning of Choris. Similarly, Old Whaling in defined by averaging several dates possibly contaminated with sea mammal carbon.

Modified from Anderson, 1984, 1988; Giddings and Anderson, 1986.

index fossils for time stratigraphic purposes. The earliest archaeological materials found on northwest Alaska beach-ridge complexes date between 4000 and 3500 B.P. and possibly as early as 4500 B.P. (based on Anderson, 1988). This cultural phenomenon, the Denbigh Flint Complex of the Arctic Small Tool tradition (ASTt), is marked by a distinctive pressure-flaked microlithic technology, cobble hearths, circular dwellings, and an absence of ceramics. After the initial occupation by ASTt peoples, more localized archaeological cultures are distinguished on the basis of decorative elements on harpoon heads and/or ceramics and the shape of houses, the number of rooms, and the length of entry passages. Through time, lithic artifacts are made increasingly on ground slate, with less use of exotic materials such as obsidian. People of the Choris culture, dated from 2800 to 2400 B.P., constructed large, oval houses above ground, used labrets, oil lamps, linear stamp pottery, and bifaces inset into sockets for land hunting or barbed harpoons for pursuing marine mammals. The Norton culture, about 2500 to 1500 B.P., is characterized by square, semi-subterranean houses with short entryways, but shares some artifactual similarities with Choris, including pottery styles, lithic tools, and the use of lamps and labrets. People of the Ipiutak culture, dated at 1600 to 1200 B.P., constructed square houses above ground, left thin occupational deposits, used labrets, had stone insets for arrowheads, and lacked pottery and oil lamps. Ipiutak is best known from the elaborate burial cult at Point Hope (Larsen and Rainey, 1948), a possibly unrepresentative view of its interior caribou hunting focus. The Thule and late-prehistoric cultures, about 1000 B.P. to the present, witnessed the development of a widespread whaling economy in northern Alaska, population increase, and the building of multiroomed houses with long entryways (Giddings, 1952; Giddings and Anderson, 1986).

CRITERIA FOR DELINEATING DEPOSITIONAL UNITS

Aerial Photographs

I used aerial photographs to delineate depositional boundaries by examining geomorphic features (frost cracks, lakes, etc.), slope breaks, and vegetational differences (see below). As is well known, color differences in the infrared portion of the spectrum clearly record differences in moisture content, plant cover, and/or lithology. Three sets of photos were used: (1) the standard false color infrared U-2 overflight imagery photographed at 1:60,000 scale, (2) the National Ocean Survey (NOS) natural color infrared imagery at 1:30,000 scale, and a (3) a false color series shot at 1:8,000 scale from fixed-wing aircraft, commissioned by the National Park Service (Anchorage Regional Office). The two smaller-scale photosets are archived by the GeoData Center of the Geophysical Institute, University of Alaska.

Granulometric Analysis

Variations in grain size statistical parameters such as mean, skewness, and kurtosis are widely considered to provide a basis for inferences about current strength, depositional environment, and the provenance of sediments (Reineck and Singh, 1980; Shepard and Young, 1961; Visher, 1969).

Grain size samples ($n = 105$) were collected from depth of 20 to 30 cm on beach or dune ridges from the Cape Espenberg spit (Mason, 1990). Three types of sample strategy were employed: (1) samples were obtained longitudinally, parallel to the coast, every 250 m and in two traverses perpendicular to the coast, across all the depositional units, (2) several ridges were intensively sampled at differing slope orientations on a ridge, and (3) deeper samples were collected from cutbank and/or archaeological exposures. Laboratory procedures followed Lewis (1983), using nested sieves at 0.25 ϕ intervals, mechanically shaken for 15 min. Statistic parameters were calculated using the procedure of Folk and Ward (1957).

The granulometric analyses indicate that Cape Espenberg is composed predominantly of well-sorted fine sand ($x = 2.33 \pm 19$ ϕ), with a range of 1.46 to 2.59 ϕ. Coarser medium sand (1 to 2 ϕ) is found in disconformable cross-beds of a storm facies, marine in origin. The height of beach facies across the Espenberg spit varies from 1.0 to 1.5 m above mean sea level (MSL)* and provides evidence of temporary sea level elevations caused by barometric effects. In most cases, the granulometric analyses fail to assist in differentiating beach from dune facies, leading to the conclusion that the modern beach is largely composed of sand-eroded updrift in the western A and B complexes and transported and redeposited downdrift in the eastern E complex.

Primary Sedimentary Structures: Beach Ridge, Dune, and Blowout Facies

The recognition of beach and dune facies is essential in describing the formation and evolution of ridges at Espenberg. To further this analyis, I inspected cutbank exposures and archaeological test pits, and dug a series of additional trenches across the modern beach, low ridges, and older ridges. Sedimentary characteristics are especially important in defining storm events on the modern beach and in distinguishing primary from secondary deposition.

To examine beach stratigraphy, several trenches were placed perpendicular to the modern beach. Beach depositional units vary between sets of planar, subhorizontal laminae (0 to 3° landward) and sets of laminae obliquely inclined seaward (up to 10°). The seaward-inclined units are often traceable for several meters and include thick (10 to 20 cm) sequences of discrete laminae (each 1 mm in thickness) composed of coarser, dark mineral sand and shell. The alternation in stratigraphic units reflects an angular disconformity produced during storm erosion, followed by deposition during waning phases of the storm. The succeeding horizontal bedding reflects the recovery after the storm. During calm-weather periods, ridge-runnel features (with 30° beds) and swash bars (''berm ridges'' in the sense of Carter, 1988) of variable width and height are added atop the post-storm horizontal beds. The net product of beach processes is a *beach ridge,* defined by Reineck and Singh (1980): ''a continuous linear mound of rather coarser sediment near the high water line . . . develop[ing] mainly during storms and exceptionally high waters.'' At Espenberg, it is rarely possible to observe primary beach-ridge structures from surface exposures or by excavation, due to the close proximity of the water table.

Coarse clasts, such as drift logs, bones, shell, etc., are deposited in the backbeach during or after storms, providing a nucleus for coastal dune growth. Dune growth is furthered by the presence of lyme grass (*Elymus arenarius*) and the rapid deflation of swash bars during the winter (cf., Carter, 1986, 1988). The cumulative effects of eolian processes on the beach lead to the formation of two types of sedimentary deposit at Espenberg.

*Mean sea level (MSL) is approximately that of mean tidal level, as estimated in the field; no gauge data available are from the region.

Dune ridges are accretional features, bell-shaped with slopes up to 30°. Formed as sand from the backbeach trapped by lyme grass, dune ridges extend up to 20 m in height, but are generally only 5 to 6 m above sea level. Primary dune ridges contain numerous 1 to 2-cm-thick beds of grass separated by sand beds. The constructure of a dune ridge involves nucleation about a beach ridge and the accumulation of successive vertical increments atop the ridge, with upper facies often cross-bedded (10 to 20°).

Blowout ridges form as the vegetative cover on dune ridges is disrupted due to a combination of storm-surge attack, trampling, fire, plant mortality, or other factors (Mason, 1990). With time, considerable topographic complexity is produced on blowout ridges, as eolian erosion produces nested blowouts, coalescing deflation basins, and residual dunes similar to the forms in eastern Australia described by Hesp (1988). Blowout ridges, containing numerous buried surfaces or organic soil horizons, record several periods of erosion and redeposition.

The three types of ridges correspond to different depositional environments as beach-ridge and dune-building processes operate over several seasons. The beach is modified most drastically during storm surges, which occur at present with greatest frequency from September to November, before and occasionally after freeze-up (Wise et al., 1981). Sand movement and concentration in the backbeach occurs during winter, especially late winter, as has been documented in Ireland (Carter, 1986). Beach and low dune ridges, by contrast, may indicate comparatively less windy conditions, saturation of sediments by rain or groundwater, and an absence of grasses (if ridges remain too saline) (Godfrey et al., 1979). Fair-weather conditions prevailing in July/August lead to a significant amount of beach-ridge accretion, especially in the horizontal direction (Davis et al., 1972). The addition of the vertical component, above the tidal range, occurs in the recovery phase after storm events (Davis and Fox, 1975).

Secondary, Cryogenic Modifications in Arctic Dune Ridges

Primary structures in arctic coastal dunes are susceptible to cryogenic processes that deform beds and impart a secondary imprint. Distinctive features such as wedges, loading structures, and convoluted beds occur primarily on well-drained ridges where a difference exists between an impermeable lower, frozen layer and an upper, seasonally thawing/refreezing zone at 1 to 2 m depth (Vandeberghe, 1988). Troughs and wedges occur on ridges 750 years and older, providing information useful in climatic reconstructions. Wedges are produced if snow cover is low, contraction cracks form in mid-winter, and the cracks are infilled by blowing snow or sand (Mackay, 1974). In low-lying areas, intersecting frost cracks produce a polygonal network in plan view, and cracks may cross dune ridges. The density and pattern of frost cracks is useful in defining and correlating depositional units, as seen in the comparison below between Espenberg and Choris (and at Cape Krusenstern; cf., Mason, 1990).

Vegetational Changes

The topography of the Cape Espenberg dune and beach ridges is largely governed by plant ecology. Dune ridges build due to the high tolerance of beach lyme grass (*Elymus* spp.) for burial by sand (30 cm per annum) and its ability to root deeply (several meters) and expand laterally by rhizomes (Chapman, 1978; Ranwell, 1972). The vertical component imparted by beach grass is matched by the horizontal nature of the root systems of prostrate shrubs such as crowberry (*Empetrum nigrum*) and willow (*Salix* spp.). The latter species are a surface-stabilizing influence, favored as additional ridges build seaward, isolating dunes from the renewable sand source on the beach. In water-saturated swales, there are specialized biotic assemblages of sedges, willows, blueberries, sphagnum moss, and tussocks. The contrasts between the different plant communities are readily observable on infrared aerial photos and provide a means to delinate depositional units.

Lakes and Palsas

Lakes and ponds form readily within interridge swales on the Espenberg spit. Lakes range in size from less than 0.1 ha to over several km^2 in area. The source of the water is ultimately from precipitation, with seasonally high evaporation leading to fluctuations in lake level and complete desiccation of some ponds in blowout basins. The development of lakes in swales is complicated by the formation of peat and the presence of permafrost, only about 70 to 80 cm below the surface in low-lying areas. After peat forms, a series of ice mounds, or palsas, are forced up by the expansion associated with the phase change of ice formation. The formation of palsas reflects circumstances with little snow cover and substantial penetration of frost, which produces a perched water table. The eventual decay of peat and the melting of palsa ice also leads to lake expansion (Seppälä and Koutaniemi, 1985). The morphologic outlines of lakes across the Espenberg spit are variable, providing a relative chronological device in correlating depositional facies from beach-ridge complexes that are undated. Lakes between the oldest ridges are polygonal in oultine, bounded by straight banks produced by frost-crack peat ramparts, while lakes between younger ridges are more oval or round in outline. Similarly, polygonal lake outlines occur in the oldest portions of the beach-ridge plains at Cape Krusenstern and Cape Prince of Wales. This allows the age of the earliest Wales and Krusenstern ridges to be estimated at 4000 to 3000 B.P., comparable to the dating of the polygonal lakes at Espenberg (cf., Mason, 1990).

Blowout Evolution

With increasing age, the coastal dunes at Espenberg undergo several changes in morphology as blowouts develop the economy of the dunes shifts from one of sand surplus to one of sand deficit and the dominant plant cover shifts from grasses to prostrate shrubs intolerant of sand burial. On the youngest, most seaward ridge, blowouts are singular and widely separated laterally, while a more complex topography results by the third ridge as smaller blowouts are added like cellular additions to the larger ones. Older ridges show multiple generations of blowouts, revealed by exposed paleosols and nested basins (Mason, 1990). The evolution of blowouts provides a means of relative dating, as well as a methodological aid in locating archaeological sites.

Soil Development

Soil profile development in sandy sediments varies considerably with the stability of the surface. Thick surficial organic O-horizons are common in areas not subject to active eolian deflation, which are covered by crowberry and other low shrubs with horizontal root systems. In areas subject to seasonal ponding and desiccation, oxidation of iron grains produces mottling and pronounced alteration of the sediments by the coating of individual grains (Pye, 1983). Episodic ground-water percolation further results in the downward displacement of heavier grains and produces distinctive tongue-like features. On the oldest ridges, the down-profile movement of iron leads to spodosolization, i.e., B-soil horizon development (Soil Survey, 1975). In areas undergoing active sand transport, soil formation is less developed, consisting of thin silt laminae, the accumulation of finer sediments eluviated from the surface (Soil Survey Staff, 1975). Thus, in broad outline, the intensity of soil formation at Espenberg provides a measure of relative age and of the relative stability of the ground surface during human occupations, which correlate closely with the paleosols.

Summary of Techniques for Correlation of Ridges

Depositional units are defined at Espenberg using variations in vegetation cover, soil development, lake and palsa formation, and granulometric parameters. Absolute age estimates are obtained from archaeological and geological radiocarbon dates and by references to tephra marker beds. Using these methods, I provide the following reconstruction of the depositional history of Cape Espenberg and then offer a correlation with the depositional history on Choris Peninsula.

DEFINING DEPOSITIONAL UNITS AT CAPE ESPENBERG

I delineate four depositional units at Cape Espenberg (Figure 3), which reflect an alternation between periods dominated by high-intensity storms occurring in rapid succession and periods with less intense, infrequent storms. I follow conventions set by Giddings (1963) at Cape Krusenstern to number ridges sequentially older landward from the modern beach (i.e., 1 = youngest). Unit designations follow geological practice, using Roman numerals (I, II . . .) in an oldest-to-youngest ascending numeration. Based on the available C-14 dates (Figures 5 and 6). Unit I formed between 4000 and 3300 B.P., Unit II from 3300 to 2000 B.P., Unit III between 2000 and 1200 B.P., and Unit IV after 1200 B.P. to the present.

Grain-size parameters indicate that Cape Espenberg is divided into only two depositional units. The most recent, higher dune ridges (Unit IV) are coarser than the older units (I to III), implying either a different sediment source or the effects of higher storm energies and deeper wave base on offshore sediments. The coarse fraction is composed of dark mineral sands found in bluffs updrift from Espenberg and is connected with increased bluff erosion during the dune-building interval (Mason, 1990).

Depositional units can be distinguished on aerial photos due to differences in moisture content and total percentage of vegetation cover. The most seaward unit, Unit IV, shows a greater albedo, reflecting more light due to a less complete vegetation cover, predominantly of grasses on actively building dune ridges and modified blowouts. Conversely, a darker, less reflective region in the middle of Espenberg spit, Unit III, is a series of marsh-covered, low-lying beach ridges with wide swales. Blowout ridges in Unit II contain extensive areas of unvegetated sand and concentric blowout basins. Unit I ridges are reflective, being comparatively well drained, with localized areas of blowouts on the oldest ridges.

Topographically, ridges alternate between higher dune and/ or blowout ridges and lower beach ridges. Dune or blowout ridges are 4 to 6 m above MSL in elevation, but individual ridges are up to 20 m above MSL (Units II and IV). Beach ridges, lacking dunes, are commonly only 1 to 2 m above MSL (Unit III, most of Unit I). Dune ridges vary in height across the Espenberg spit: a series of up to six principal dune ridges lies within the youngest portion of the spit, less than 0.5 km from the sea (Unit IV), while isolated dune ridges are at 0.68 to 0.74, 1.5 to 1.53 (Unit III), and 1.7 to 1.9 km from the sea (Unit II). The middle portion of the spit, 1 km in width, is formed by a series of low beach ridges separated by wide (up to 200 m) swales (Unit III).

THE HOLOCENE DEPOSITIONAL HISTORY OF CAPE ESPENBERG

Unit I consists of up to six low beach ridges (E-20 to E-15) attached to the mainland on the west and bordered by a shallow estuary of Kotzebue Sound on the east. Unit I contains about one quarter of the total progradation at Espenberg, with up to 0.65 km in width. There are large, polygonally shaped lakes within and cross-cutting the wide swales of Unit I. The youngest, most landward portion of Unit I was scarped and transgressed by dunes during Unit II deposition. While most ridges are low in elevation, the earliest Unit I ridges are higher, up to 5 m above MSL, the result of intermittent eolian activity throughout the last 4000 years. Sand in Unit I is often heavily oxidized, and cryogenic phenomena such as wedges and convolutions disrupt primary sedimentary structures and paleosols. Two buried spodosolic B-horizons extend laterally for over 300 m on the oldest Unit I ridge, E-20. The lowermost paleosol contains 4000 to 3400-B.P.-old Aniakchak tephra, an airfall deposit transported from the Alaska Peninsula over 1200 km to the south (Riehle et al., 1987). The age of the tephra fall at Espenberg is established by a C-14 date of 3700 ± 90 B.P. (β-23170) on buried grass from a more recent portion of the same E-20 ridge. Archaeologically, only small encampments of ASTt-related cultures are recorded on Unit I ridges; one occupation dates at 3750 ± 80 B.P. (β-33758) (Harritt, 1990). Based on the limiting ages cited above, beach-ridge deposition in Unit I started before 3800 B.P., probably at about 4000 B.P., and then reflects two additional processes: (1) an initial period of dune-ridge formation and comparative stasis in shorelines followed by a (2) period of rapid beach-ridge progradation.

Unit II describes a prominent blowout ridge (E-14), laterally continuous for the entirety of the Espenberg spit. The Unit II ridge widens to a maximum of about 150 m and extends up to 9 m above MSL. In places, the Unit II dune ridges cover older beach ridges of Unit I, but a distinct, uneroded 10 to 15° slope distinguishes the younger margin of the unit. Frost cracks cross-cut the Unit II ridges, forming deep 10 to 20-cm troughs, although no active ice wedges are found on the ridge at present. At least three cycles of blowout formation occurred on Unit II ridges, bracketed by two buried paleosols, the lower dated at 2800 to 2200 B.P. by archaeological C-14 dates (Schaaf, 1988; Harritt, 1990). Archaeological remains are sparse lithic or ceramic scatters often associated with charred sand impregnated with sea mammal oil, the remains of seal-hunting Choris and Norton cultures, widespread throughout western Alaska in the third millennium B.P. (Giddings and Anderson, 1986; Schaaf, 1988). Dates from Choris and Norton sites at Espenberg allow an upper-age assignment on dune construction during Unit II: before 2790 ± 80 B.P. (β-33759) and extending to 2285 ± 90 (β-17968) B.P. An episode of marine erosion that scarped the older Unit I occurred during 3300 to 3000 B.P., prior to the construction of the dunes during Unit II. On the whole, Unit II remained adjacent to the shore for a considerable period of time and reflects only limited progradation and a net erosional regime. The millennium between 3000 and 2000 B.P. is a cold interval of glacial expansion throughout Alaska (Calkin, 1988) and world-wide (Porter and Denton, 1967). The buried organic horizons at Espenberg may record a stasis in wind activity or a brief warm interval between 2800 and 2500 B.P. (Mason, 1990).

Unit III includes a region of low ridges (E-13 to E-6) separated by wide swales; the ridges are often discontinuous and extend only several hundred meters in length. Unit III, about 1 km in width, accounts for nearly half of the progradation at Cape Espenberg. Permafrost is only 70 cm below the surface, and polygonal networks of frost cracks, bordered by ramparts, form below the active layer. In addition, rows of palsas are aligned across the swales, perpendicular to the southeastward ground-water flow. Poorly preserved primary sedimentary structures include seaward-dipping laminae, but no shell, which tentatively suggests a marine origin for these ridges. Organic horizons of 2 to 5 cm thickness cover stable ridge surfaces, but distinct subsurface soil horizons are rare, although grains may be slightly stained or oxidized. Two of the Unit III ridges, E-12 and E-8, are higher in elevation (up to 3 m above MSL) and possess a cover of low, 1 to 2-m deflated dunes and two paleosols. Few archaeological sites were located during surveys on Unit III ridges, limiting the assessment of ridge age(s). Archaeological excavation in several Ipiutak sites provides an upper-limiting age of 1358 ± 41 B.P. (average of three dates) for ridge E-8, one of the low dune ridges in Unit III (Harritt, 1989; cf., Mason, 1990). The other low dune ridge, E-12, probably dates to 1800 B.P. based on interpolations from neighboring ridges. Its lack of appreciable dunes, low elevation, and the great width between ridges suggest that Unit III was a time of comparatively infrequent or low-intensity storm surges. However, brief periods of stronger winds led to the building of low dunes on two of the Unit III ridges. The evidence for human occupation, despite the low elevation, also implies that massive storms were infrequent during the period between 2000 and 1200 B.P., during the rapid progradation at Espenberg. Across northern and western Alaska, this millennium is marked by warmer conditions, as evidenced by soil and peat formation atop alluvial terraces (Hamilton, 1981; Kaufman et al., 1989).

Unit IV consists of five prominent dune ridges (E-5 to E-1) up to 20 m above MSL. On the seaward aspect, the grass-covered dune ridges receive sand from the backbeach, but active dune building decreases and plant species diversity increases with distance from the beach, as the exclusionary effects of salt-spray lessen. Blowouts develop in a wide variety of forms, ranging from isolated oval or round blowouts on the most seaward, recent dune to secondary, smaller blowouts along the margins of larger ones by the second or third ridges. Stratigraphic exposures in cutbanks show that a single ridge consists of several smaller beach (= berm) ridges and incipient dunes welded together by overlying tabular sets of cross-bedded eolian deposits. The lowermost beach-ridge facies are capped by shell and coarse sands and resemble ridge/runnel features on the modern beach. Paleosols do not occur in the youngest ridges, although distinct, organic-stained, silty horizons appear on the E-5 ridge, the oldest ridge of Unit IV. The higher dunes of Unit IV choke off several of the cross-cutting channels in the low-lying Unit III ridges. Dune formation in Unit IV started before 1010 ± 90 B.P. (β-17969) with the addition of the E-5 ridge and has continued, in places, until the present. Radiometric determinations from buried archaeological components associated with the western Thule and old Kotzebue cultures reveal that most of the dune building occurred before the occupations dated at 800 to 600 and 300 to 200 B.P. (Harritt, 1989). The construction of the first dune ridges in Unit IV correlates with a brief Northern Hemisphere glacial expansion about 900 to 1000 A.D. (Porter, 1986), yet prior to the early Medieval climatic optimum (Lamb, 1982), as evidenced by a comparison of the Espenberg cultural chronology with tree-ring chronologies from the lower Kobuk River (Giddings, 1952). The seaward aspect of the Unit IV ridges is currently eroding, as indicated by a prominent scarp and measurements of 8 to 13 m of retreat during the 26-year period from 1949 to 1976 (Jordan, 1988).

Summary of the Cape Espenberg Chronology

Using archaeological C-14 dates and other geomorphic inferences, I have established a chronology for the evolution of the sandy Cape Espenberg spit. In summary, the four units record alternations between rapid horizontal progradation from 4000 to 3300 B.P. and 2000 to 1200 B.P. and vertical dune building from 3300 to 2000 and from 1200 B.P. to the present, a difference which translates into a proxy record of storm intensity and frequency. The gravel ridges at Choris reveal a similar picture of storm intensity related only to the high wave energies capable of moving gravel upslope. In this way, the Choris record provides a test case for the interpretations offered for Espenberg.

THE CHORIS PENINSULA BEACH-RIDGE SEQUENCE

The 300-m-wide Choris A subcomplex is comprised of nine ridges and extends only about 2 km north to south (Figure 4). Vegetational differences allow the delineation of the most seaward ridge, sparsely vegetated with grass, from the remainder of the ridges covered with crowberry. Still subject to occasional storm overwash, the most recent Choris (A-1) ridge is over 4 m above MSL and 100 m in width, almost twice as high and wide as any of the other Choris A ridges. The vertical stratigraphy of individual Choris ridges shows that a single ridge forms as a composite of numerous overwash events graded in response to waning currents. Thin grass laminae separate discrete storm events, representing former surfaces and, hence, hiatuses in storm deposition. Choris ridges older than the A-1 ridge are covered by crowberry and cinquefoil (*Potentilla* spp.) and are crossed by polygonal patterns produced by cryogenic processes. Organic soil horizons, more than 5 cm thick, lie just below the surface of the oldest A-8 and A-9 ridges and in swales between other ridges, but evidence of soil profile development is lacking.

Narrow, lower ridges (2.0 to 2.5 m above MSL) and wide swales (20 to 30 m) occur in the middle of the Choris A complex. However, only two Choris ridges are comparatively wide, the eighth (A-8) and first ridges (A-1), which also contain the most significant archaeological remains: Choris and late-prehistoric cultures, respectively. The cultural features (oval houses, cache pits, and artifacts) on the A-8 ridge form the type site of the Choris culture (Giddings and Anderson, 1986). Isolated diagnostic lithic artifacts (end-blades, etc.) of the Norton culture lie on the surface of the A-4 ridge, and undated Thule culture and late-prehistoric house depressions and graves on ridge A-1.

The depositional sequence of Choris A complex ridges (Figure 4) parallels that of Espenberg, both in plan view and in relative height. Four depositional periods are traceable in the west Choris A complex; wider, higher ridges occur only within two intervals. *Unit I* includes only a single ridge at the base of the bluff. *Unit II* consists of the 8th ridge, about 30 m wide and 2.75 m above MSL. *Unit III* defines six narrow (less than 20 m wide) ridges about 2.0 to 2.5 m above MSL. *Unit IV* is the most seaward ridge, the youngest and the most massive of the Choris ridges. Date assignments are available (Figure 6) from Unit II based on the Choris occupation at 2646 ± 177 B.P. (P-203) (Giddings and Anderson, 1986). Norton-aged cache pits on the A-5 ridge in Unit III are dated at 2190 ± 51 B.P. (P-611) (Giddings and Anderson, 1986). Units II and III at Choris, containing Choris and Norton artifacts, correlate with the Choris/Norton Unit II blowout ridge at Cape Espenberg, which dates from 3300 to 2000 B.P. After about 2100 B.P., the accretion rate of gravel ridges at Choris decreased in terms of horizontal distance, resulting in lower ridges with wider swales, implying less intense storms, as at Espenberg. During comparatively recent times—probably the last 500 years—a single ridge has formed. Though undated radiometrically, the modern ridge is the widest and highest of all Choris ridges and contains Thule, late-prehistoric, and modern settlements. In summary, because gravel ridges respond primarily to storm events, it may be argued that storminess increased after 2000 B.P.— and this event must be correlated with the most recent Unit IV, less than 1200 years old, at Espenberg. The alignment of Choris to the west allows the impact of only a limited number of intense storms—from the west-northwest, accounting for the infrequency of preserved storm events there (Figure 1).

SUMMARY AND CONCLUSIONS

To study the history of prograding beach complexes in Kotzebue Sound, I have used several lines of inference, relying on radiometric dates from archaeological sites as chronological markers. The delineation of chrono- and lithostratigraphic units relies on the surficial record of geomorphic and ecologic processes. These are reflected by changes in plant communities, lake development, paleosol formation, and the appearance of cryogenic features such as convolute bedding, frost cracks, and palsas. The wider applicability of the geoarchaeological method employed at Cape Espenberg is illustrated by a correlation with a partially dated beach-ridge complex at the nearby Choris Peninsula.

In the last 4000 years, the Cape Espenberg sand spit has undergone episodic alternations in the amount of horizontal progradation vs. vertical dune building. Alternations in horizontal progradation and vertical dune growth may be explained by changes in the cumulative dune/beach sediment budget, a concept advanced by Psuty (1988). As described at Cape Espenberg, vertical dune growth is greatest in areas experiencing net beach erosion, while progradation is marked by the addition of low beach ridges occasionally capped by low dunes, but separated by wide swales. The progradational regime is linked to stable, perhaps warm, climatic conditions interrupted by few storms, coupled with low winter winds. Rapid progradation prevailed from about 4000 until 3300 B.P. and from 2000 until 1200 B.P. Composite dunes, the vertical component, possess a beach-ridge nucleus topped by several generations of dunes, reactivation surfaces, and blowouts infilled by planar beds, as revealed in cutbank exposures. Dune-building activity is concentrated in three periods at Espenberg: (1) 3300 to 2000 B.P., (2) 1200 to 600 B.P., and, with lessened intensity, (3) 250 B.P. to the present. Less intense dune activity occurred before 3700, 1800 (estimated), and about 1400 to 1300 B.P. If dune building is correlative with increased storminess, as, for example, in the North Sea (Jelgersma et al., 1970; Lamb, 1988) and Australia (Thom, 1978, 1984), then the ridges at Espenberg are a proxy climatic indicator. On the gravel ridges of the Choris Peninsula A subcomplex, higher, wider composite ridges formed during periods of greater storm frequency, as in the last few centuries and 2900 to 2200 B.P., while fewer, lower ridges built during the intervening period.

The correlation of the Espenberg and Choris depositional histories with other climatic indicators requires an extended treatment, not possible in the present context (cf., Mason, 1990). To summarize, briefly, it is quite significant that dune building episodes cooccur with widespread alluviation in west and north Alaska (Hamilton, 1981) and glacial expansion throughout Alaska (Calkin, 1988). By contrast, the progradational phase correlates with peat formation atop alluvial terraces (Hamilton, 1981; Kaufman et al., 1989) and expansion of the tree line into higher altitudes in eastern Alaska (Denton and Karlén, 1977).

In conclusion, by using a geoarchaeological methodology, I present depositional records from two Kotzebue Sound beach-ridge complexes as a proxy record of climatic fluctuations. Hopefully, researchers working in other prograding systems will find the geoarchaeological method readily applicable, making modifications for local geomorphic conditions. It is not a substitute for the standard, deep-coring method, but, rather, is a complementary approach. However, to use the method, researchers will need to assess the cultural factors which influence settlement pattern and the context of archaeological radiocarbon samples within geologic strata.

ACKNOWLEDGMENTS

The National Park Service (Alaska Regional Office and Bering Land Bridge National Preserve) provided logistic and financial support, as well as monies for radiocarbon dating. Additional support was provided by the Geist Fund of the University of Alaska Museum and the Alaska Quaternary Center. The cooperative efforts of Jeanne Schaaf and Dr. Roger K. Harritt were essential in undertaking the research. Dr. David M. Hopkins provided insight, his well-known intensive editorial review, and intellectual support. James Jordan, Dale Vinson, Mark Moore, and Stefanie Ludwig assisted in the field and lab research.

REFERENCES

Ahlbrandt, T.S. and Andrews, S. (1978) Distinctive sedimentary features of cold-climate Eolian deposits, North Park, Colorado. *Paleogeogr. Paleoclimatol. Paleoecol.* 25: 327.

Anderson, D.D. (1984) Prehistory of North Alaska. In: *Handbook of North American Indians,* vol. 5, Arctic, pp. 80–93. Damas, D., Ed. Smithsonian Institution, Washington, DC.

Anderson, D.D. (1988) Onion Portage: The Archaeology of a Stratified Site from the Kobuk River, Northwest Alaska. *Anthropol. Pap. Univ. Alaska.* 22: 1.

Bird, E.C.F. and Schwartz, M.L. (1985) *The World's Coastline.* Van Nostrand Reinhold, New York.

Bockstoce, J. (1979) *The Archaeology of Cape Nome.* Univ. Mus. Monogr. University of Pennsylvania, Philadelphia.

Calkin, P. (1988) Holocene glaciation of Alaska (and adjoining Yukon Territory, Canada). *Quat. Sci. Rev.* 7: 159.

Carter, R.W.G. (1986) The morphodynamics of beach-ridge formation: Magilligan, northern Ireland. *Mar. Geol.* 73: 191.

Carter, R.W.G. (1988) *Coastal Environments: An Introduction to the Physical Ecological and Cultural Systems of Coastlines.* Academic Press, New York.

Carter, R.W.G. and Orford, J.D. (1981) Overwash processes along a gravel beach in south-east Ireland. *Earth Surf. Process. Landforms.* 6: 413.

Chapman, V.J. (1978) *Coastal Vegetation.* Pergamon Press, London.

Clark, J.A. and Lingle, C.S. (1979) Predicted sea level changes 18,000 years B.P. to the present caused by the Late Glacial Retreat of the Anatarctic ice sheet. *Quat. Res.* 11(3): 279.

Collins, H.B. (1937) Archaeology of St. Lawrence Island, Alaska. *Smithson. Misc. Collect.* 96(1).

Creager, J.S. and McManus, D.A. (1966) Geology of the Southeastern Chukchi Sea, In: *Environment of the Cape Thompson Region, Alaska,* pp. 755–786. Wilimovsky, N.J. and Wolfe, J.N., Eds. U.S. Atomic Energy Commission, Oakridge, TN.

Curray, J.R. (1964) Transgressions and regressions. In: *Papers in Marine Geology: Shepard Commemorative Volume,* pp. 175–183. Miller, R.L., Ed. Macmillan, New York.

Curray, J.R., Emmel, F.J., and Crampton, P.J.S. (1969) Holocene history of a strand plain, lagoonal coast, Nayarit, Mexico. In: *Coastal Lagoons: a Symposium,* pp. 63–100. Castanares, A.A. Ed. Universidad Nacional Autonoma de México, México, D.F.

Dansgaard, W. et al. (1984) North Atlantic climatic oscillations revealed by deep Greenland ice cores. *Am. Geophys. Union Monogr.* 29: 288–298.

Davis R.A. (1985) Beach and nearshore zone. In: *Coastal Sedimentary Environments,* pp 379–444. Davis, R.A., Ed. Springer-Verlag, New York.

Davis, R.A. and Fox, W.T. (1975) Process-response patterns in beach and nearshore sedimentation. I. Mustang Island, Texas. *J. Sediment. Petrol.* 45(4): 852.

Davis, R.A., Fox, W.T., Hayes, M.O., and Boothroyd, J.C. (1972) Comparison of ridge and runnel systems in tidal and non-tidal environments. *J. Sediment. Petrol.* 42(2): 413.

Denton, G.H. and Karlén, W. (1977) Holocene glacial and tree-line variations in White River and Skolai Pass, Alaska and Yukon Territory. *Quat. Res.* 7: 63.

DePratter, C.B. and Howard, J.D. (1977) History of shoreline changes determined by archaeological dating: Georgia coast, U.S.A. *Trans Gulf Coast Assoc. Geol. Soc.* 27: 252.

Dumond, D.E. (1987) A re-examination of Eskimo-Aleut prehistory. *Am. Anthropol.* 89: 32.

Eittreim, S., Grantz, A. and Whitney, O.T. (1977) Tectonic imprints on sedimentary deposits in Hope basin. In: *The U.S. Geological Survey in Alaska: Accomplishments during 1976,* pp. 100–103. Blean, K.M., Ed. U.S. Geol. Surv. Circ. 751-b.

Fisher, J.J. (1967) Development Pattern of Relict Beach Ridges of the Outer Banks, North Carolina. Ph.D. thesis, University of North Carolina, Chapel Hill.

Folk, R.L. and Ward, W.C. (1957) Brazos River bar: a study in the significance of grain size parameters. *J. Sediment. Petrol.* 27(1): 3.

Giddings, J.L., Jr. (1952) *Arctic Woodland Culture of the Kobuk River.* University Museum, Philadelphia.

Giddings, J.L., Jr. (1963) Some arctic spear points and their counterparts. *Anthropol. Pap. Univ. Alask.* 10(2): 1.

Giddings, J.L., Jr. (1966) Cross-dating the archeology of northwestern Alaska. *Science.* 153: 127.

Giddings, J.L., Jr. (1967) *Ancient Men of the Arctic.* Alfred Knopf, New York.

Giddings, J.L., Jr. and Anderson, D.D. (1986) Beach Ridge Archaeology of Cape Krusenstern: Eskimo and Pre-Eskimo Settlements around Kotzebue Sound. Publ. in Archaeol. 20, National Park Service, Washington, D.C.

Godfrey, P. J., Leatherman, S.F., and Zaremba, R. (1979) A geobotanical approach to classification of barrier beach systems. In: *Barrier Islands from the Gulf of St. Lawrence to the Gulf of Mexico,* pp 99–126. Leatherman, S.P., Ed. Academic Press, New York.

Hamilton, T.D. (1981) Episodic alluviation in the Central Brooks Range—chronology, correlations and climatic implications. In: *United States Geological Survey in Alaska: Accomplishments during 1979,* pp. B21–B24. Albert, N.R.D. and Hudson, T., Eds. U.S. Geol. Surv. Circ. 823 B.

Harritt, R.K. (1989) Recent archaeology in Bering Land Bridge: National Preserve: the 1988 field season at Cape Espenberg. Paper presented at the 16th Annu. Meet. Alaska Anthropology Association, Anchorage.

Harritt, R.K. (1990) Recent archaeology in Bering Land Bridge National Preserve: the 1989 field season at Cape Espenberg and in the Ikpek area. Paper presented at the 17th Annu. Meet. Alaska Anthropology, Association, Fairbanks.

Hayes, M.O. (1979) Barrier island morphology as a function of tidal and wave regime, In: *Barrier Islands from the Gulf of St. Lawrence to the Gulf of Mexico,* pp. 1–28. Leatherman, S.P., Ed. Academic Press, New York.

Hesp, P.A. (1988) Morphology, dynamics and internal stratification of some established foredunes in southeast Australia. *Sediment. Geol.* 55: 17.

Hopkins, D.M. (1967) Quarternary marine transgressions in Alaska. In: *The Bering Land Bridge,* pp. 47–90. Hopkins, D.M., Ed. Stanford University Press, Stanford.

Hopkins, D.M. (1988) The Espenberg Maars: a Record of Explosive Volcanic Activity in the Devil Mt.-Cape Espenberg, Seward Peninsula. Alaska. In: *The Bering Land National Preserve: An Archaeological Survey,* pp. 262–321. Schaaf, J., Ed. Res./Resour. Managem. Rep. 14, National Park Service, Anchorage.

Hosley, E. (1972) Archaeological Evaluation of Ancient Habitation site, Point Hope, Alaska. In: *Point Hope Beach Erosion, Point Hope, Alaska,* pp. 1–47. Survey Report, Alaska District, Corps of Engineers, Archorage.

Jelgersma, S., de John, J., Zagwijn, W. H., and van Regteren Altena, J.F. (1970) The coastal dunes of western Netherlands: geology, vegetational history and archaeology. *Med. Rijks Geol. Dienst. N. Ser.* 21: 93.

Jordan, J.W. (1988) Erosion Characteristics and Retreat Rates along the North Coast of Seward Peninsula, In: *The Bering Land National Preserve: An Archaeological Survey,* pp. 322–362. Schaaf, J., Ed. Res./Resour. Manage. Rep. 14, National Park Service, Anchorage.

Jordan, J.W. (1990). Late Holocene Evolution of Barrier Islands in the Southern Chukchi Sea, Alaska. M.S. thesis, Quaternary Studies, University of Alaska, Fairbanks.

Kaufman, D.S., Calkin, P.E., Whitford, W.B., Przybyl, B.J., Hopkins, D.M., Peck, B.J., and Nelson, R.E. (1989) Surficial Geologic Map of the Kigluaik Mountains Area, Seward Peninsula, Alaska. U.S. Geol. Surv. Misc. Field Studies Map MF 2074.

Koster, E.A. and Dijkmans, J.W.A. (1988) Niveo-aeolian deposits and denivation forms, with special reference to the Great Kobuk Sands Dunes, northwestern Alaska. *Earth Surf. Process. Landforms.* 13: 153.

Kraft, J.C. and Chrzastowski, M.J. (1985) Coastal stratigraphic sequences. In: *Coastal Sedimentary Environments,* pp. 625–664. Davis, R.A., Springer Verlag, New York.

Kraft, J.C., Kayan, I., and Aschenbrenner, S.E. (1985) Geological Studies of Coastal Change applied to Archaeological Settings. In: *Archaeological Geology,* pp. 57–84. Rapp, G., Jr. and Gifford, J.A., Eds. Yale University Press, New Haven.

La Belle, J.C., Wise, J.L., Voelker, R.P., Schulze, R.H., and Whol, G.M. (1983) *Alaska Marine Ice Atlas.* Arctic Environmental Information and Data Center, University of Alaska, Anchorage.

Lamb, H.H. (1982) *Climate, History and the Modern World.* Methuen, London.

Lamb, H.H. (1988) Climate in historical times and transgressions of the sea, storm floods and other coastal changes. In: *Weather, Climate and Human Affairs,* pp. 78–103. Lamb, H.H., Ed. Routledge, London.

Larsen, H. and Rainey, F. (1948) Ipiutak and the Arctic whale hunting culture. *Anthropol. Pap. Am. Mus. Nat. Hist.* 42.

Larson, L.H. (1980) *Aboriginal Subsistence Technology on the Southeastern Coastal Plain during the Late Prehistoric Period,* pp. 6–23. University Presses of Florida, Gainesville.

Lewis, D.W. (1983) *Practical Sedimentology.* Hutchinson Ross, Stroudsburg, PA.

Mackay, J.R. (1974) Ice-wedge cracks, Garry Island, Northwest Territories. *Can. J. Earth Sci.* 11: 1366.

Mason, O.K. (1988a) The sand ridge stratigraphy of northern Seward Peninsula. In: *The Bering Land National Preserve: An Archaeological Survey,* pp. 364–409. Schaaf, J.J., Ed. Res./Resour. Manage. Rep. 14, National Park Service, Anchorage.

Mason, O.K. (1988b) Geoarchaeological correlations at NW Alaska beach ridge complexes. Paper presented at the 54th Annu. Meet. Society for American Archaeology, Phoenix.

Mason, O.K. (1990) Beach Ridge Geomorphology of Kotzebue Sound: Implications for Paleoclimatology and Archaeology. Ph.D. thesis, Quaternary Science, University of Alaska, Fairbanks.

Mason, O.K. and Ludwig, S.L. (1990) Resurrecting beach ridge archaeology: parallel depositional histories from St. Lawrence Island and Cape Krusenstern, Alaska. *Geoarchaeology.* 5(4): 349.

McCullough, D.S., Taylor, W., and Rubin, M. (1965) Stratigraphy, non-marine mollusks and radiometric dates from quaternary deposits in Kotzebue Sound, western Alaska. *J. Geol.* 73(3): 442.

McManus, D.A., Creager, J.S., Echols, R.J., and Holmes, M.L. (1983) The Holocene transgression on the Arctic flank of Beringia: Chukchi Valley to Chukchi Estuary to Chukchi Sea. In: *Quaternary Coastlines and Marine Archaeology: Towards a Prehistory of Landbridges and Continents,* pp. 365–388. Masters, P.M. and Fleming, N.C. Eds. Academic Press, New York.

Meehan B. (1982) *Shell Bed to Shell Midden.* Australian Institute for Aboriginal Studies, Canberra.

Miller, T.P. and Smith, R.L. (1987) Late quaternary caldera-forming eruptions in the Eastern Aleutian arc, Alaska. *Geology.* 15: 434.

Moore, G.W. (1966) Arctic beach sedimentation. In: *Environment of the Cape Thompson Region, Alaska,* pp. 587–608. Wilimovsky, N. and Wolfe, J.N., Eds. Atomic Energy Commission, Oak Ridge, TN.

Moore, G.W. and Giddings, J.L., Jr. (1961) Record of 5000 years of Arctic Wind Direction Recorded by Alaskan Beach Ridges. Abstract, Spec. Geol. Soc. Am. Pap. 68.

Nelson, E.W. (1899) *Eskimos About Bering Strait.* Bureau of American Ethnology Annual Report. 18(1).

North American Commission on Stratigraphic Nomenclature. (1983) North American Stratigraphic Code. *Am. Assoc. Petrol. Geol. Bull.* 67(5): 841.

Orford, J.D. (1987) Coastal processes: the coastal response to sea level variation. In: *Sea Surface Studies: A Global View,* pp. 415–463. Devoy, R.V.N., Ed. Croom Helm, New York.

Patton, W.W. and Miller, T.P. (1968) Regional Geologic Map of the Selawik and Southeast Baird Mountains Quadrangle, Alaska. U.S. Geol. Surv. Misc. Geol. Invest. Map I-530.

Peratrovich & Nottingham, Inc. (1982) Shishmaref Erosion Control Engineering Studies. Alaska Department of Transportation and Public Facilites, Anchorage.

Péwé, T.L. (1975) Quaternary Geology of Alaska. U.S. Geol. Surv. Prof. Pap. 835.

Porter, S.C. (1986) Pattern and forcing of Northern Hemisphere glacier variations during the last millennium. *Quat. Res.* 26: 27.

Porter, S.C. and Denton, G.H. (1967) Chronology of neoglaciation in the North American Cordillera. *Am. J. Sci.* 265: 177.

Psuty, N.P. (1988) Sediment budget and dune/beach interaction. *J. Coast. Res. Spec. Issue.* 3: 1.

Pye, K. (1983). Early Post-depositional Modification of Aeolian Dune Sands. In: *Eolian Sediments and Processes,* pp. 197–221. Brookfield, M.E. and Ahlbrandt, T.S., Eds. Elsevier, Amsterdam.

Ranwell, D. (1972) *Ecology of Salt Marshes and Sand Dunes.* Chapman and Hill, London.

Reineck, H. and Singh, I. (1980) *Depositional Sedimentary Environments,* pp. 352–356. Springer-Verlag, New York.

Reinson, G. (1984) Barrier island and associated strand-plain systems. In: *Facies Models,* pp. 119–140, 2nd ed., Walker, R.G., Ed. Geological Association of Canada, Toronto.

Riehle, J.R., Meyer, C.E., Ager, T.A., Kaufman, D.S., and Ackerman, R.E. (1987) The Aniakchak Tephra deposit, a late Holocene marker horizon in Alaska. In: *Geologic Studies in Alaska 1986,* pp. 19–24. Hamilton T.D. and Galloway, J.P., Eds. U.S. Geol. Surv. Circ. 998.

Ritchie, W. and Penland, S. (1988) Rapid dune changes associated with overwash processes on the deltaic coast of south Louisiana. *Mar. Geol.* 81: 97.

Roep, T.B. (1986) Sea level markers in coastal barrier sands: examples from the North Sea coast. In: *Sea Level Research: A Manual for the Collection and Evaluation of Data,* pp. 97–128. van de Plassche, O., Ed. Geo-Books, Norwich, England.

Sandweiss, D.H. (1986) The beach ridges at Santa, Peru: El Nino, uplift and prehistory. Geoarchaeology. 1(1): 17.

Schaaf, J. (1988) The Bering Land National Preserve: An Archaeological Survey, Vol. I (unclassified) and II (classified site reports), Res./Resour. Manage. Rep. 14, National Park Service, Anchorage.

Searle, D.J. and Woods, P.J. (1986) Detailed documentation of a Holocene sea level record in the Perth region, southernwestern Australia. *Quat. Res.* 26: 299.

Seppälä, M. and Koutaniemi, L. (1985) Formation of a string and pool topography as expressed by morphology, stratigraphy and current processes on a mire in Kuusamo, Finland. *Boreas.* 14: 287.

Shepard, F. (1973) *Submarine Geology,* 3rd Ed. Harper & Row, New York.

Shepard, F.P. and Wanless, H.R. (1971) *Our Changing Coastlines.* McGraw-Hill, New York.

Shepard, F.P. and Young, R. (1961) Distinguishing between beach and dune sands. *J. Sediment. Petrol.* 31(2): 196.

Shinkwin, A.D. (1977) *Excavations at Point Hope, Alaska.* Report to the National Park Service and North Slope Borough, University of Alaska, Fairbanks.

Soil Survey Staff. (1975) Soil Taxonomy: A Basic System of Soil Classification for Making and Interpreting Soil Surveys, pp. 25, 202, 203. Soil Conservation Service, U.S. Department of Agriculture Handbook No. 436, Washington, D.C.

Stuiver, M. and Pearson, G.W. (1986) High precision calibration of radiocarbon time scale AD 1950–500 BC. In: Proc. 12th Int. ^{14}C Conference, Radiocarbon 28(2B), pp. 805–838. Stuiver, M. and Kra, R.S., Eds.

Stuiver, M., Pearson, G.W., and Brazunias, T. (1986) Radiocarbon age calibration of marine samples back to 9000 cal yr B.P. In: Proc. 12th Int. ^{14}C Conference, Radiocarbon 28(2B), pp. 980–1021. Stuiver, M. and Kra, R.S., Eds.

Tanner, W.F. (1988) Beach ridge data and sea level history from the Americas. *J. Coast. Res.* 4(1): 81.

Thom, B.G. (1978) Coastal sand deposition in southeast Australia during the Holocene. In: *Landform Evolution in Australasia,* pp. 197–214. Davies, J.L. and Williams, M.A.J., Eds. Australian National University Press, Canberra.

Thom, B.G. (1984) Transgressive and regressive stratigraphies of coastal sand barriers in southeast Australia. *Mar. Geol.* 56: 137.

Vandenberghe, J. (1988) Cryoturbations, In: *Advances in Periglacial Geomorphology,* pp. 179–198. Clark, M.J. Ed. John Wiley & Sons, New York.

Visher, G.S. (1969) Grain-size distributions and depositional processes. *J. Sediment. Petrol.* 39: 1074.

Westgate, J.A. and Briggs, N.D. (1985) Tephrochronology and fission track dating. In: *Dating Methods of Pleistocene Deposits and Their Problems,* pp. 31–38. Rutter, N.W., Ed. Geosci. Can. Reprint Ser. 2.

Wise, J.L., Comiskey, A.L., and Becker, R. (1981) *Storm Surge Climatology and Forecasting in Alaska.* Arctic Environmental Information and Data Center, Anchorage.

Wright, L.D., Chappell, J., Thom, B.G., Bradshaw, M.P., and Cowell, P. (1979) Morphodynamics of reflective and dissipative beach and inshore systems: southeastern Australia. *Mar. Geol.* 32: 105.

Chapter

4

Paleoshorelines and the Sambaquis of Brazil

Kenitiro Suguio
Instituto de Geociências
University of São Paulo
São Paulo, Brazil

Louis Martin
ORSTOM/ON-CNPq
Departmento de Geofísica
Rio de Janeiro, Brazil

Jean-Marie Flexor
ON-CNPq
Departmento de Geofísica
Rio de Janeiro, Brazil

INTRODUCTION

The paleoshorelines of the past 7000 years have been systematically studied since 1974 by scientists from the University of São Paulo, the Federal University of Bahia, and the National Observatory (CNPq) in association with the Institute of Scientific Research for Development in Cooperation (ORSTOM), in France. Studies have been completed on the Quaternary coastal formations of the States of São Paulo, Rio de Janeiro, Bahia, Sergipe, Alagoas, northern Espírito Santo, Paraná, and Santa Catarina (Figure 1). Systematic surveys of sedimentary deposits at the mouths of the Paraíba do Sul, Doce, Jequitinhonha, and São Francisco Rivers have also been undertaken in order to determine the role played by sea level variations and by littoral drift in the construction of these coastal plains (Figure 1). The Quaternary coastal marine deposits of Rio Grande do Sul have been investigated by a team from the federal university of that state (Villwock et al., 1986). The most important conclusions from these studies have been reviewed in Suguio et al. (1985) and Martin et al. (1987).

To reconstruct a paleoshoreline, it is necessary to define a marker for it in time and space. The latter can be established by determining its present position with respect to the shoreline at the time of its formation. The age of formation or deposition of the marker must also be determined. A marker thus defined gives the position of an ancient shoreline at a certain time in the past relative to the present shoreline.

Evidence for Quaternary paleoshorelines comes from several sources. Sandy littoral

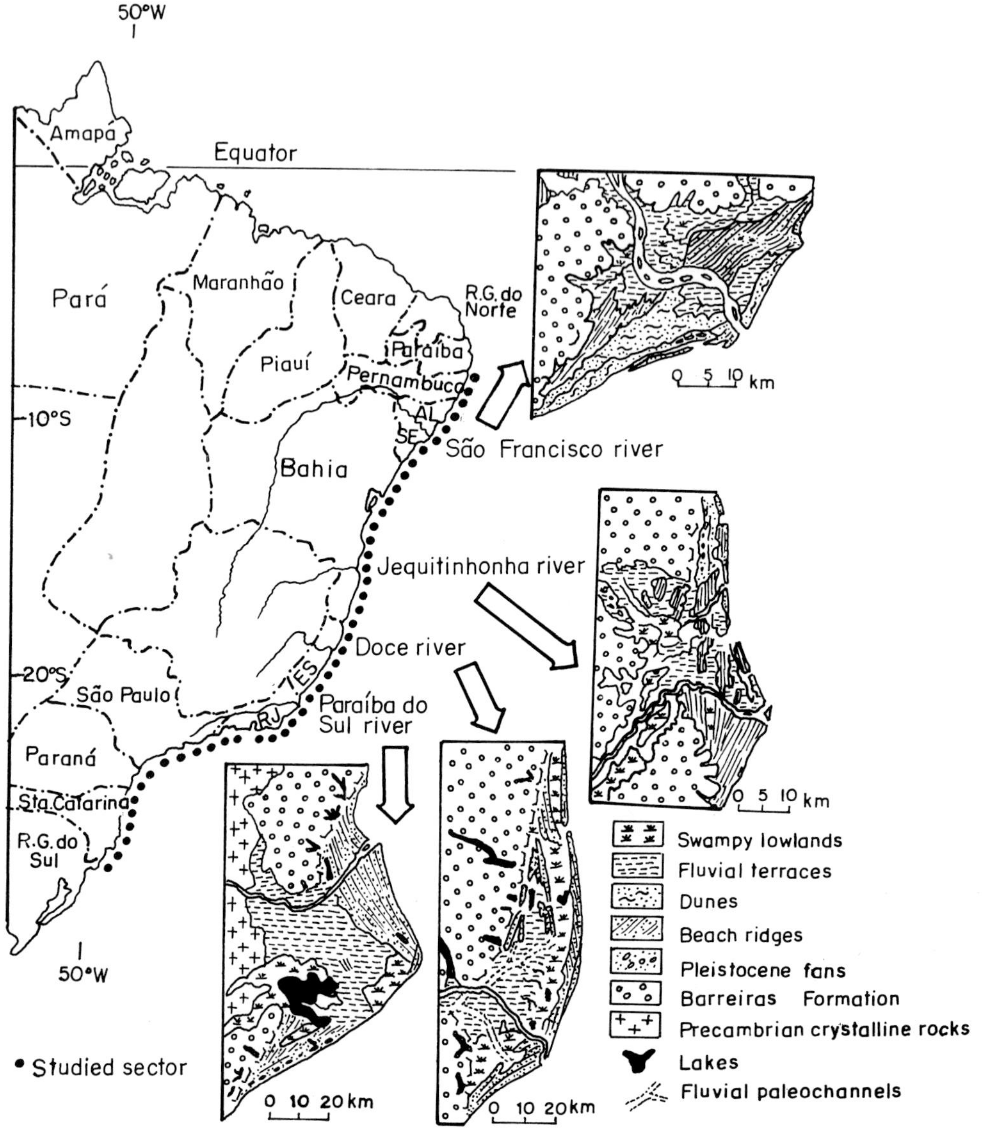

Figure 1. Coastal zones of Brazil with indication of the studied areas, including the most important Quaternary deltaic features.

deposits elevated high above present, equivalent deposition zones provide indisputable evidence for former raised sea levels. Detailed mapping together with about 800 absolute age determinations has led to the recognition of two principal generations of sandy terraces, recording two periods of high Quaternary marine stands.

Along most of the rocky portion of the Brazilian coast, there is biological evidence for marine levels higher than at present. This generally consists of oyster and limpet (gastropod) deposits and sea urchin holes situated above the present living zone of these organisms. In sandy terraces, one can also find fossilized burrows of *Callichirus major,* a marine arthropod, above their present living zone (Suguio and Martin, 1976a).

In coastal areas of Brazil, particularly between the States of Espírito Santo and Santa Catarina, several hundred giant sambaquis (shell middens) have been built-up from the food remains of ancient human inhabitants (Figure 2) (Löfgren, 1893; Bigarella, 1951a, 1951b; Empéraire and Laming, 1956; Hurt, 1974; Martin and Suguio, 1976b; Martin et al., 1981/1982). The location and species content of these sambaquis can also provide evidence of past shorelines.

PALEOSHORELINES AND RELATIVE SEA LEVEL CHANGES DURING THE LAST 7000 YEARS ALONG THE BRAZILIAN COAST

Papers by Martin and Suguio (1975, 1978), Martin et al. (1979a,b, 1980), Suguio and Martin (1976b, 1978), Suguio et al. (1980), and Bittencourt et al. (1979) and others, recently reviewed by Suguio et al. (1985) and Martin et al. (1987), provide valuable information about paleoshorelines and relative sea level changes along the Brazilian coast.

An evolutionary history of these regions during the Quaternary has been reconstructed from this information. About 123,000 years ago, the relative sea level stood about 8 ± 2 m above the present level. Between the end of this transgression (Cananéia transgression) and the beginning of the following regression, a first generation of sandy beach ridges was deposited. During the last glacial when the sea level was low, these deposits were carved by rivers, and a system of valleys, some very large and deep, developed. The original wave-built terrace surfaces, frequently marked by beach-ridge alignments, have been preserved only in interfluvial zones. When, during the Holocene, the sea level reached its

Figure 2. Partial view of a huge sambaqui, located beside a sand dune, near the present strandline. It is composed almost exclusively of *Anomalocardia brasiliana* shells (Jaguaruna, State of Santa Catarina). For scale, compare with the vehicle used during the field surveys.

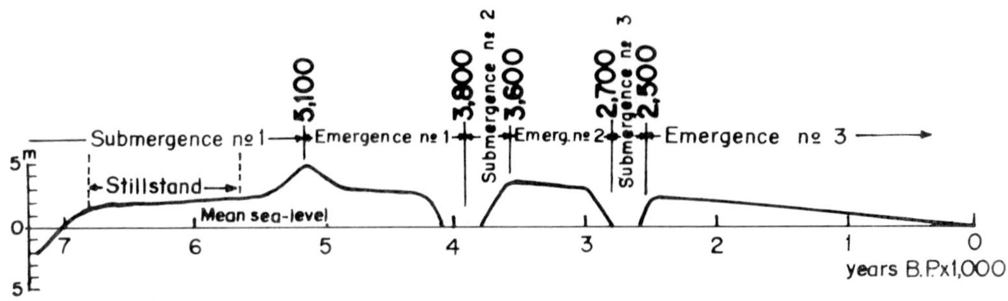

Figure 3. Schematic relative sea level fluctuation curve during the last 7000 years for the northern area of Salvador (State of Bahia). From Martin, L., Flexor, J.-M., Vilas-Boas, G.S., Bittencourt, A.C.S.P., and Guimarães, M.M.M. (1979a). In: Proc. 1978 Int. Symp. Coastal Evolution in the Quaternary, pp. 264–274. Suguio, K., Fairchild, T.R., Martin, L., and Flexor, J.-M., Eds., São Paulo. Used by permission.

present position (about 7000 to 6500 C-14 years ago; Santos transgression), the low-lying parts of the ancient terraces were invaded by the sea, forming extensive lagoonal systems.

Relative sea level fluctuation curves for the last 7000 years have been constructed for several sectors of the Brazilian coast, which indicate that:

- The present-day mean sea level was surpassed for the first time about 6500 to 7000 B.P.
- By about 5100 B.P., the sea level had risen to about 4 to 5 m above today's mean sea level.
- At about 3800 B.P., the sea level experienced a low stand and was positioned slightly below today's mean sea level.
- At 3600 B.P., the sea level rose to 2.5 to 3.5 m above today's level.
- At 2700 B.P., the sea level again experienced a low stand, dropping slightly below the present-day mean sea level.
- By about 2500 B.P., a third high stand was reached. At this time, the sea level rose 2 to 2.5 m above today's mean sea level, and since then it has been gradually dropping.

In summary (Figure 3), three main phases of submergence (7000 to 5100 B.P., 3800 to 3600 B.P., and 2700 to 2500 B.P.) and three main phases of emergence (5100 to 3800 B.P., 3600 to 2700 B.P., and 2500 B.P. to the present) can be recognized along the central Brazilian coast during the Holocene.

INFORMATION ABOUT PALEOSHORELINES TO BE DERIVED FROM THE SAMBAQUIS

Geographic Position

In coastal areas subjected to an emergence period following a submergence phase, such as the central coast of Brazil, prehistoric settlement sites located further inland were probably associated with a period of maximum lagoonal extent during a period of maximum sea level. Some central coast sambaquis occur as far as 30 to 35 km inland from the present shoreline.

It may be assumed that the location of settlements is decided primarily by proximity to shellfish beds of sufficient richness to provide nourishment over a long period of time (Fairbridge, 1976). Obviously, shallow and quiet water bays, lagoons, and estuarine zones, with muddy surfaces exposed during low tides, would be more favorable locations than sandy areas at the open-ocean margin. In the former case, biological productivity is much higher. It is probable that the prehistoric occupants chose campsites above high-tide level, immediately adjacent to favorable collecting grounds where conditions of comfort and safety were available. If this is true, one may assume a close relationship between the positions of the sambaquis and the paleoshorelines, and the position of some sambaquis well inland can be explained only by a lagoon extension significantly greater, and hence a sea level higher, than at present.

While it is relatively easy to establish the geographic relationship between the position of a sambaqui and a nearby ancient lagoon, estuary, or bay, it is much more difficult to establish the vertical relationship between the altitude of the sambaqui's substrate and the position of the sea level during its construction. We can only assume that initially it was above the local high-water spring tide (HWST) level, a very important assumption for interpreting sambaquis whose substrates are now located beneath the present HWST. It is possible that some of the latter sambaquis originated as wastedumps below houses which were built on poles standing in shallow water during high tides. This possibility can be eliminated by a detailed investigation of the sambaqui's substrate and composition. The assumption that sambaquis were built-up close to mollusk-collecting areas allows the nearest paleoshoreline to the site to be cited as the location of shellfishing areas at the time of the site's occupation.

Size of the Sambaquis

The sambaquis exhibit a variety of shapes and sizes, ranging from huge, elongated ($300 \times 60 \times 60$ m) or ellipsoidal ($86 \times 40 \times 25$ m) mounds to small, circular (10×1 to 1.5 m) accumulations. One sambaqui examined by Fairbridge (1976) in the State of Santa Catarina was approximately 20 m high and 100 m in diameter, representing about 2.5 billion shellfish, that is, 100 shellfish per day for a group of 100 people for 500 years.

In general, the more inland sambaquis are smaller than those situated near the present strandline. This is to be expected as maximum lagoon extent periods were very short. Thus, only in the outer lagoon zones was there sufficient time for the buildup of huge sambaquis.

Faunal Composition

About 50 different species of mollusks have been identified within the sambaquis of the Brazilian coast (Bigarella, 1949), but there are only five dominant species:

- *Anomalocardia brasiliana* Gmelin
- *Ostrea brasiliana* sp.
- *Ostrea arborea* Chemnitz
- *Lucina jamaicensis* Chemnitz
- *Modiolus brasiliensis* Chemnitz

Except for *O. arborea,* which lives fixed on aerial roots of mangrove trees, these bivalves live within sandy or clayey-sandy sediments deposited in shallow-water lagoons and bays. Shell remains of other organisms are generally very scarce.

TABLE I. Variation of $\delta^{13}C_{PDB}$ as a Function of Position of the Samples Within a Paleolagoon, Coastal Plain of the State of São Paulo

Sample	Radiocarbon Age (B.P.)	$\delta^{13}C_{PDB}$ (o/oo)	Distance from Open Sea (km)
Itapoã III	5245 ± 125	−0.63	5
Jataituba	5235 ± 150	−3.26	20
Vapumaúva II	5080 ± 60	−0.06	1
Pariqüera-Açu	5035 ± 140	−3.66	24
Vapumaúva I	4680 ± 115	−0.12	2
A196	4636 ± 100	−6.17	18
Ubatuba	3870 ± 100	−0.02	?
A132	3775 ± 130	−5.69	33

From Flexor, J.-M., Martin, L., and Suguio, K. (1979). In: Proc. 1978 Int. Symp. Coastal Evolution in the Quaternary, pp. 356–375. Suguio, K., Fairchild, T.R., Martin, L., and Flexor, J.-M. Eds. São Paulo. Used by permission.

C-13/C-12 Ratios of the Carbonate Shells

$\delta^{13}C_{PDB}$ measurements of the carbonate shells show a spectrum of values changing as a function of the influence of continental environmental conditions during the carbonate formation. Shells from lagoonal organisms show $\delta^{13}C_{PDB}$ values between those for freshwater (about 13 o/oo) and those for marine organisms (about 0 o/oo), varying as a function of their geographic position within the lagoon (Flexor et al., 1979). Shells from outer zones (nearer the open-sea edge of a lagoon) show only slightly negative $\delta^{13}C_{PDB}$ values, while those from inner zones are clearly characterized by more negative values (Table I). In the latter cases, carbon derived from the decomposition of land plants has been incorporated by the mollusks within their carbonate shells. Deeper and inner portions of the lagoons are characterized by poor water circulation, promoting the accumulation of large quantities of plant-derived organic material. This is independent of water salinity and explains why the carbonate shells of inner lagoon organisms exhibit $\delta^{13}C_{PDB}$ values characterized by a strong land influence. Conversely, according to the increase or decrease of the lagoonal area, shells from the same locality will have distinct $\delta^{13}C_{PDB}$ values. If such shells are found in stratified sambaquis, their $\delta^{13}C_{PDB}$ values will be good indicators of lagoon oscillations and thus indirectly of sea level fluctuations (Martin et al., 1985).

For example, looking at Figure 4, if the sambaquis located at A, B, and C are of a similar age, they will be expected to show quite different $\delta^{13}C_{PDB}$ values for their carbonate shells, with A being quite negative and C close to zero. On the other hand, if B and C show several periods of occupation, the least-negative $\delta^{13}C_{PDB}$ values will correspond to periods of greatest lagoonal extent and the most negative to periods of lagoon contraction. These data cannot provide information on paleosea-level position with respect to the present level. However, they can indicate falling and rising lagoon levels and the approximate time of maximum and minimum still stands.

Dating of the Sambaqui

The usefulness of dates from sambaquis depends on the position of the dated sample within the mound. If a sample from the basal portion has been dated, it should correspond to the beginning of the site's occupation and thus provide a minimum date for the establishment of the neighboring shoreline. On the other hand, if the sample comes from the surface of the sambaqui and thus dates the end of occupation, it should indicate the approximate date at which shellfish were no longer available in the vicinity.

If central portions of inland sambaquis are dated, particularly periods of desertion and reoccupation, the departures may be tied to periods of lagoonal desiccation, or low-sea-level periods, and the reoccupations to periods of resurgent high sea level.

Types of Substrates

Sambaquis are located on four distinctive types of substrates (Figure 5).

Type I: Pleistocene marine terraces. These terraces are found inland from the maximum extent of Holocene lagoons. Therefore, the sambaquis found on them should have been built during periods of highest Holocene sea level.

Type II: Holocene marine terraces. These terraces were deposited as beach ridges during the regressive phases that followed the sea level maxima, particularly the 5100-B.P. Santos transgression. Sambaquis on these terraces have necessarily been constructed after high Holocene-sea-level stands, and in some cases their positions on the terraces can provide additional data on sea levels.

Type III: Ancient lagoonal deposits. These lagoon deposits are located in front of sandy marine terraces. Hence, the sambaquis found on them must have been constructed after a period of high sea level, probably during the ensuing lowering of the sea level. One

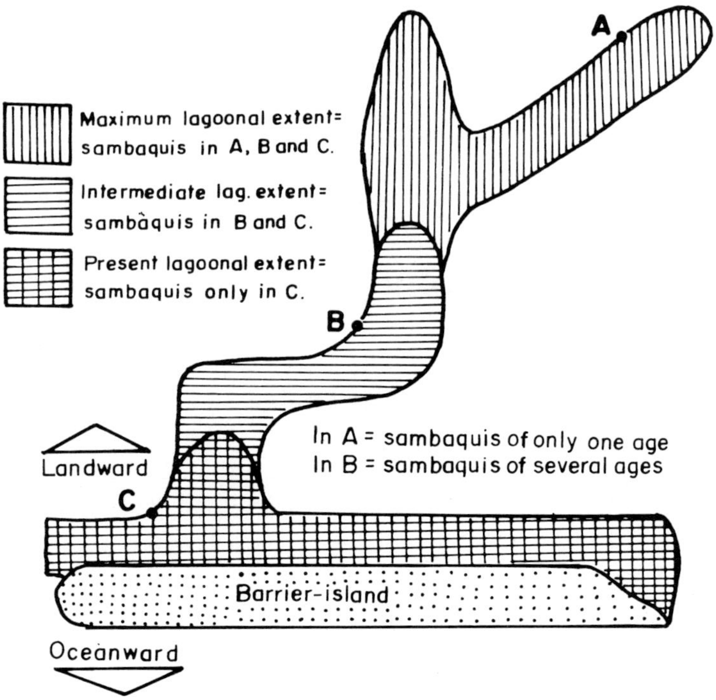

Figure 4. Age variations of the sambaquis as a function of changes in lagoonal extension along the Brazilian coast during the Holocene.

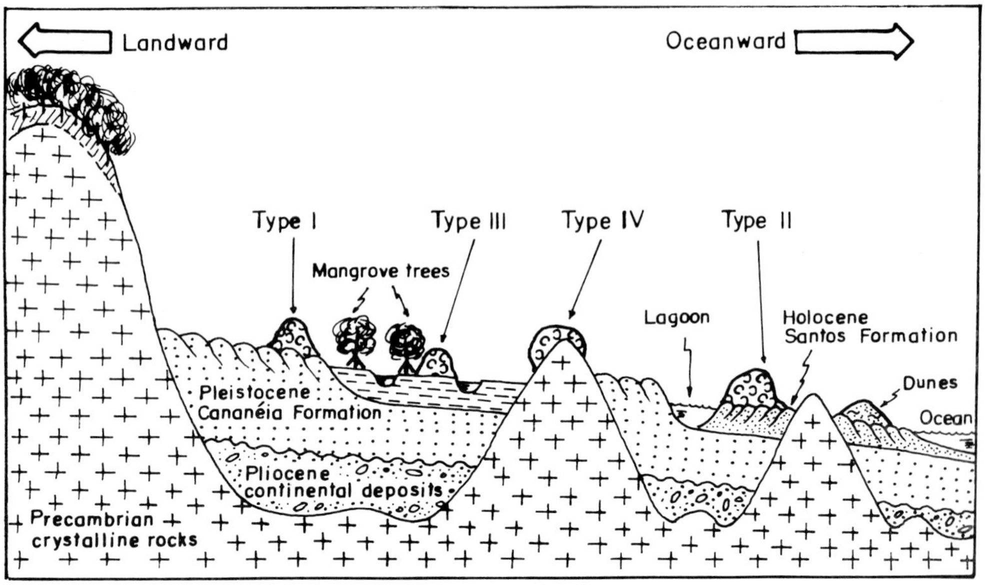

Figure 5. Distinctive types of substrates of the sambaquis along the Brazilian coast. (Modified from Fairbridge, R.W. (1976) *Science.* 191: 353–359.)

must be cautious in interpreting their elevation because the lagoonal organic clay is subject to compaction under the weight of the shell midden, thus diminishing the height of the substrate.

Type IV: Crystalline rocky hills. These hills are located at the margins or centers of ancient lagoons or bays. As they are never covered by high sea stands, they may have been occupied at any time and are thus not sensitive time markers. However, those situated more inland may have been settled when the lagoon reached its greatest extent.

COMPARISON OF GEOLOGICAL AND ARCHAEOLOGICAL DATA ON SEA LEVEL

Cananéia-Iguape Region (State of São Paulo)

The Cananéia-Iguape sedimentary plain has the shape of a large crescent and covers an area of about 2500 km^2. Its outer portion is presently drained by a lagoonal system and tidal estuaries (Figure 6). A large part of the plain is occupied by remnants of more or less dissected Pleistocene terraces and by dried-out ancient lagoons, at the far inland margins of which several sambaquis have been found.

Shell debris and wood fragments from littoral marine or lagoonal deposits have been dated, and this information, when associated with the nature of the sediments, indicates that (Figure 7A):

- About 6600 to 6500 B.P., the relative sea level reached approximately its present level and was rising.

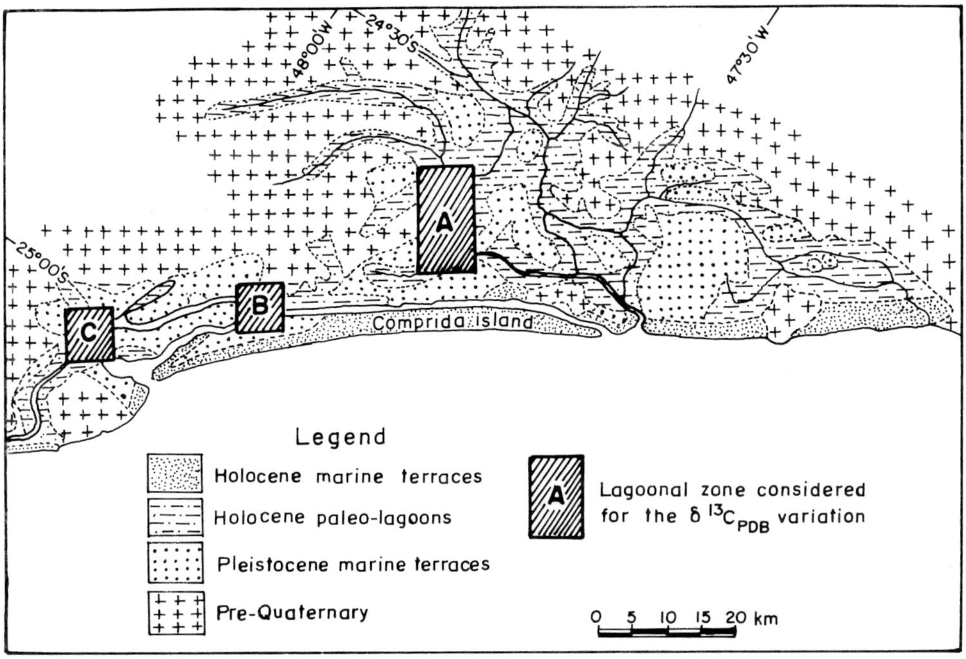

Figure 6. Schematic geologic map of the Cananéia-Iguape coastal plain (State of São Paulo).

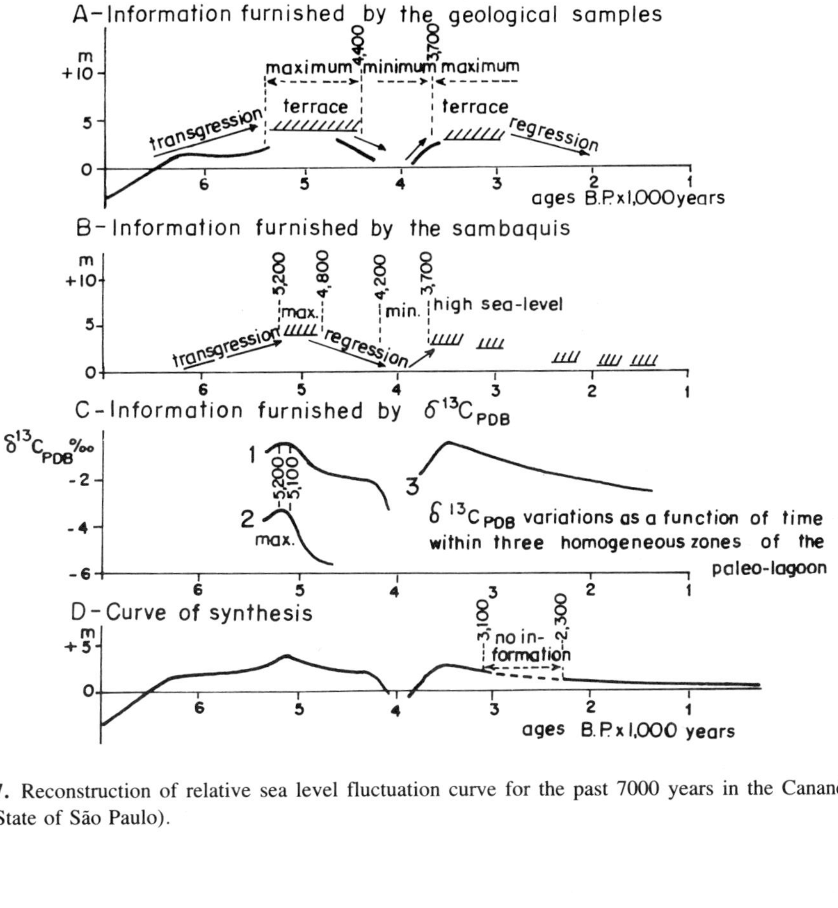

Figure 7. Reconstruction of relative sea level fluctuation curve for the past 7000 years in the Cananéia-Iguape region (State of São Paulo).

- About 5400 B.P., the relative sea level was about 1.5 m above its present level.
- About 4400 B.P., the relative sea level was about 2 m higher than its present level, but was falling. Hence, between 5400 and 4400 B.P., the relative sea level passed through a maximum height which, according to morphological data, must have been 3.5 to 4 m above the present one.
- Between 4400 and 3700 B.P., the relative sea level passed through a minimum.
- Between 3800 and 3700 B.P., the relative sea level rose very rapidly, reaching a second maximum of about 3 m above its present level.
- After the second maximum, the relative sea level returned more or less steadily to its present level. Unfortunately, we have insufficient data for detailing this return.

Dating of sambaquis in this area confirms the conclusions on relative sea level reached on the basis of geological evidence. As indicated above, between 5400 and 4400 B.P., the relative sea level was higher than at present, attaining a maximum of 3.5 to 4 m above its current level. The maximum lagoonal extent which would accompany this high stand should result in the establishment of inland sambaquis (Tables II and III). The location of these sambaquis only makes sense if the sea level is higher than at present (Figure 7B) and the ages of these shell middens lie between 5200 and 4800 B.P. These dates narrow the estimate of the time for maximum lagoonal extent, which we now estimate at about 5100 B.P. Moreover, the location of later sites suggests that by 4100 B.P., the sea level was no more than 0.5 m above the present level. This confirms the conclusion reached on the basis of geological evidence that the sea level reached a minimum between 4100 and 3800 B.P.

Table IV shows that the oldest of the late-Holocene sambaquis (Estaleiro, 3690 ± 80 B.P.) is situated on a Pleistocene terrace, which should indicate a high lagoon level. This hypothesis seems to be confirmed by the $\delta^{13}C_{PDB}$ value of its shell carbonate. This value is much less negative than is that of the shells of Sambaquinho, which was built in the same area when the sea level could have been 0.5 m above sea level at the most. This

TABLE II. General Characteristics of the Sambaquis Constructed Between 5400 and 4400 Years B.P.

Sambaqui	Radiocarbon Age (B.P.)	Laboratory Number	Nature of Substrate	Position in the Paleolagoon	$\delta^{13}C_{PDB}$ (o/oo)
Itapoã III	5245 ± 125	Bah 365	P.T.	Outer	−0.63
Jataituba	5240 ± 150	Bah 346	P.T.	Inner	−3.26
Guaxixi	5110 ± 100	Bah 370	P.T.	Outer	−0.48
Vapumaúva	5080 ± 60	Bah 365	P.T.	Outer	−0.06
Pariqüera-Açu	5035 ± 140	Bah 295	P.T.	Inner	−3.88
Juruvaúva I	5010 ± 115	Bah 359	P.T.	Outer	−0.76
Juruvaúva	4970 ± 110	Bah 361	P.T.	Outer	−1.34
Batatal	4920 ± 100	I 9186	P.T.	Outer	—
R. das Pedras I	4860 ± 100	Bah 343	C.B.	Inner	−3.17
Momuna	4790 ± 100	Bah 308	P.T.	Inner	−5.39
R. das Pedras III	4750 ± 110	Gif 3641	P.T.	Inner	−4.67
	4710 ± 145	Bah 300			
Rio Comprido	4560 ± 110	Gif 3646	C.B.	Outer	—
Cananéia	4340 ± 110	Gif 3435	P.T.	Outer	−1.28
	4300 ± 140	Bah 302			

Note: P.T. = Pleistocene Terrace; C.B. = Crystalline Basement; Bah = Laboratório de Física Nuclear Aplicada, Universidade Federal da Bahia, Brazil; Gif = Laboratoire de Faibles Radioactivités, Gif-Sur-Yvette, France; I = Isotopes, U.S.A.

TABLE III. Sambaquis Constructed Between 5000 and 3700 Years B.P. on Holocene Terraces and Paleolagoonal Deposits

Sambaqui	Radiocarbon Age (B.P.)	Laboratory Number	Position of Sea Level	$\delta^{13}C_{PDB}$ (o/oo)
Vapumaúva II	4680 ± 110	Bah 362	< +3.5 m	−0.72
Nóbrega	4380 ± 160	SPC 21	< +3.5 m	—
Boguaçu II	4160 ± 100	Bah 303	< +1.5 m	−1.39
	4120 ± 100	Gif 3436		
Ararapira II	4175 ± 100	Bah 290	< +0.5 m	−1.67
Ubatuba	3870 ± 100	Bah 294	< +2.0 m	−0.02
Ararapira I	3790 ± 110	Gif 3437	<0 m	—

Note: Bah = Laboratório de Física Nuclear Aplicada, Universidade Federal da Bahia, Brazil; SPC = Centro de Pesquisas Geocronológicas, Instituto de Geociências, USP, Brazil; Gif = Laboratoire de Faibles Radioactivités, Gif-Sur-Yvette, France.

TABLE IV. Sambaquis Constructed Since 3700 Years B.P.

Sambaqui	Radiocarbon Age (B.P.)	Laboratory Number	Nature of Substrate	Position of Sea Level	$\delta^{13}C_{PDB}$ (o/oo)
Estaleiro	3690 ± 80	Bah 367	P.T.	>0 m	−0.65
Fosfasa I	3350 ± 135	Bah 340	P.T.	>0 m	−1.80
Pereirinha	3330 ± 125	Bah 286	H.T.	>0 m	−0.09
Boguaçu III	3220 ± 90	Bah 307	H.T.	< +3 m	−3.25
	3090 ± 110	Gif 3645			
Pindu	3090 ± 120	Bah 348	C.B.	>0 m	−4.31
Guarapari	2285 ± 45	Bah 368	L	< +1 m	−2.24
R. das Minas	1850 ± 100	Gif 3643	H.T.	< +0.5 m	—
Sambaquinho	1500 ± 120	Bah 292	L	< +0.5 m	−2.61
Itapitangui	1490 ± 120	Bah 293	H.T.	< +1.5 m	−4.45
Vamiranga	1015 ± 70	Bah 369	P.T.	Indefinite	−3.81

Note: H.T. = Holocene Terrace and L = lagoonal.

confirms the geological conclusion that between 3800 and 3700 B.P., the relative sea level reached +3 m maximum. The sambaqui Boguacu I (3080 ± 55 B.P.) was constructed when the sea level could not have been more than 2.5 m above the present level, indicating that by 3100 B.P. the maximum of 3 m was passed. Finally, the information from the sambaquis Guarapari and Sambaquinho indicates that the relative sea level at about 2300 and 1500 B.P. was no more than 1 and 0.5 m above present level respectively. This provides data on the rate of sea level fall in the latest period not available from geological data.

As discussed above, $\delta^{13}C_{PDB}$ values can be used as an index to marine vs. continental influences and as indirect evidence of relative sea level change (Figure 7C). Three groups of sambaquis in the Cananéia-Iguape coastal plain can be used to assess the value of $\delta^{13}C_{PDB}$ values in estimating sea level change (Figure 4: A, B, and C). The sambaquis in Group A were inhabited between 5200 and 4800 B.P., those in Group B between 5200 and 4800 B.P., and those in Group C between 3800 and 1500 B.P.

Group A consists of seven sambaquis on the Cananeia coastal plain (Figure 4). The first six have ages ranging from 5200 to 4200 B.P. during the first Holocene transgression (Table V). It would be valuable to have information on a sambaqui constructed about 4100 B.P., when the sea level could not have been higher than 0.5 m above its present level. We could then compare the $\delta^{13}C_{PDB}$ values of carbonate shells formed in the same geographic position between 5200 and 4100 B.P., encompassing the first Holocene regressive phase. Lacking this, we have considered a sambaqui dated to about 1000 B.P., when the sea level was quite similar to that at 4100 B.P., assuming that the $\delta^{13}C_{PDB}$ values will also be comparable.

When plotted as a function of time (curve 1 in Figure 7C), the $\delta^{13}C_{PDB}$ values show a maximum between 5200 and 5100 B.P., decrease rapidly until about 4800 B.P., and then decrease slowly until 4200 B.P. Thereafter, the interpolated data suggest another rapid decrease to 4100 B.P. These $\delta^{13}C_{PDB}$ differences can be interpreted both in terms of changing lagoonal extent and, assuming constant local tide range, in terms of changing relative sea level as follows:

- Maximum lagoonal extent, and sea level maximum, occurred between 5200 and 5100 B.P.
- The maximum was followed by a rapid decrease in lagoonal extent, indicating a quick sea level lowering between 5100 and 4800 B.P.
- Lagoonal extent and, hence, relative sea level were stationary between 4800 and 4200 B.P.
- Lagoonal extent reduced rapidly after 4200 B.P., suggesting an equally rapid sea level lowering.

Group B consists of three sambaquis from an area of the inland portion of the Iguape coastal plain (Figure 4, area B). The data derived from these sambaquis (Table VI) allows

TABLE V. Variation of $\delta^{13}C_{PDB}$ as a Function of Age Within the Same Portion of a Paleolagoon, Coastal Plain of the State of São Paulo, Brazil

Sample	Radiocarbon Age (B.P.)	$\delta^{13}C_{PDB}$ (o/oo)
Itapoã III	5245 ± 125	−0.63
Guaxixi	5110 ± 70	−0.48
Juruvaúva I	5010 ± 115	−0.78
Juruvaúva III	4970 ± 150	−1.34
Juruvaúva II	4305 ± 140	−2.07
Itapoã II	4215 ± 140	−2.27
Vamiranga	1015 ± 70	−3.81

From Flexor, J.-M., Martin, L., and Suguio, K. (1979). In: Proc. 1978 Int. Symp. Coastal Evolution in the Quaternary, pp. 365–375. Suguio, K., Fairchild, T.R., Martin, L., and Flexor, J.-M., Eds. São Paulo. Used by permission.

TABLE VI. $\delta^{13}C_{PDB}$ Values of Carbonate Shells from Three Sambaquis Constructed in the Same Inner Zone of the Iguape Coastal Plain

Sambaqui	Radiocarbon Age (B.P.)	$\delta^{13}C_{PDB}$ (o/oo)
Jataituba	5250 ± 150	−3.26
Pariqüera-Açu	5040 ± 140	−3.88
Momuna	4790 ± 115	−5.81

TABLE VII. $\delta^{13}C_{PDB}$ Values of Carbonate Shells from Four Sambaquis of the Cananéia Coastal Plain

Sambaqui	Radiocarbon Age (B.P.)	$\delta^{13}C_{PDB}$ (o/oo)
Fosfasa II	3790 ± 110	−2.63
Estaleiro	3490 ± 60	−0.67
Guarapari	2285 ± 45	−2.24
Sambaquinho	1500 ± 120	−2.61

the creation of a curve of variation of $\delta^{13}C_{PDB}$ which shows that the maximum was well established between 5200 and 5100 B.P. and that the fluctuation was very rapid between 5100 and 4800 B.P. (curve 2 in Figure 7C).

Group C consists of four sambaquis located on the Cananéia coastal plain (Figure 4, area C) which date from 3800 to 1500 B.P. (Table VII). The $\delta^{13}C_{PDB}$ variation curve for area C is given in Figure 7C (curve 3). Between 3800 and 3500 B.P., values increase very rapidly and then decrease. Put in terms of relative sea level, it appears that there was a rapid rise between 3800 and 3500 B.P., followed by a drop.

The results of $\delta^{13}C_{PDB}$ analyses provide additional confirmation of both the geological and archaeological conclusions adumbrate above. Thus, radiocarbon dating of shell debris and fossil wood fragments from both sedimentary deposits and sambaquis, as well as information on the $\delta^{13}C_{PDB}$ of carbonate shells from sambaquis, have allowed us to establish a fluctuating relative sea level curve for the Cananéia-Iguape region (Figure 7D). The accuracy of this curve is supported by the agreement between the three independent lines of information used to create it.

Santos-Itanhaém region (State of São Paulo)

The Santos-Itanhaém coastal plain has the same characteristics as the Cananéia-Iguape area. A lagoonal system originated during each high-sea-level period and sambaquis, some of which have been dated, were established at the margins of paleolagoons. The data from this area will be summarized rapidly as it confirms the results from the Cananéia-Iguape area.

A large number of radiocarbon dates of shell debris and fossil wood fragments contained in lagoonal and shallow-marine deposits, as well as of vermetid incrustations, have allowed us to delineate a rather detailed curve of relative sea level fluctuations for this region (Figure 8A). This curve is quite similar to that for Cananéia-Iguape, although some differences in amplitude are apparent.

Dated sambaquis are much less numerous here than in the Cananéia-Iguape area, but they are important for corroborating the information from dated geological samples (Figure 8B; Table VIII).

The sambaquis Rio Preto, Ararаú, and Mundo Novo are situated in an inner portion of the Itanhaém paleolagoon. Hence, they should have been inhabited during maximum lagoonal extent. However, if this hypothesis is correct, the conditions during maximum lagoonal extent were different in Itanhaém than in the Cananéia-Iguape region. The $\delta^{13}C_{PDB}$ values of the shells from these sambaquis are much more negative than those of the samples dating the high-sea-level period of Cananéia-Iguape. The $\delta^{13}C_{PDB}$ values of the sambaquis from Itanahaém are greater than that of the sambaqui Momuna (Table II), which is the

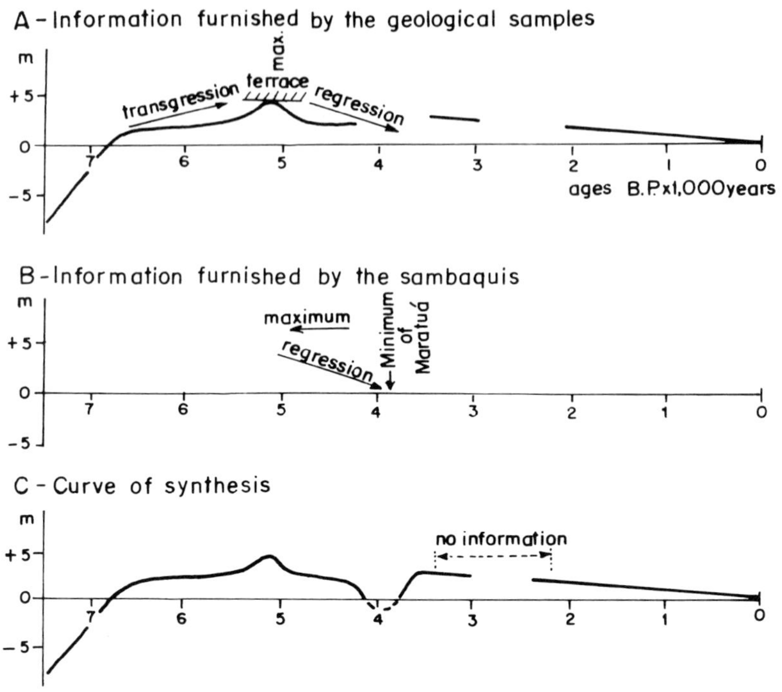

Figure 8. Reconstruction of relative sea level fluctuation curve for the past 7000 years in the Santos-Itanhaém region (State of São Paulo).

TABLE VIII. General Characteristics of the Dated Sambaquis of the Santos-Itanhaém Coastal Plain

Sambaqui	Radiocarbon Age (B.P.)	Laboratory Number	Nature of Substrate	Position of Sea Level	$\delta^{13}C_{PDB}$ (o/oo)
Piaçaguera	4930 ± 100	I 4491	C.B.	>0 m	—
Rio Preto	4635 ± 100	Bah 331	P.T.	>0 m	−6.07
Ararαú	4630 ±	Bah 296	C.B.	>0 m	−5.29
Mundo Novo	4575 ± 110	Bah 446	C.B.	>0 m	−8.91
A229	4520 ± 130	Bah 328	P.T.	< +3.5 m	−2.98
Mar Casado	4400 ± 130	Gif 1194	H.T.	< +3.5 m	—
Maratuá	3925 ± 145	Bah 382	L	<0 m	—
	3865 ± 95	Gif 9185			

Note: I = Isotopes, U.S.; Bah = Laboratório de Física Nuclear Aplicada, Universidade Federal da Bahia, Brazil; Gif = Laboratoire de Faibles Radioactivités, Gif-Sur-Yvette, France.

most negative of the Cananéia-Iguape sambaquis from this 4600-B.P. period. This suggests that the continental influence in the inner portion of the Itanhaém paleolagoon was very strong, perhaps because the maximum had passed and the paleolagoon was beginning to dry out.

The sambaquis A229 and Mar Casado (Table VIII), located on a Holocene terrace, obviously were constructed after the transgression maximum, when the sea level was probably less than their 3.5 and 3-m elevations above the present level. This is supported by the moderate $\delta^{13}C_{PDB}$ value of sambaqui A229.

The sambaqui Maratuá, which has, unfortunately, been completely destroyed, has furnished the most interesting information (Figure 8B). The base of this sambaqui was located below the present sea level. This fact cannot be attributed to substrate subsidence due to the sambaqui's weight because the sambaqui's layers were not deformed in any way. If the sambaqui had sunk under its own weight, the center would be expected to have lowered farther than the margins and the layers would be deformed. Laming-Empéraire (1968) submitted two samples from this sambaqui to be dated at the beginning of radiocarbon investigations. The results, 7330 ± 1300 B.P. (Gif-15) and 7800 ± 1300 B.P. (Gif-16), were initially accepted because the dates did coincide with a period when relative sea level was definitely lower than today. However, archaeological remains within the sambaqui of Maratuá suggested much later dates. Redating of Laming-Emperaire's samples produced ages of 3925 ± 145 B.P. (Bah-382) and 3865 ± 95 B.P. (Gif-9185), which are more in keeping with the artifactual evidence. They also confirm the hypothesis adduced from the Cananéia-Iguape region that relative sea level about 3800 B.P. was lower than today.

FINAL CONSIDERATIONS

Sambaquis cannot be used directly to reconstruct, in space and time, the positions of former sea levels as it is not possible to directly establish a vertical relationship between the base of a sambaqui and the sea level. The only thing of which we are more or less certain is that the base of the sambaqui was above the local HWST level at the beginning of its occupation. If we postulate that the paleoinhabitants established their campsites near the mollusk-harvesting place, it is easy to establish the geographical relationship between a sambaqui and an ancient lagoonal, estuarine, or shallow-marine zone. Sambaquis located further inland were probably associated with a period of maximum lagoonal extent, correlated with a sea level maximum. However, it is necessary to ascertain that coastal progradation was unrelated to intense continental sedimentation, but, rather, to a relative sea level change. If one has a series of dates from a number of sambaquis, they can be used to confirm or adjust the dates derived from geological materials. Thus, for example, sambaquis whose substrates are located beneath the present high-tide level suggest a period of lower sea level. Values of $\delta^{13}C_{PDB}$ for carbonate shells from dated sambaquis give us additional information about the high- or low-sea-level positions, as well as about the trend of sea level change.

Archaeologically, we have confirmed the close association of sambaquis with past shorelines. This supports our hypothesis that the ancient inhabitants did not carry their shellfish dinners long distances, but, rather, established their settlements in close proximity to their molluscan resources. In studies of sambaquis in other zones, it will be possible not only to reconstruct the paleoenvironment of the site from its location relative to features of the paleoshoreline, but also to hypothesize a range of possible dates from a sambaqui located in its situation based on our sea level-fluctuation curves. Excavation and dating of the sambaqui will further test and refine the curves, so that the two areas of research continue to support and amplify one another.

REFERENCES

Bigarella, J.J. (1949) Nota prévia sobre a composição dos sambaquis do Paraná e Santa Catarina. *Bol. Arq. Biol. Tecnol.* 4: 95–106.

Bigarella, J.J. (1951a) Contribuição ao estudo dos sambaquis no Estado do Paraná. I. Regiões adjacentes às baías de Paranaguá e Antonina. *Bol. Arq. Biol. Tecnol.* 5/6: 231–292.

Bigarella, J.J. (1951b) Contribuição ao estudo dos sambaquis no Estado do Paraná. II. Regiões adjacentes à baía de Guaratuba. *Bol. Arq. Biol. Tecnol.* 5/6: 293–314.

Bittencourt, A.C.S.P., Martin, L., Vilas-Boas, G.S., and Flexor, J.-M. (1979) Quaternary marine formations of the coast of the State of Bahia, Brazil. In: Proc. 1978 Int. Symp. Coastal Evolution in the Quaternary, pp. 232–253. Suguio, K., Fairchild, T.R., Martin, L., and Flexor, J.-M., Eds. São Paulo.

Empéraire, J. and Laming, A. (1956) Les sambaquis de la côte méridionale du Brésil. Campagne de Fouilles (1954–1956). *J. Soc. Am. Mus. de l'Homme.* 45: 5–163.

Fairbridge, R.W. (1976) Shellfish-eating preceramic indians in coastal Brazil. *Science.* 191: 353–359.

Flexor, J.-M., Martin, L., and Suguio, K. (1979) Utilisation du rapport isotopique $^{13}C/^{12}C$ comme indicateur d'oscillations lagunaires. In: Proc. 1978 Int. Symp. Coastal Evolution in the Quaternary, pp. 356–375. Suguio, K., Fairchild, T.R., Martin, L., and Flexor, J.-M., Eds. São Paulo.

Hurt, W.R. (1974) The interrelationships between the natural environment and four sambaquis, coast of Santa Catarina, Brazil. Occas. Pap. Monogr. 1, 23 pp. Indiana University Museum, Bloomington.

Laming-Empéraire, A. (1968) Missions archéologiques françaises au Chili et au Brésil. *J. Soc. Am.* 57: 77–99.

Löfgren, A. (1893) *Os Sambaquis de São Paulo.* Comiss. Geogr. Geol. Estado São Paulo, Bol. 9, São Paulo.

Martin, L. and Suguio, K. (1975) The state of São Paulo coastal marine Quaternary geology: the ancient strandlines. *An. Acad. Bras. Ciênc* 47: 249–263.

Martin, L. and Suguio, K. (1976) Les variations du niveau de la mer au Quaternaire Récent dans le sud de l'État de São Paulo (Brésil): utilisation de "sambaquis" (kjokkenmodings) dans la détermination des anciennes lignes de rivage holocènes. *Actes 42nd Congr. Int. Am.* 9A: 73–83.

Martin, L. and Suguio, K. (1978) Excursion route along the coastline between the town of Cananéia (State of São Paulo) and Guaratiba outlet (State of Rio de Janeiro). In: *1978 Int. Symp. Coastal Evolution in the Quaternary.* Spec. Publ. 2, São Paulo.

Martin, L., Flexor, J.-M., Vilas-Boas, G.S., Bittencourt, A.C.S.P., and Guimarães, M.M.M. (1979a) Courbe de variation du niveau rélatif de la mer au cours des 7000 dernières années sur un secteur homogène du littoral brésilien (nord de Salvador, Bahia). In: Proc. 1978 Int. Symp. Coastal Evolution in the Quaternary, pp. 264–274. Suguio, K., Fairchild, T.R., Martin, L., and Flexor, J.-M., Eds. São Paulo.

Martin, L., Suguio, K., and Flexor, J.-M. (1979b) Le Quaternaire marin du littoral brésilien entre Cananéia (SP) et Barra de Guaratiba (RJ). In: Proc. 1978 Int. Symp. Coastal Evolution in the Quaternary, pp. 296–331. Suguio, K., Fairchild, T.R., Martin, L., and Flexor, J.-M., Eds. São Paulo.

Martin, L., Suguio, K., Flexor, J.-M., Bittencourt, A.C.S.P., and Vilas-Boas, G.S. (1980) Le Quaternaire marin brésilien (littoral pauliste, sud-fluminense et bahianais). *Cah. O.R.S.T.O.M. Sér. Géol.* 9(1): 95–124.

Martin, L., Suguio, K., and Flexor, J.-M., (1981/1982) Utilisation des amas coquilliers artificiels dans les réconstructions des anciennes lignes de rivage: exemples brésiliens. *Cah. O.R.S.T.O.M. Sér. Géol.* 12(2): 135–146.

Martin, L., Flexor, J.-M., Blitzkow, D., and Suguio, K. (1985) Geoid change indications along the Brazilian coast during the last 7000 years. In: Proc. 5th Int. Coral Reef Congr., Vol. 3, 85–90. Tahiti.

Martin, L., Suguio, K., Flexor, J.-M., Dominguez, J.M.L., and Bittencourt, A.C.S.P. (1987) Quaternary evolution of the central part of the Brazilian coast: the role of relative sea level variations and the shoreline drift. In: *UNESCO Rep. Mar. Sci.* 43: 97–145.

Suguio, K. and Martin, L. (1976a) Presença de tubos fósseis de Callianassa nas formações quaternárias do litoral do Estado de São Paulo e sua utilização na reconstrução paleoambiental. Bol. IG, Instituto de Geociências, USP, vol. 7, p. 17–26.

Suguio, K. and Martin, L. (1976b) Brazilian coastline Quaternary formations: the states of São Paulo and Bahia littoral zone evolutive schemes. *An. Acad. Bras. Ciênc.* 48: 325–334.

Suguio, K. and Martin, L. (1978) Quaternary marine formations of the states of São Paulo and southern Rio de Janeiro. In: *1978 Int. Symp. Coastal Evolution in the Quaternary.* Spec. Publ. 1, São Paulo.

Suguio, K., Martin, L., and Flexor, J.-M. (1980) Sea level fluctuations during the past 6000 years along the coast of the State of São Paulo. In: *Earth Rheology, Isostasy and Eustasy,* pp. 471–486. Ed. Mörner N.-A., John Wiley & Sons, New York.

Suguio, K., Martin, L., Bittencourt, A.C.S.P., Dominguez, J.M.L., Flexor, J.M., and de Azevedo, A.E.G. (1985) Flutuações do nível relativo do mar durante o Quaternário Superior ao longo do litoral brasileiro e suas implicações na sedimentação costeira. *Rev. Bras. Geociênc.* 15(4): 273–286.

Villwock, J.A., Tomazelli, L.J., Loss, E.L., Dehnhardt, E.A., Horn Filho, N.O., Bachi, F.A., and Dehnhardt, B.A. (1986) Geology of the Rio Grande do Sul coastal province. *Quat. South Am. Antarct. Penin.* 4: 79–97.

Underwater

Lucille Lewis Johnson
Department of Anthropology
Vassar College
Poughkeepsie, New York

INTRODUCTION

In working underwater, archaeologists have pursued two strategies, and here we have examples of both. The first uses high-technology methods to derive general predictions about past subaerial landscapes. These have so far proven most useful in encouraging the avoidance of these regions in sub-marine resource extraction activities. The second uses low-technology scuba exploration to look for sites in shallow water where their presence can be predicted from land-based surveys. In discussing both strategies, Masters and Flemming (1983) have pointed out that

> Diving teams equipped with simple equipment and small budgets are at present revealing a steady flow of information about submarine prehistoric archaeology: no other technique is producing prehistoric archaeological data from any parts of the continental shelf.
>
> The OCS deep water surveys are preparing the background for future archaeological excavations. As shallow water sites are discovered and excavated in greater numbers, and in deep[er] water, there will be a progressively greater need for the background data obtained by remote sensing instruments. The two approaches are therefore convergent through time.

Both of the chapters here which examine underwater environments discuss the Gulf of Mexico. The Gulf of Mexico is particularly important for developing methods of underwater investigation because it is a warm-water, shallow coastal plain which presents optimal conditions for the burial of early-Holocene sites. Garrison shows how sophisticated geological methods can be used to identify large relict surface features at great depth. Dunbar, Webb, and Faught, on the other hand, examine nearshore environments where archaeological methods and relatively simple equipment can be used to examine submerged sites. The survey reported by Garrison is most important in indicating the methods that can be used in locating past shoreline features far offshore. The actual features located by Garrison are probably too far offshore to be pertinent to archaeological inquiry, although the −73-m shoreline will prove of interest if pre-Clovis occupation of the Americas is proven. In their analysis, Dunbar et al. use the locations of known subaerial sites in northwestern Florida to predict the location of submerged sites which still would have been inland at their time of occupation. While their results are preliminary, their predictive model has resulted in the location of a number of submerged archaeological sites. The validity of their method has been proven by the significant results which they attained following very little field research.

REFERENCE

Masters, P.M. and Fleming, N.C., Eds. (1983) *Quaternary Coastlines and Marine Archaeology.* Academic Press, New York.

Chapter

5

Recent Archaeogeophysical Studies of Paleoshorelines of the Gulf of Mexico

Ervan G. Garrison
NOAA Ocean and Coastal Resource Management
Washington, D.C.

ABSTRACT

Geological mapping cruises conducted in 1985 and 1987–1988 in the eastern Gulf of Mexico obtained detailed sidescan sonar and high-resolution subbottom profiler data on paleoshorelines and sinkhole features. Two relict shorelines have been tentatively identified at 20-fathom (36 m) and 40-fathom (73 m) depths below present sea level. In addition, buried stream channels have been identified entrenched within identifiable escarpments that still exist above the present sea floor in some areas. Sinkholes of great size, with little fill sediment, were seen on the Florida platform. Preliminary analyses indicate a Holocene age and origin for the paleoshorelines, together with evidence of a deeper stillstand based on sidescan and bathymetric data from the Florida platform. A relationship to prehistoric settlement of the then-subaerial continental shelf is suggested.

INTRODUCTION

This chapter presents the results of three geological mapping cruises conducted in September 1985, October-November 1987, and March-April 1988. The archaeogeophysical data discussed herein were originally gathered by the author as principal investigator for the mapping components of larger marine studies sponsored by the National Oceanic and Atmospheric Administration (NOAA) and the Minerals Management Service (MMS). It was evident that these data could be utilized for a suite of interpretative goals, one of which includes the detection and study of paleoshorelines and subaerially exposed geologic features, now submerged. The order of presentation will start with the latest date—those of the submerged paleoshorelines mapped during the 1987 study off Mississippi-Alabama.

CHARACTERIZATION OF PALEOSHORELINE FEATURES, MISSISSIPPI-ALABAMA CONTINENTAL SHELF

In fall 1987 and spring 1988, a geological mapping cruise was carried out 172 km off the Mississippi-Alabama coasts in the eastern Gulf of Mexico (Figure 1). The survey area was 51×18 km, or almost 1000 km^2. The objective was to map the area using high-resolution acoustical instrumentation such as a 100-kHz sidescan sonar and a 4-7kHz subbottom profiler. Navigation was accomplished using the STARFIX satellite-positioning system, which provides a dynamic positional accuracy of ± 5 m.

The survey area includes the edge of the continental shelf south of Mississippi and Alabama. Subbottom profiler records indicate that the shelf edge is located upon delta-front foreset beds that were truncated by erosion during the last low stand of sea level in the Pleistocene. Holocene sediments 0 to 15 m thick cap the erosional surface, and the topographic features of primary interest to this study were constructed on top of these sediments. The Holocene sediments are thickest in the central part of the survey area, perhaps indicating a small delta lobe that was deposited in that area (Rezak, 1988).

Interesting geologic features were found throughout the survey area. In general, the sediment reflectivity at the sea floor displayed complex variations. High and low reflectivity were found, and the variations between high and low reflectivity occurred in waves and patches of varying size and complexity. Topographic features were of three types: (1) pinnacles, (2) linear ridges, and (3) banks (Figures 2, 3, and 4). The first two categories account for most of the observed features, and many of these are located along an isobath approximately 40 fathoms (73 m) deep. This line is believed to be related to a stillstand in the recent postglacial rise in sea level (Figure 5).

Following the acquisition of sidescan and subbottom data from two cruises, the data were analyzed in a preliminary manner in order to determine which areas would be visited using the remotely operated vehicle (ROV) during a September 1988 cruise. This preliminary analysis was completed prior to the compilation of a photomosaic constructed using the sidescan sonar records.

CHARACTERIZATION OF THE 73-M SHORELINE, MISSISSIPPI-ALABAMA CONTINENTAL SHELF

Sediment Facies

Ludwick (1964) described the following five sediment facies on the Mississippi-Alabama continental shelf (Figure 6):

- Chandeleur facies
- St. Bernard prodelta facies
- Mississippi-Alabama sand facies
- Mississippi-Alabama reef and interreef facies
- Mississippi prodelta facies

The Chandeleur facies is a fine-grained, well-sorted quartz sand lying just seaward of the island arc. The St. Bernard prodelta facies lies as a broad arc on the western part of the Mississippi-Alabama shelf. The sediment consists of a Chandeleur facies and other remnants of the abandoned St. Bernard delta lobe. The Mississippi-Alabama sand facies

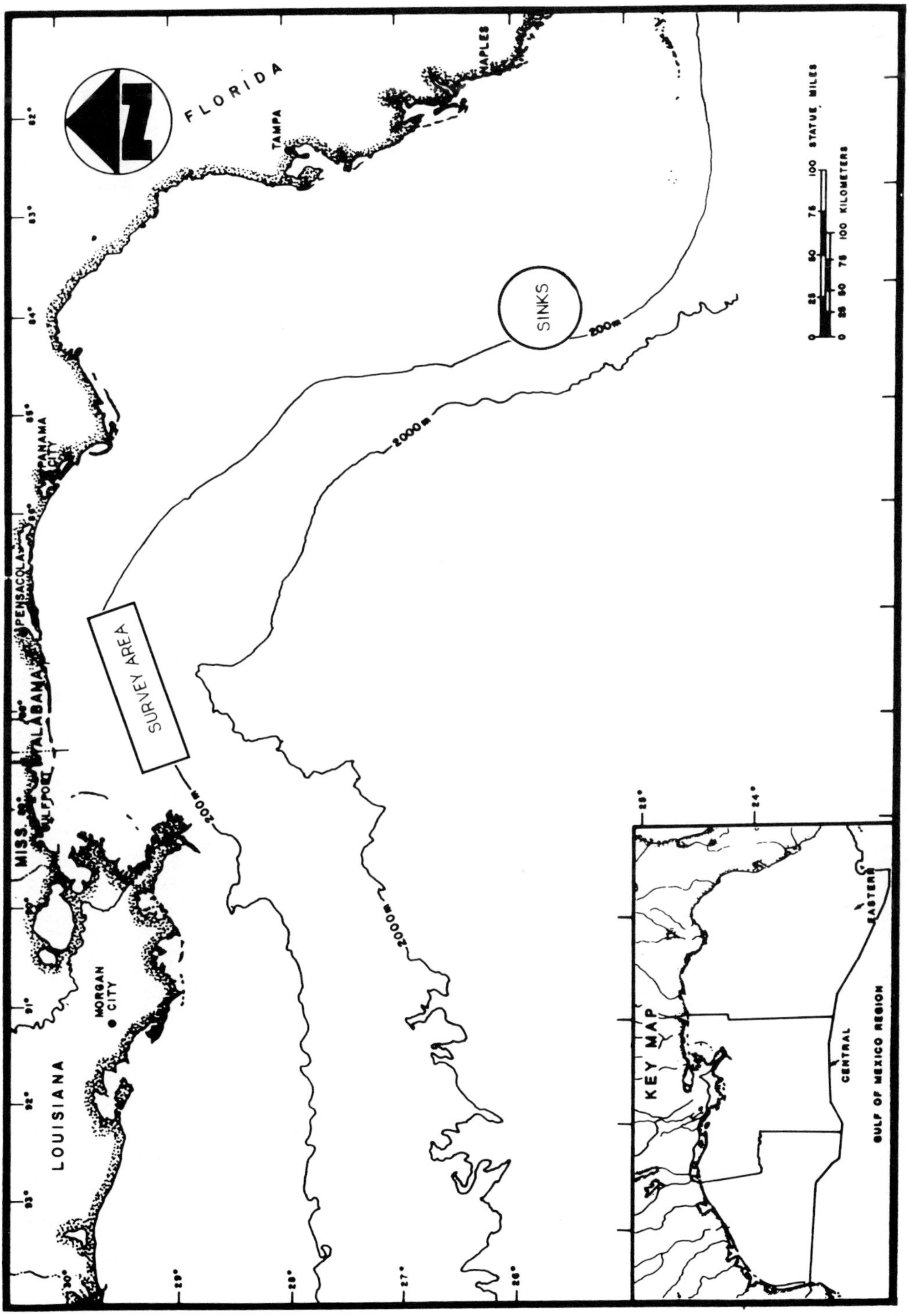

Figure 1. Gulf of Mexico showing the locations of paleoshoreline, survey area, and sinks.

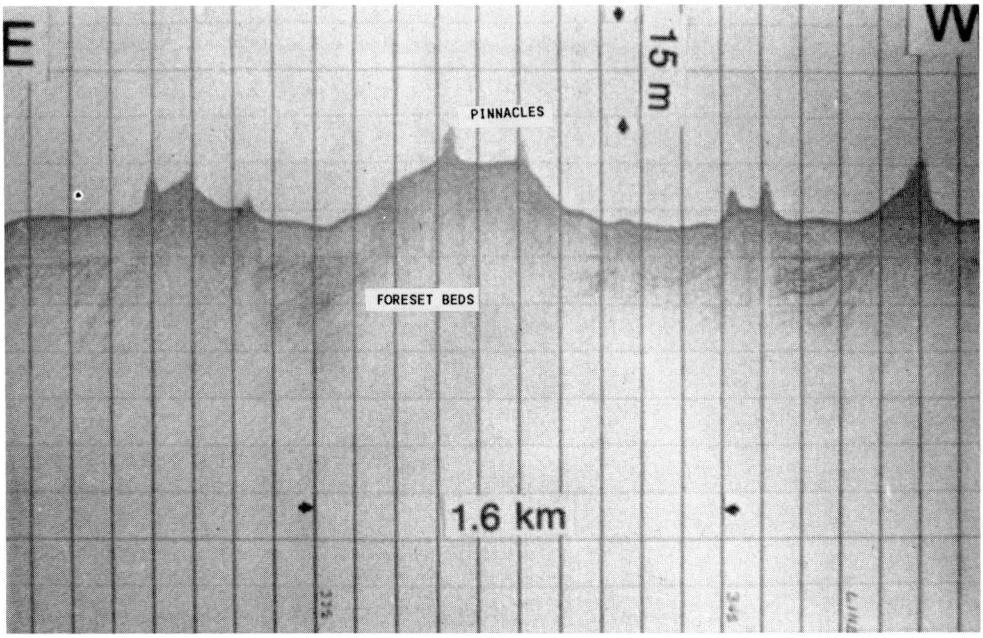

Figure 2. Topographic features of the paleoshoreline: pinnacles (7-kHz subbottom profiler record).

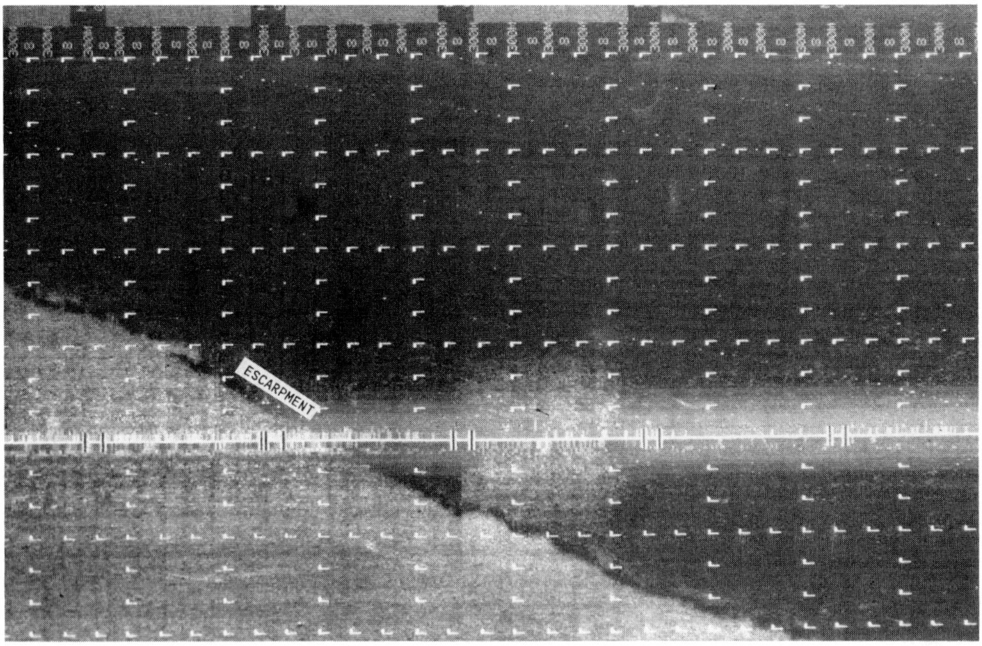

Figure 3. Topographic features of paleoshorelines: linear ridges (100-kHz sidescan sonar record).

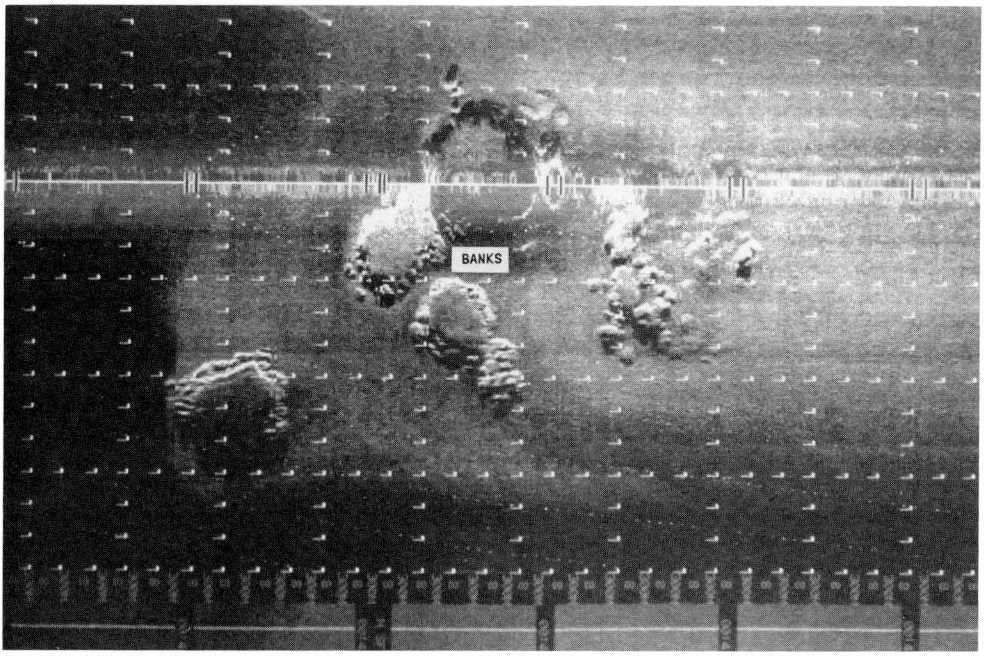

Figure 4. Topographic features of paleoshorelines: banks (100-kHz sidescan sonar).

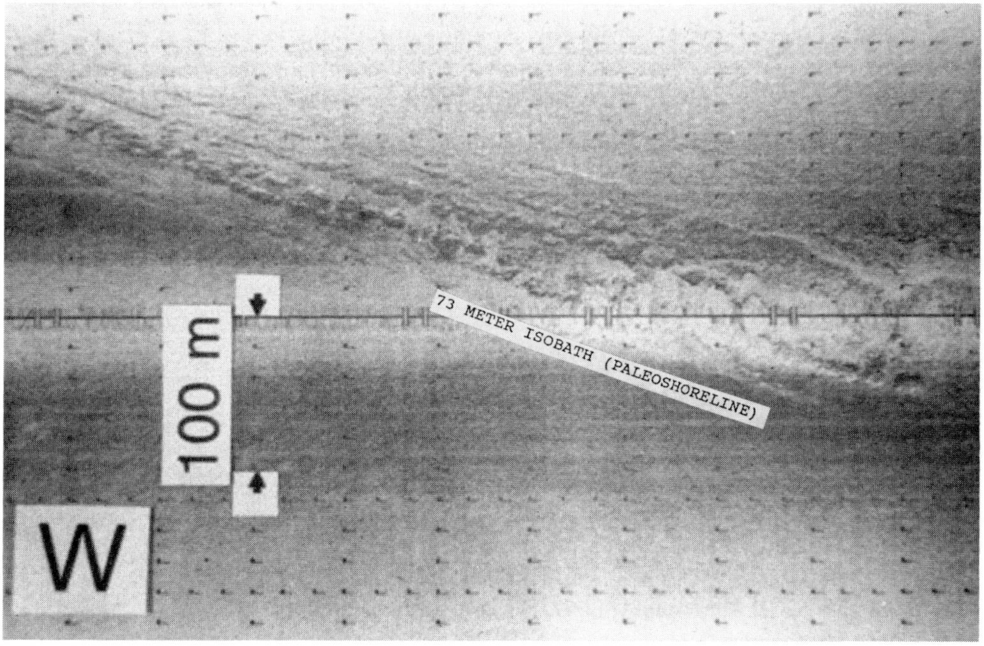

Figure 5. 40-Fathom paleoshoreline (100-kHz sidescan sonar).

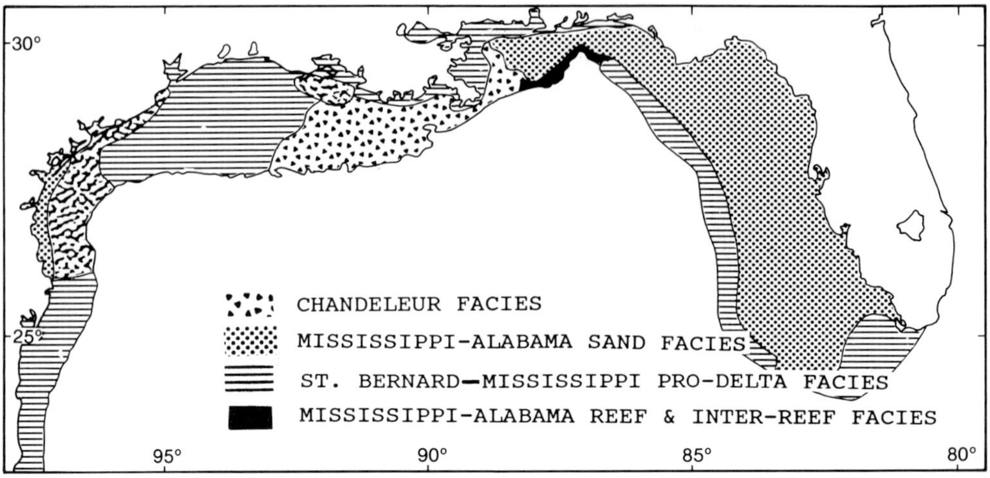

Figure 6. Sediment distribution, Gulf of Mexico continental shelf.

consists predominantly of a well-sorted, fine-grained, clean quartz sand. The Mississippi-Alabama reef and interreef facies lies to the southeast of the relict quartz sand facies. Beginning adjacent to DeSoto Canyon and running to the west along the shelf break, the reefs described by Ludwick and Walton (1957) include the Pinnacles area and the Quaternary shorelines. Sediments in the Mississippi prodelta facies are clay and silt-sized material that are either deposited very rapidly at the shelf edge or carried westward by the prevailing surface currents.

Ludwick's (1964) description of the occurrence of corroded and coated disarticulated oyster shells close to the southern margin of the Mississippi-Alabama sand facies, together with our own sediment analyses, mapping, and sidescan data, the direct observations of the sea floor by W. W. Schroeder (1988, personal communication), and observations of the sediments in the vicinity of Southwest Rock off the mouth of Mobile Bay (Rezak, 1988) are convincing evidence that the sediments on the Mississippi-Alabama continental shelf are relict Holocene to recent deposits that have been modified in varying degrees by the rise in sea level since they were originally deposited. The presence of stained mollusk shells, including blackened and brownish oysters, and the occurrences of calcite-cemented and siderite-cemented quartz sands found at several different water depths in this facies suggests, according to W. W. Schroeder (1988, personal communication), that the post-Pleistocene history of this shelf is punctuated by a series of sea level stillstands that created barrier islands and protected brackish-water lagoonal areas similar to the modern-day Mississippi Sound. The variation of sediment texture in this facies reported by Pyle et al. (1975) is most probably due to the fact that it is a relict deposit representing several different environments. The lagoonal deposits, being typically muddy, as are the modern Mississippi Sound sediments, were easily eroded as the sea level rose, leaving a lag deposit of blackened shell hash, with more resistant, exhumed, blackened oyster reefs standing as shoals on the otherwise featureless sea floor. The sands of the old barrier islands were then smeared out over the sea floor, covering the old lagoon with a thin veneer of sand as new barrier islands were formed by some of the same older sands plus newly arrived quartz sands from the east. However, not enough sand was available to completely bury the high-relief oyster reefs, leaving them as low-relief pinnacles or flat areas of high reflectance as seen on the sidescan records.

This hypothesis can be tested by taking rock dredges on some of the low-relief pinnacles that have been identified in the sand facies and vibracoring in their vicinity. Rock dredges should bring up large, blackened oyster shells. Vibracores should show a clean quartz sand overlying a lag of small shell hash that, in turn, overlies a muddy lagoonal deposit.

CHARACTERIZATION OF SUBMERGED SINKHOLES, FLORIDA CONTINENTAL SHELF

In the fall of 1985, a similar geological mapping cruise took place on the outer Florida platform of the eastern Gulf of Mexico's continental shelf. On a joint research cruise with the NOAA ship Oregon II and the Harbor Branch Foundation's Edwin Link in September, it was possible to conduct extensive instrument and subsequent submersible inspections of features detected by the instruments off the western coast of Florida in the general area shown in Figure 1.

Insonification by sidescan sonar of an area roughly 300 km southwest of St. Petersburg detected two large sinkhole features 3.5 km apart. They lie in 175 m (96 and 97 fathoms) of water. Both sinks were clearly seen by the sidescan sonar, with the sink termed "96 Fathom" being the larger of the two features in terms of width and depth (Figure 7). This sink was chosen for an exploratory dive using the JOHNSON-SEA LINK I four-man

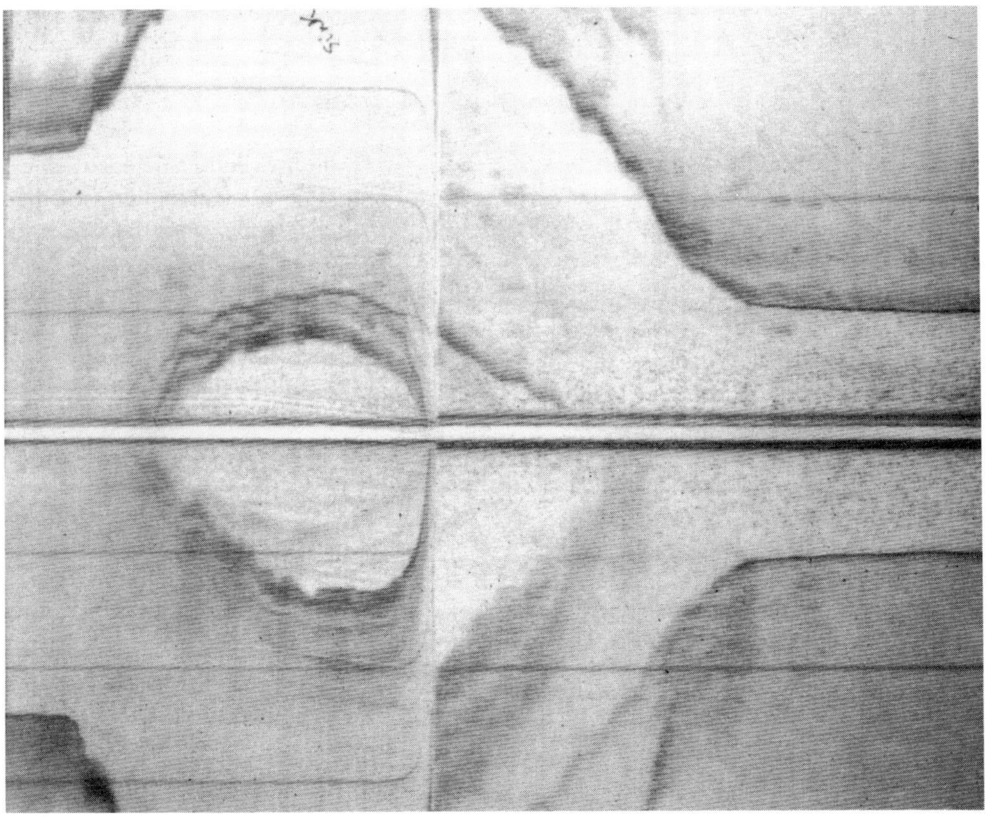

Figure 7. 96-Fathom sink (100-kHz sidescan sonar, 50 and 100-m scales).

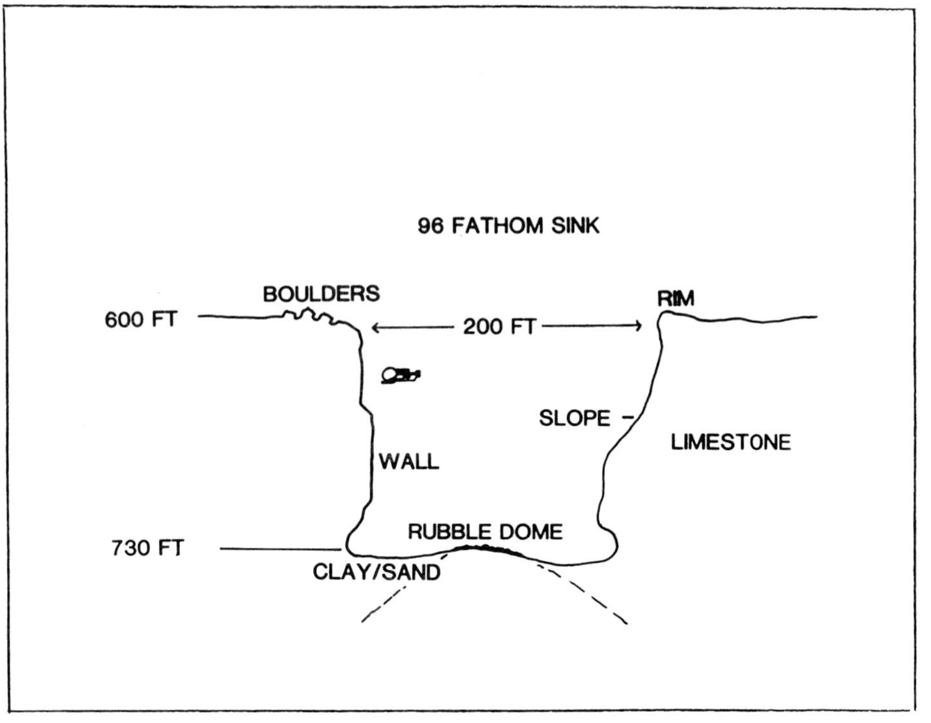

Figure 8. Cross-section of 96-fathom sink.

submersible. A 3.5-h dive was made into the sink to a depth of over 222 m. Instrumental measurements were made of (1) conductivity, (2) currents, and (3) temperature. Shallow core samples were taken of the bottom sediments and salient geological and biological phenomena were documented by color video cameras.

Shown in cross-section (Figure 8), the sink's rim, with a surrounding boulder field, begins at roughly 96 fathoms or 175 m. The upper walls of the sink were relatively sheer, with no visible geological strata. The walls were bioeroded vertically, with the fissures and crevasses containing both mobile and sessile fauna. Near the base of the wall, an undercut extends around the perimeter of the sink. The sediments of the sink's floor were composed of organic debris such as coral and sponge. These formed a dome-like feature in the sink's bottom. This "dome" is probably underlain by a limestone roof fall. Anaerobic soil features were observed in the sediments, suggesting a lack of dissolved oxygen. Current and conductivity readings do not indicate the presence of freshwater inflow, but this cannot be ruled out based on the limited duration of the sampling dive. No archaeological or paleontological materials were directly observed, but this was to be expected given the extensive biological matting of all surficial hard-bottom and wall features. The depth of this particular sink, 175 m, also militates against the high probability of human utilization of the sink in the Pleistocene, given current ideas on seastand chronologies, i.e., its being outside the commonly accepted −130 m maximum low seastand of the late Pleistocene.

ARCHAEOLOGICAL SIGNIFICANCE OF THE 73-M (40 FATHOM) PALEOSHORELINE AND THE 175-M (100 FATHOM) SINKHOLES

The shoreline and karstic sinkhole features are direct evidence of subaerial exposure of the present Gulf of Mexico continental shelf during lowered sea levels. Their archaeological importance hinges upon the linkage of (1) age and (2) association with human groups.

Recent thinking on sea levels for the Quaternary shows general agreement on the trends in sea level and its fluctuation (Figures 9 and 10). Whatever controversy exists centers on

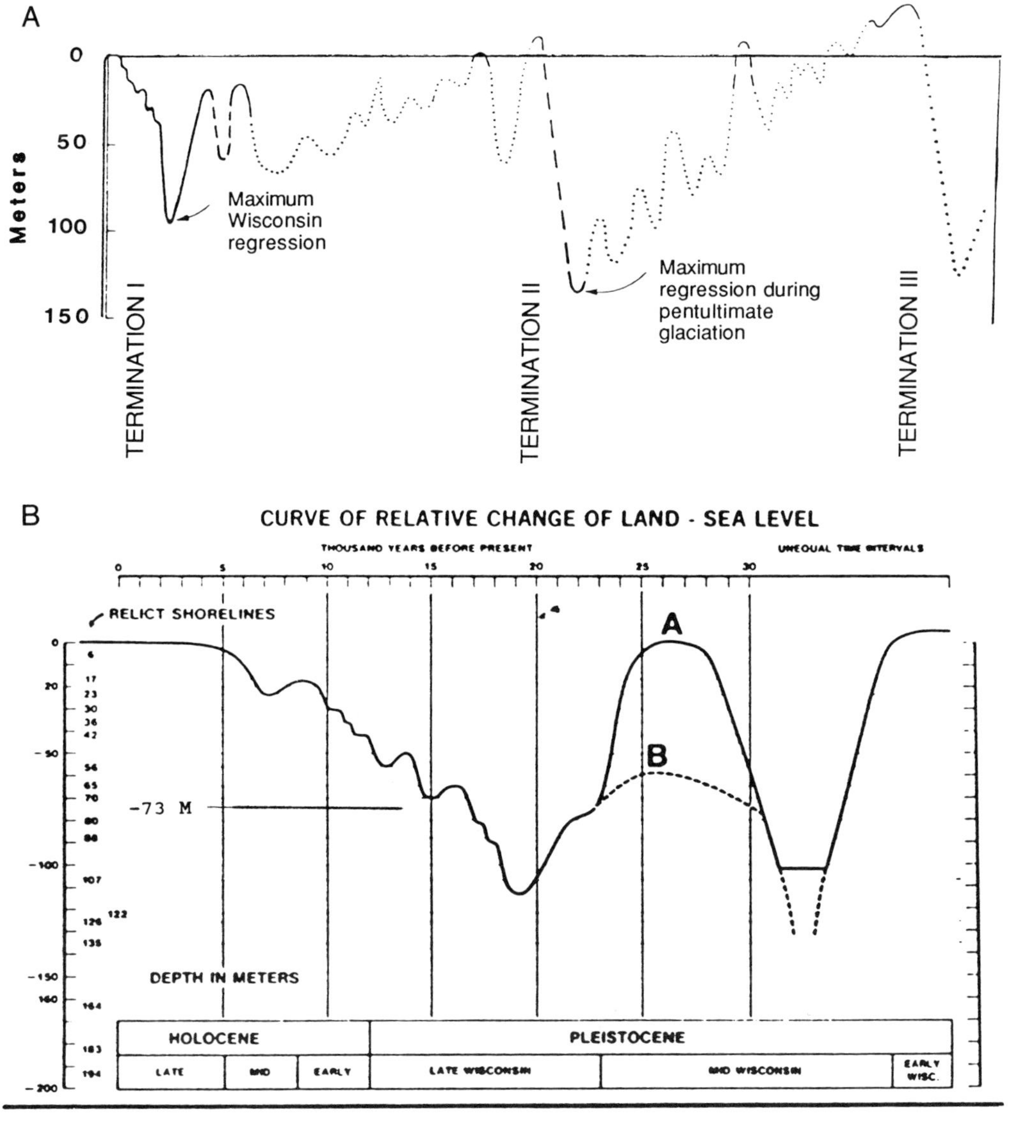

Figure 9. (A) Sea level history of Beringia during the last 250,000 years. Solid portion of the curve is based on firm data. (B) Curve of relative land-sea level for the northern Gulf of Mexico. Curve derived from various published sources, which include: Coleman and Smith, 1964; Schnable and Goodell, 1968; Rehkemper, 1969; Nelson and Bray, 1970; Frazier, 1974; and Coastal Environments, Inc., 1980.

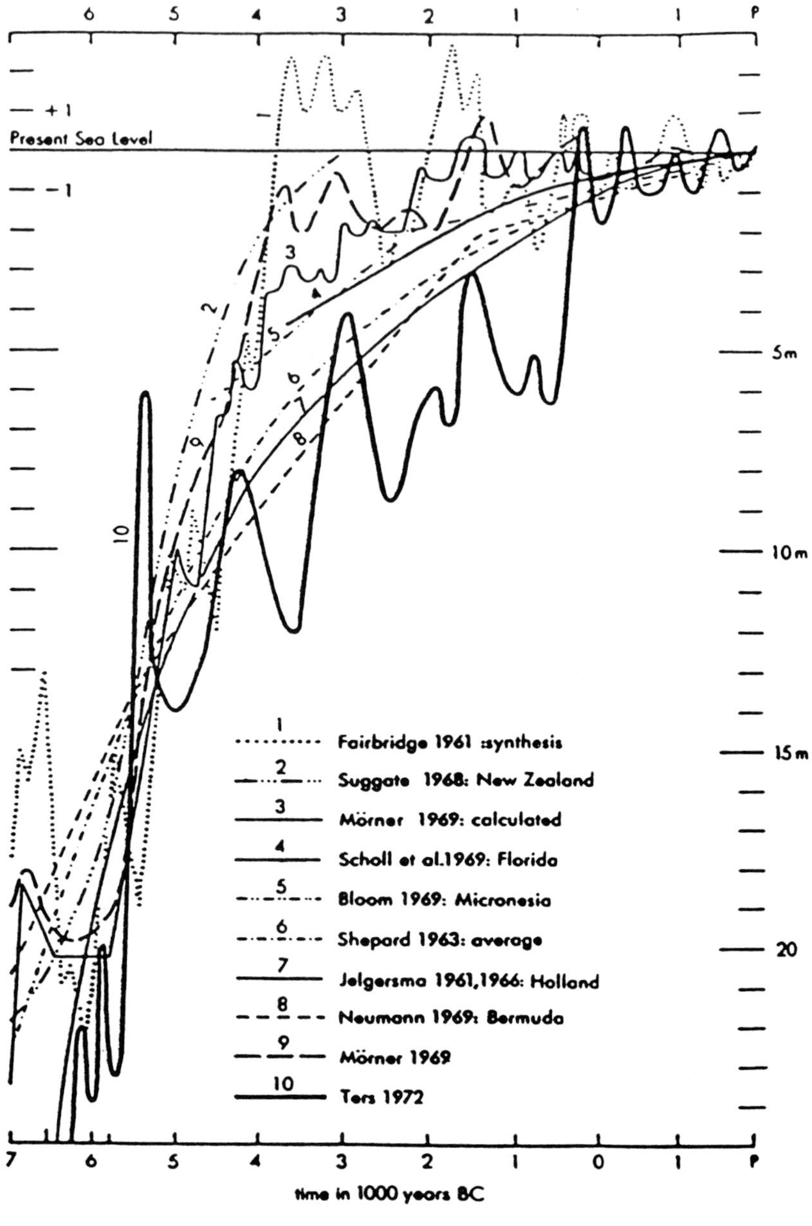

Figure 10. Graphic portrayal of interpretation of the changes in sea level over the past 8000 years. (From Everard, C.E. (1980) *Archaeology and Coastal Change,* pp. 1–23. Thompson, F.H., Ed. Society of Antiquities, London. Used by permission.)

the issue of ''continual'' vs. ''episodic'' curves which relate to more complex issues of glacial processes.

The episodic model, particularly for the postulated fluctuations of the last glaciation (Wisconsin), has gained in currency over the last few years (John Anderson, 1989, personal communication). This is due, in large part, to the work of Thomas and Bentley (1978), who have proposed a catastrophic ice-sheet decoupling mechanism of rapid, episodic sea level rise.

The sea level curve for the northern Gulf of Mexico by Gagliano et al. (1982) suggests ages for a series of late-Pleistocene stillstands, as shown in Figure 11. If their curve for the northwestern Gulf area is applicable to the eastern Gulf, then the 73-m shoreline was formed sometime between 21,300 and 17,500 B.P., with the older date representing the shoreline position as the sea was regressing to its maximum lowstand and the younger date that of the shoreline during the sea's later transgression of the shelf (Stright, 1986).

The dates for the deeper sinkhole features are more problematical. Their depth places them beyond the generally accepted values for the late-Pleistocene lowstand of −100 to −130 m (Hopkins, 1973; Saucier, 1977; Komar, 1976). Komar (1976) points out that not much information is available on lowstands. Indications of lower levels come from "nickpoints", terraces, deltaic deposits, and drowned reefs observed near the outer shelf edge. In the Gulf of Mexico, reliable indicators fail below 100 m.

Terraces at depths comparable to those of the sinkholes have been found in northern Australia (Jongsma, 1970) at −175 and −200 m. Van Andel and Veevers (1967) and Veeh and Veevers (1970) date the −175-m terrace to 13,000 to 17,000 years B.P. and Jongsma (1970) dates the −200-m terraces at 170,000 B.P. or earlier. Coleman et al. (1980) summarize that four to five major lowstands occurred in the Pleistocene, with the lowest stand at 150 to 180 m below present sea level. Due to deformation of the geoid that is the Earth, it is difficult to extrapolate or compare lowstands from different parts of the world.

Of the two geologic features discussed in this paper, the 73-m shoreline existed in a period when prehistoric cultures such as pre-Clovis or early Paleoindian are believed to have been established in the New World (Figure 12). Relic landforms along late-Pleistocene seashores and associated estuaries have reasonable probabilities for the occurrence of their campsites (Ruppe, 1983; Stright, 1986). The deep sinkholes fit less comfortably into the

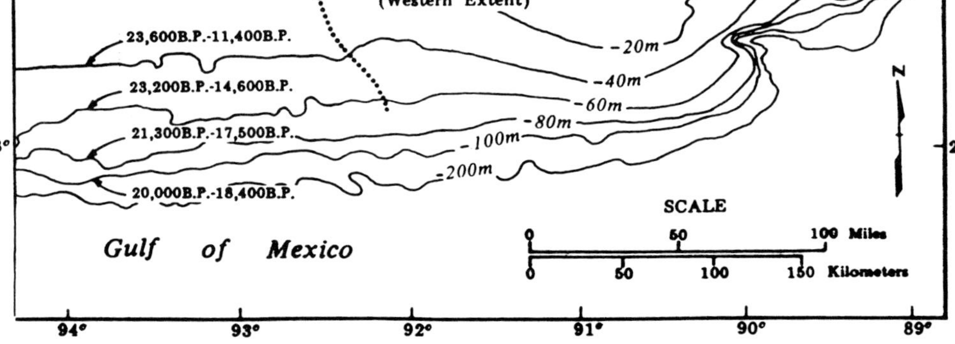

Figure 11. Late-Pleistocene stillstands. (From Stright, M.J. (1986) *Geoarchaeology.* 1(4): 347–364. Used by permission.)

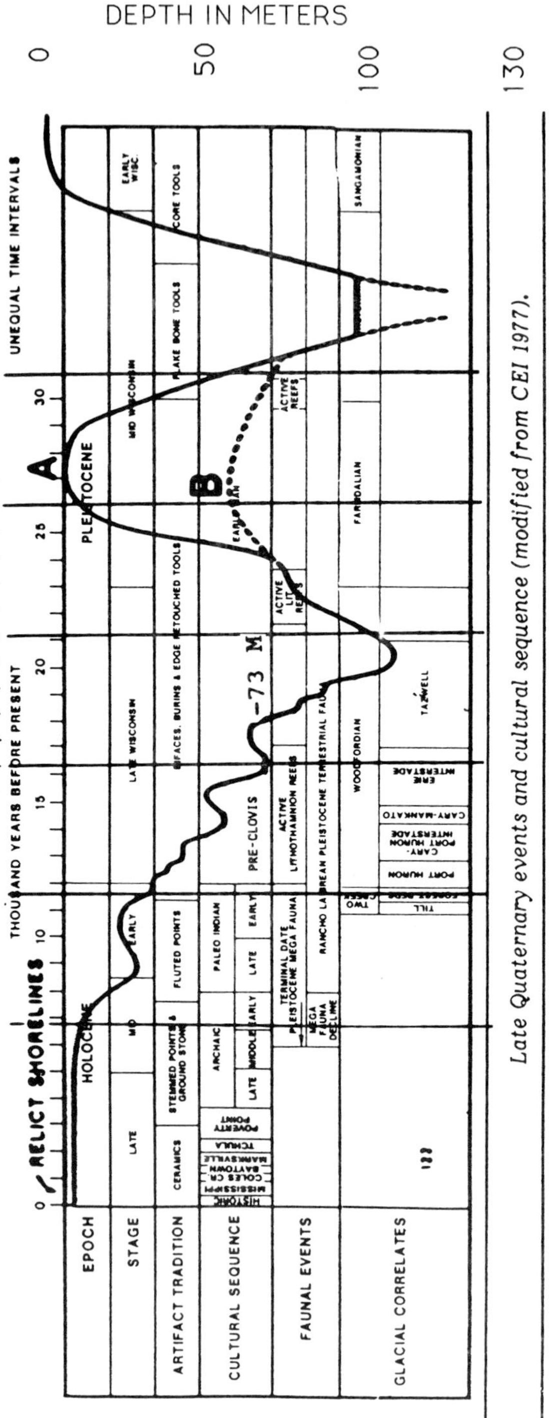

Figure 12. Late-Quaternary events, cultural sequence, and superimposed sea level curve. While the position of the paleoshoreline (−73 m) falls primarily within the pre-Clovis portion of the early Paleoindian stage, it is entirely possible for the curve to fall within later portions of this early stage of man's occupation of the New World due to uncertainties in our understanding of both the sea level curve and the early cultural sequence for the continental shelf. (Modified from Coastal Environments, Inc. (1980) Cultural Resources of the Northern Gulf of Mexico Continental Shelf. Vol. 1–3. National Park Service, Washington, D.C.)

chronological models for sea level and the concomitant occurrence of human groups. Recent archaeological research has established sinks as a locus for human settlement in Florida (Neil, 1958, 1964; Cockrell, 1973; 1975 cf. Dunbar; taught this work). The possibility does exist that these drowned features may have archaeological significance in our studies of the late Pleistocene.

REFERENCES

Anderson, J.B. Department of Geology and Geophysics, P.O. Box 1892, Rice University, Houston, TX 77251. Phone: 713-527-4884.

Coastal Environments, Inc. (1980) Cultural Resources of the Northern Gulf of Mexico Continental Shelf. Vol. 1–3. National Park Service, Washington, D.C.

Cockrell, W.A. (1973) Remains of early man recovered from spring cave. *Arch. Hist. News.* 4(2).

Cockrell, W.A. (1975) Warm Mineral Springs, 1975. A multi-disciplinary approach to a 10,000 B.P. archaeological site (Abstr.). *Fla. Sci.* 38(Suppl. 1): 19.

Coleman, J.M. and Smith, W.G. (1964) Late recent rise in sea level. *Geol. Soc. Am. Bull.* 75: 833–840.

Coleman, J.M., Prior, D.B., and Garrison, L.E. (1980) Subaqueous Sediment Instabilities in the Offshore Mississippi River Delta. BLM Open File Rep. 80-02, U.S. Department of the Interior, New Orleans, La.

Everard, C.E. (1980) On sea level changes. In: *Archaeology and Coastal Change,* pp. 1–23. Thompson, F.H., Ed. Society of Antiquities, London.

Frazier, D.E. (1974) *Depositional Episodes: Their Relationship to the Quaternary Stratigraphic Framework in the Northwestern Portion of the Gulf Basin. Geol. Circ.* 74-1, Buerau of Economic Geology, University of Texas, Austin.

Hopkins, D.M. (1973) Sea level in Beringia during the past 250,000 years. *Quat. Res.* 3: 520–540.

Gagliano, S.N., Pearson, C.E., Weinstein, R.A., Wiseman, V.E., and McClendon, C.M. (1982) *Sedimentary Studies of Prehistoric Archaeological Sites.* Coastal Environments Inc., Baton Rouge, La.

Jongsma, D. (1970) Eustatic sea level changes in the Arafura Sea. *Nature.* 228(5267): 150–151.

Komar, P.D. (1976) *Beach Processes and Sedimentation.* Prentice-Hall, Englewood Cliffs, New Jersey.

Ludwick, J.C. (1964) Sediments in the northeastern Gulf of Mexico. In: *Papers in Marine Geology, Shepard Commemorative Volume.* MacMillan, New York.

Ludwick, J.C. and Walton, W.R. (1957) Shelf edge calcareous prominences in the northwestern Gulf of Mexico. *Am. Petrol. Geol. Bull.* 41: 2054–2101.

Neil, W.T. (1958) A stratified early site at Silver Springs, Florida. *Fla. Anthropol.* 16(4): 99–104.

Neil, W.T. (1964) The association of Suwanee points and extinct animals in Florida. *Fla. Anthropol.* 17(1): 17.

Nelson, H.F. and Bray, E.E. (1970) Stratigraphy and history of the Holocene sediments in the Sabine-High Island area, Gulf of Mexico. In: *Deltaic Sedimentation: Modern and Ancient,* pp. 48–72. Morgan, J.P., Ed. Spec. Publ. 15, Society of Economic Paleontologists and Mineralogists.

Pratt, R.M. and Dill, R.F. (1974) Deep eustatic terrace levels: further speculations. *Geology.* 2(3): 155–159.

Pyle, T.E., Henry, V.J., McCarthy, J.C., Giles, R.A., and Neurauter, T.W. (1975) Baseline monitoring studies, Mississippi, Alabama, Florida outer continental shelf, 1975–1976. In: *Geophysical Investigations for Biolithologic Mapping of the MAFLA Lease Area.* Vol. 5. BLM/ST-78/34, Bureau of Land Management, Washington, D.C.

Rehkemper, L.J. (1969) Sedimentology of Holocene estuarine deposits, Galveston Bay. In: *Holocene Geology of the Galveston Bay,* pp. 12–52. Lankford, R.R. and Rofers, J.J.W., Eds. Houston Geological Society, Houston, Tx.

Rezak, R.J. (1988) Sediment analyses. In: Mississippi-Alabama Marine Ecosystem Study, Year 1 Annual Report. Vol. 1. Brooks, J.M. and Giammona, C.P., Eds. Department of the Interior, Mineral Management Serive, Gulf of Mexico Regional Office, New Orleans, La.

Ruppe, R.J. (1983) Predictive models for locating underwater cultural resources on the outer continental shelves. In: *Proc. Alaskan Marine Archaeology Workshop,* Langdon, S.J., Ed. Alaska Sea Grant Rep. 83–9.

Saucier, R.T. (1977) The Northern Gulf Coast During the Farmdalian Substage: A search for Evidence. Prepared for the U.S. Army Engineer District, New Orleans, La.

Schnable, J.E. and Goodel, G.H. (1968) *Pleistocene Recent Stratigraphy, Evolution and Development of the Apalachicola Coast, Florida.* Spec. Pap. 112, Geological Society of America.

Stright, M.J. (1986) Human occupation of the outer continental shelf during the late Pleistocene/ early Holocene: methods for site location. *Geoarchaeology.* 1(4): 347–364.

Thomas, R.H. and Bentley, (1978) A model for Holocene retreat of the West Antartic ice sheet. *Quat. Res.* 2: 150–170.

Van Andel, T.H. (1960) Sources and dispersion of Holocene sediments, northern Gulf of Mexico. In: *Recent Sediments, Northwestern Gulf of Mexico,* pp. 35–55. Shepard, F.P., Phleger, F.B., and Van Andel, T.H., Eds. American Association of Petroleum Geologists, Tulsa, Ok.

Van Andel, T.H. and Poole. D.M. (1960) Sources of recent seidments in the northern Gulf of Mexico. *J. Sediment. Petrol.* 30: 91–122.

Van Andel, T.H. and Veevers, J.J. (1967) Morphology and sediments of the Timor Sea. *Bull. Bur. Miner. Resour. Geol. Geophys. Aust.* 83: 173.

Veeh, H.H. and Veevers, J.J. (1970) Sea level at −175 m off the Great Barrier Reef 13,600 to 17,000 years ago. *Nature.* 226(536): 536–537.

Chapter

6

Inundated Prehistoric Sites in Apalachee Bay, Florida, and the Search for the Clovis Shoreline

James S. Dunbar
Florida Bureau of Archaeological Research
Division of Historical Resources
Department of State
Tallahassee, Florida

S. David Webb
Department of Natural Sciences
Florida Museum of Natural History
University of Florida
Gainesville, Florida

Michael Faught
Department of Anthropology
University of Arizona
Tucson, Arizona

INTRODUCTION

Florida's Gulf Coast has long been hypothesized to be a likely source of submerged early-prehistoric sites, but only in recent years have significant discoveries begun to reinforce such views. During the 1950s and 1960s, the marked increase in Paleoindian artifacts raised by scuba divers from the bottoms of inland waterways led Neill (1964), Lazarus (1965), and Gluckman (1967) to project the search onto the shelf, proposing that similar resources would be found under marine waters of the Gulf of Mexico. The existence of inundated marine sites was demonstrated when artifacts and fossils were removed by shell-mining operations in Tampa Bay (Warren, 1970; Goodyear and Warren, 1972). During the next 2 decades, marine archaeology was vastly augmented by the application of seismic profiling, remote sensing, and other geological exploration methods, as well as by an immense leap forward in the sophistication of oceanographic research (e.g., Ballard and Uchupi, 1970). In this enriched context, marine archaeologists have significantly refined search methods for Paleoindian and Archaic sites in the Gulf of Mexico (Coastal Environments, 1977; Stright, 1986; Dunbar et al., 1989).

Our research into underwater archaeology and paleontology in the Gulf Coastal Plain of Florida has benefited from these advances. For reasons to be explored below, we have concentrated on the "Big Bend" area in the north-central corner of Florida where the north/south peninsula joins the east/west panhandle. During the past 5 years, we have established a rough transect running westward from Half-Mile Rise on the Aucilla River into Apalachee Bay (Figure 1). In this paper, we provide some preliminary results from our archaeological and paleontological surveys beneath marine and fresh waters in this area. On the basis of this and other work, we tentatively interpret some of the environmental conditions encountered by the first Floridians. We wish to stress, however, that the scope of our interpretations vastly exceeds the reach of our present data. The principal justification for discussion at this juncture is to encourage and to help set a sound course for further underwater investigations.

REGIONAL SETTING

Many of Florida's richest archaeological resources, from the treasures of the Spanish Plate Fleet back to the carved ivory implements of the earliest Paleoindians, are distinguished by their exquisite preservation due to underwater entombment. Florida's fresh and marine waters often provide the ideal range of depositional conditions that produce gentle, yet rapid burial, and thus provide incalculable dividends in organic preservation and environmental context. Purdy (1988) illustrates the potential range of such wet-site archaeology. Whereas 3 decades ago the idea of preserved human brain tissue from earliest Holocene seemed so preposterous (Goggin, 1962) that the first authentic specimens from Warm Mineral Springs were allowed to decompose, now such finds from the Windover site are quick frozen and their DNA analyzed (Doran and Dickel, 1988). Wooden artifacts also have been reported from a Paleoindian context in rich organic sediments within Warm Mineral Springs and Little Salt Springs in southwestern Florida (Clausen et al., 1975, 1979; Cockrell, 1980; Skow, 1986).

For several decades, Florida's inland waters have been widely recognized as an important source of Paleoindian artifacts as well as late Pleistocene megafaunal remains (Jenks and Simpson, 1941; Simpson, 1948; Goggin, 1962; Neill, 1964; Waller, 1969, 1970; Webb, 1974; Dunbar and Waller, 1983; Webb et al., 1984; Dunbar et al., 1988; Dunbar, 1991). Furthermore, the association of early hunters with the megafauna represented in Florida underwater sites has been directly demonstrated by a butchered mammoth vertebra (Bullen et al., 1970), butchered giant tortoise shells (Clausen et al., 1979; Clayton, 1981, 1983), and by a *Bison antiquus* skull bearing an embedded spearpoint (Webb et al., 1984). Dunbar and Waller (1983) show that the general distribution as well as major clustering of Paleoindian artifacts are closely correlated with chert outcrops and the surface exposure of Paleogene limestone (Ocala, Suwannee, and St. Marks) formations in the northwestern peninsula of Florida. Dunbar (1991) proposes that the site distribution correlates not only with chert sources, but also with the occurrence of karstic terrain and with the paleohydrology.

The primary purpose of our underwater archaeological and paleontological field work in Florida has been to develop a reliable stratigraphic and environmental framework in which to place the rich Paleoindian and megafaunal evidence. Much of our work in Half-Mile Rise, establishing a local network of stratigraphic sections, is being reported more fully elsewhere. Early in this work, however, it became evident that a secondary goal

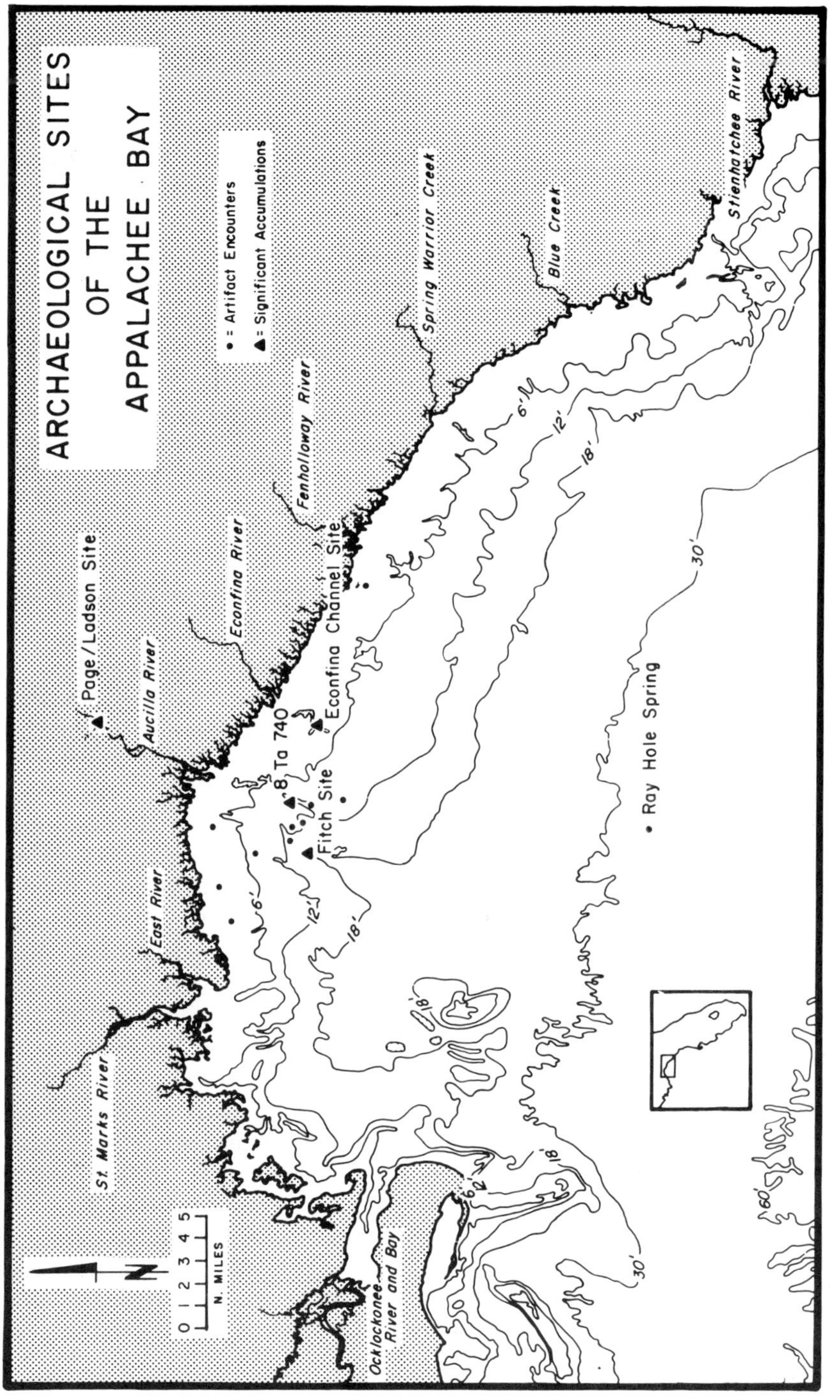

Figure 1. Map of Apalachee Bay showing the inland Page/Ladson site and offshore site locations. Depth contours are shown at 6, 12, 18, 30, and 60 ft below mean sea level.

should be to extend our network of controlled underwater excavations westward into the Gulf of Mexico.

The two fundamental reasons for concentrating archaeological and paleontological survey efforts along the north-central part of the Florida Gulf Coast are related to the karst lowland physiography that is so clearly developed in that region. First, as noted above, is the availability of chert, evidently a crucial resource in Florida Paleoindian economy (Goodyear, 1988). Chert outcrops extend outward into the Gulf of Mexico, and, as noted below, are the key to our marine exploration strategy. Second, the extensive exposures of Paleogene limestones (Ocala, Suwannee, and St. Marks Formations) produce no clastic erosional products. This explains the absence of sandy beaches, spits, and barrier islands north of Tampa and east of the Ochlockonee River (Coastal Environments Inc., 1977). Sinkholes, river channels, springs, and other former freshwater features remain exposed in the seabed of this Tertiary karst region of the shelf and have not become sediment filled subsequent to the mid-Holocene transgression. Thus, topographic targets that might otherwise only be detected with the use of sophisticated remote-sensing equipment can be identified by simple visual or bathymetric techniques. The combined absence of clastic load and high-energy currents, which limit exploration along the Atlantic Seaboard (Kraft, 1978), offer an ideal situation for recovering preserved samples from relatively undisturbed archaeological and paleontological sites.

GEOLOGY, HYDROLOGY, AND SEA LEVEL FLUCTUATIONS

In Florida, physiographic regions underlain by similar geologic formations and sharing similar hydrologic potentials (geohydrology) also share similar patterns of geographic site distribution. Goodyear (1988) and Anderson (1990), among others, have compared resource procurement areas with site distributions and have gathered convincing evidence for the lithic resource orientation of eastern Paleoindians. However, how site distribution patterns changed through time in relation to accessible resources has not been well studied, probably because the majority of the site-distribution analysis has taken place in noncoastal regions. The potential for dynamic geohydrologic change to alter the distribution of available resources, and thereby affect prehistoric settlement, has only begun to be addressed (Dunbar, 1981) even though it is probably central to the issue to understanding site distributions on the continental shelf.

In order to make sense of the Florida pattern, one is forced to consider sea level change, because rising Holocene sea levels drowned chert outcrops, making them inaccessible as an exploitable resource, and replaced terrestrial habitats with brackish, then marine underwater environments. In addition, the predominance of karstic limestone in Florida has caused other changes in the landform, particularly along the north-central Gulf Coast/Ocala Ridge area where Tertiary limestones occur near or at the surface. In this karst region, inland water tables were considerably lower during glacial events and, like the sea, transgressed over terrestrial lowlands with the onset of glacial meltdown (Webb, 1974). Several freshwater sites located in drowned karst features were utilized by prehistoric peoples in the late Pleistocene and early Holocene. As the Tertiary karst lowlands were transgressed by freshwater, resource types and the availability of them also changed, as did the potential for prehistoric utilization.

During the various phases of Pleistocene/Holocene transgression, there may have been interval(s) that favored prehistoric use of a location due to the temporary coalescing of multiple resources. Of the potential number of resources of interest to a prehistoric population, lithic and other mineral resources are the most geographically and temporally fixed. Other resources, oysters or megafauna, for instance, are environmentally dependent and do not occur unless the correct type of habitat has been established, but once established, the physical evidence of their existence is added, in a tangible way, to the sedimentary record. Mineral resources, such as chert outcrops, and other marked resource-gathering areas, such as oyster bars, rock outcrops, and drowned channels and sinkholes, provide large underwater targets and promise to expedite the search for inundated sites on the continental shelf.

KARST AND THE SIGNIFICANCE OF PALEOHYDROLOGY TO PALEOINDIANS

Karst is formed as limestone, which is alkaline, is exposed, and dissolved by acid in the soil, water, and atmosphere (Jennings, 1985; Sweeting, 1981). Major sources of acidity are carbonic acid, which is a product of the dissolution of atmospheric carbon dioxide into water, and organic acids in the soil. Solution features are voids dissolved into once-solid limerock. Sinkholes are not solution features, but result because the solution has had time to weaken the surrounding rock. As the processes of solution continue, subterranean voids may become too large, the weight of the overlying sediment too great, or a combination of these and other factors may cause the earth above a cavity (cave) to collapse, forming a sinkhole. The processes of karstification sculpt subterranean networks which become connected to the surface via open sinkholes, springs, and cave entrances.

Almost all of Florida is underlain by Tertiary limestones. Florida's topography, especially in the Tertiary karst region where the Ocala Uplift helped amplify solution (Opdyke et al., 1984; Schmidt and Scott, 1984), has developed into a mature karst with well-developed and complex underground channel systems. In areas where the Tertiary limestones are near the ground surface, numerous sinkholes and other openings connect to the surface. One of the nation's largest groundwater aquifers, the Floridian Aquifer, is held in these karstic limestone systems (Cooper et al., 1953; Davies and LeGrand, 1972).

In nonkarst regions, meandering river systems form extensive surface networks that drain runoff and experience increased flow and sediment transport during floods. Because of sediment transport, depositional features such as point bars and natural levees form in and along nonkarst channels and deltas build out onto adjacent continental shelves (Donoghue, 1990).

In the Tertiary karst region, surface drainage is absent except for channel systems that dip below the level of the aquifer's surface. Several, well-developed surface channels do exist in the karst regions. Some, like the Suwannee River, flow uninterrupted to the sea, while others, like the Aucilla River, are linear sets of sinkholes which connect disjointed sets of underground and surface channels. Dry channels, now elevated above the aquifer's surface, formed flowing rivers during higher aquifer and sea level stands (Stubbs, 1940; Edwards, 1948).

Of the rivers that maintain consistent flow in the karst region, few carry appreciable clastic sediment loads and, therefore, they do not contribute sediment to the adjacent continental shelf. During high-water stages, karst rivers lose much of their potential to transport sediment because currents are diminished by partial diversion of the total water volume into underground networks. Thus, similar to the karstic shelf, which displays features uncharacteristic of "typical" marine settings, karst rivers also lack sedimentary features such as natural levees, point bars, and deltas that build out to sea (Vernon, 1951).

Sinkhole deposits are alkaline, as are some karst lowland soils, because the limestone adds a calcareous content which neutralizes the acidity. Alkaline soils are rare in other settings, and acidic soils dominate. Alkaline sediment promotes bone preservation and, where it occurs, fossil sites spanning the Pleistocene are common. Deposits that were once in wetland areas and became more deeply inundated during Holocene transgression have the potential of preserving in them wood, fiber, seeds, and animal tissue.

Today, the Floridian Aquifer is near the surface throughout much of the mainland Tertiary karst region, but during the Paleoindian time frame, the aquifer, like the sea level, was lower and site clusters were concentrated in areas where numerous karst features gave access to persistent sources of water (Dunbar, 1991).

In addition to the dramatic effects of transgression, it is important to consider human adaptations from the perspective of the paleohydrology that existed on the shelf prior to inundation. Regardless of yearly precipitation, karst terrain elevated high above groundwater levels tends to be unusually dry and to support xeric habitats because the porous rock permits water to quickly drain to the level of the aquifer. Therefore, climate, which generally exerts the dominant influence on ecology, is somewhat nullified in limestone regions where karst plays an influential role in ecology (LeGrand and Stringfield, 1973).

In Florida, the late Pleistocene decline in sea level lowered the karst aquifer 26 m or more in some places (Webb, 1974; Clausen et al., 1979), and perched water levels in lakes marginal to the Tertiary karst region appear to have declined some 18 to 20 m (Watts, 1983). Shallow lake basins were dry during the last Wisconsinan glacial; for example, Lake Okeechobee was dry except for a brief interval about 12,000 years ago and did not form a perennial lake until the mid-Holocene (Brooks, 1974). During the late Pleistocene, the Tertiary karst shelf is believed to have displayed a gradation of ecological variation, depending on the elevation of the ground's surface above the water table. This, in turn, was largely affected by the level of the sea. What are now coastal or near-coastal areas would have been well inland and elevated high above Pleistocene groundwater levels in the Floridian Aquifer. Areas now far out to sea and near the Clovis shoreline would have been located where the Floridian Aquifer was near the surface.

Sediments, although minimal in most places in the Tertiary karst region, have accumulated in one noteworthy setting, karst depressions. Sinkholes and other deep depressions in the otherwise flat coastal limestones act as natural sediment traps. The types of sediment and nature of their deposition are largely dependent on a sinkhole's elevation above or below water tables (fresh or salt). Because the Floridian Aquifer has fluctuated with the sea level, both the type of sediment being deposited and the rates of deposition have changed, affected by the late-Quaternary climatic shifts. The eustatic-hydrologic cycle changed Florida, including its shelf, as sea level pulses brought about the growth and reduction of landmass throughout the Pleistocene (Webb, 1974). At least in theory, the eustatic-hydrologic cycle caused changes in the material entering sinkholes and preserved in them a Quaternary history of the karst shelf and adjoining mainland. While the sediment in one sinkhole will not provide a complete view of the Quaternary history of the karst region, samples from several sinkholes extending from the mainland onto the shelf most likely will.

SEA LEVELS AND PALEOINDIAN RECORDS ON THE SHELF

The search for Paleoindian sites on the submerged Gulf shelf of Florida is closely keyed to the history of late-Pleistocene and early-Holocene sea levels. When the last glacial epoch reached its apogee some 18,000 years ago, the sea level presumably had reached its nadir, which, according to various estimates, lay at depths of 60 to 80 m or more below present. Thanks to its shallow western slope, the Florida peninsula would have augmented its width by more than 300 km during such low sea levels. With its width thus doubled and its glacial climate quite benign, Florida's coastal plain ''approached optimal conditions for the earliest Americans'', as Edwards and Merrill (1977) recognized.

There are also several other factors that make the Gulf Coast of the Florida peninsula a promising area to stage a search and recovery program for evidence of late-Pleistocene life. The Big Bend of Florida is relatively undisturbed by tidal fluctuations and, because its surface geology consists predominantly of limestone, it has a relatively shallow mantle of clastic sediments. The fact that buried forests from the last glacial epoch are still visible in some parts of this submerged coastal terrain encourages the view that it may yield a rich record of other late Pleistocene life as well (Brooks, 1973). And of course its width is immense: by comparison, the shelf off Cape Hatteras is 25 km wide, and that off Miami is only 2 km wide.

The idealized worldwide record of eustatic sea level, even for the last deglacial hemicycle, is complicated in any given region by diverse local factors, including tectonism, isostasy, and compaction of sediments (Pirazolli et al., 1989). For these reasons, many geomorphologists, for example, Fletcher (1985) and Morner (1984), have recommended that any given coastal area develop an independent sea level trajectory on local empirical evidence, rather than arbitrarily assuming a fit to some worldwide hypothetical curve.

With regard to the last deglacial hemicycle in the Big Bend of Florida, we follow a mixed strategy. On the one hand, we fully subscribe to the many concerns about local variations from a standard eustatic curve. Furthermore, because of its favorable setting, the west coast of Florida, specifically the Big Bend area, where Neogene clastic sediments only minimally blanket the underlying Paleogene limestone surface, is an especially important area for which to construct an empirical sea level record from local evidence. Donoghue (1990) has undertaken such an investigation for the Big Bend area. And some of the sites reported in this chapter, when dates are confirmed, will help bracket that local sea level curve. On the other hand, we strongly suspect that local variations from the eustatic curve are minimized in the Big Bend area, for it is located on the trailing margin of the North American continent, south of the glacial margin, and east of sediment loading from the Mississippi Embayment. It might thus be expected to provide a relatively close approximation to the true eustatic record (Brooks, 1973).

For purposes of the present analysis, we will adapt a version of the standard eustatic curve to model the last deglacial rise of sea level on Florida's Gulf Coast, specifically in the Big Bend area. We turn first to the long tradition of studies of submerged sea levels in adjacent coastal regions of the western Atlantic. Low escarpments on the continental shelf near Hudson Canyon were recognized and named more than 50 years ago (Veatch and Smith, 1939). The classic studies by Curray (1965) and by Milliman and Emery (1968) extended these features along the shelf in eastern North America and dated them as Wisconsinan. These two classic curves, however, differed by as much as 50 m regarding postulated ocean depths during the interval from 15,000 to 10,000 years ago. Subsequent studies favored the shallower curve of Curray (1965) and suggested that some of the dated materials, possibly peats and especially oyster shells, were ''mobile'' and had been transported postdepositionally to greater depths (McIntyre et al., 1978; Dillon and Oldale, 1978). A still shallower curve developed off the Carolinas by Blackwelder et al. (1979) indicates that at 17,000 years ago the sea stood 60 m below its present level and that by 10,000 years ago it rose to a level of 22 m. These figures seem generally concordant with the younger curve derived by Scholl et al. (1969) from a large, well-dated set of paired freshwater and tidal sediment samples from the Florida Keys, which shows a gradually decreasing rate of sea level rise at about 6000 radiocarbon years ago. We have taken the somewhat conservative sea levels of Curray (1965) to model the possible Clovis shoreline in the Big Bend area of Florida. Curray estimates the levels to have been about 80 m below present sea level around 17,000 years ago and about 20 m below present around 10,000 years ago.

Because the region of possible Paleoindian occupation along Florida's submerged coast is so very broad, it is desirable to identify certain areas or zones of greater probability for preservation and discovery of late-Pleistocene life. Probably depths for stillstands during the last deglaciation are of special significance. Foraminiferal studies of deep sea cores indicate that the melting rate of the deglacial hemicycle reached a peak between 13,000 and 11,500 years ago (Ruddiman and Duplessy, 1985). Then there was a pause, followed by a second ''step'' of more rapid melting beginning at about 10,700 years ago. According to Frazier (1974) and Ballard and Uchupi (1970), there may be evidence of a significant early stillstand at about 53 m and another, younger one at 15 m. Late-Pleistocene proboscidean teeth are concentrated on the Atlantic shelf at depths of 40 to 49 m. In reviewing this evidence, Webb (1981) speculated that it might represent a particularly productive lagoonal system during the latest Pleistocene stillstand and added that ''apparent clustering of megafaunal records at certain depths on the Continental Shelf may provide the best available clues as to probable places of Paleoindian associations.'' We list these possibly more sensitive depths and equivalent distances off the coast in the Big Bend area in Figure 2.

A further strategy for locating the Clovis shoreline examines features which run more or less perpendicular to these sensitive depth zones. These are the late Pleistocene straths of present river systems. In many instances, their seaward extensions can be traced readily on bathymetric charts, and certainly they can be accurately mapped using subsurface geophysical methods. Our preliminary results demonstrate the efficacy of this approach, at least in shallow shelf terrain.

APPROACHING THE CONTINENTAL SHELF WITH TERRESTRIAL MODELING

Although our major interest focuses on the submerged continental shelf, locating sites in offshore marine environments is difficult. Therefore, we began by determining the frequency and predictability of site locations on the adjacent coastal plain in hopes that we could use this information to model site-location parameters for once similar, but now submerged, environments.

Onshore, the low-energy, often anaerobic conditions of the karst features of the northeast Florida Gulf Coast have preserved organic remains and sedimentary profiles, including at least one continuous stratigraphic profile at the Page/Ladson site which includes the Pleistocene/Holocene boundary (Dunbar et al., 1988). The preservation of stratigraphies of the Pleistocene/Holocene boundary are unusual in North American geochronological sequences in general (Haynes, 1984), so their presence here enhances the importance of the area for multidisciplinary research.

In addition, 90% of the Paleoindian sites containing diagnostic artifacts—Clovis, Suwannee, or Simpson projectile points, carved proboscidean ivory foreshafts—are located near such karst depressions, where they penetrate the Tertiary limestones to expose the Floridian Aquifer, a source of potable water (Figure 3). Tertiary limestones in this region

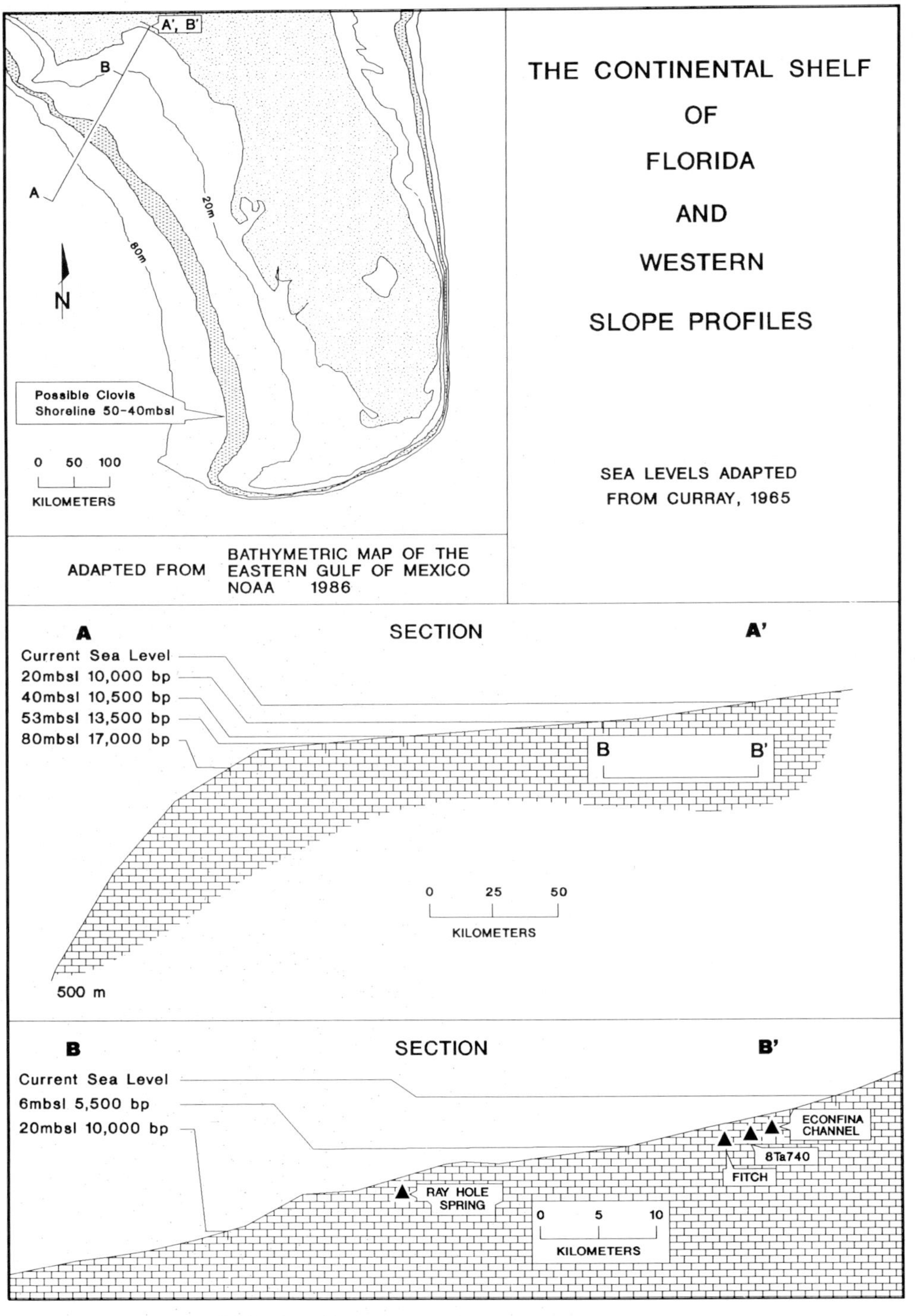

Figure 2. Map of the Florida continental shelf showing the possible Clovis shoreline and the extent of additional late-Pleistocene landmass. Profiles A-A′ and B-B′ show proposed sea level stands in times before the present and position below present sea level for the Apalachee Bay area. Profile B-B′ also shows the various sea level stands in relation to the inundated sites in Apalachee Bay.

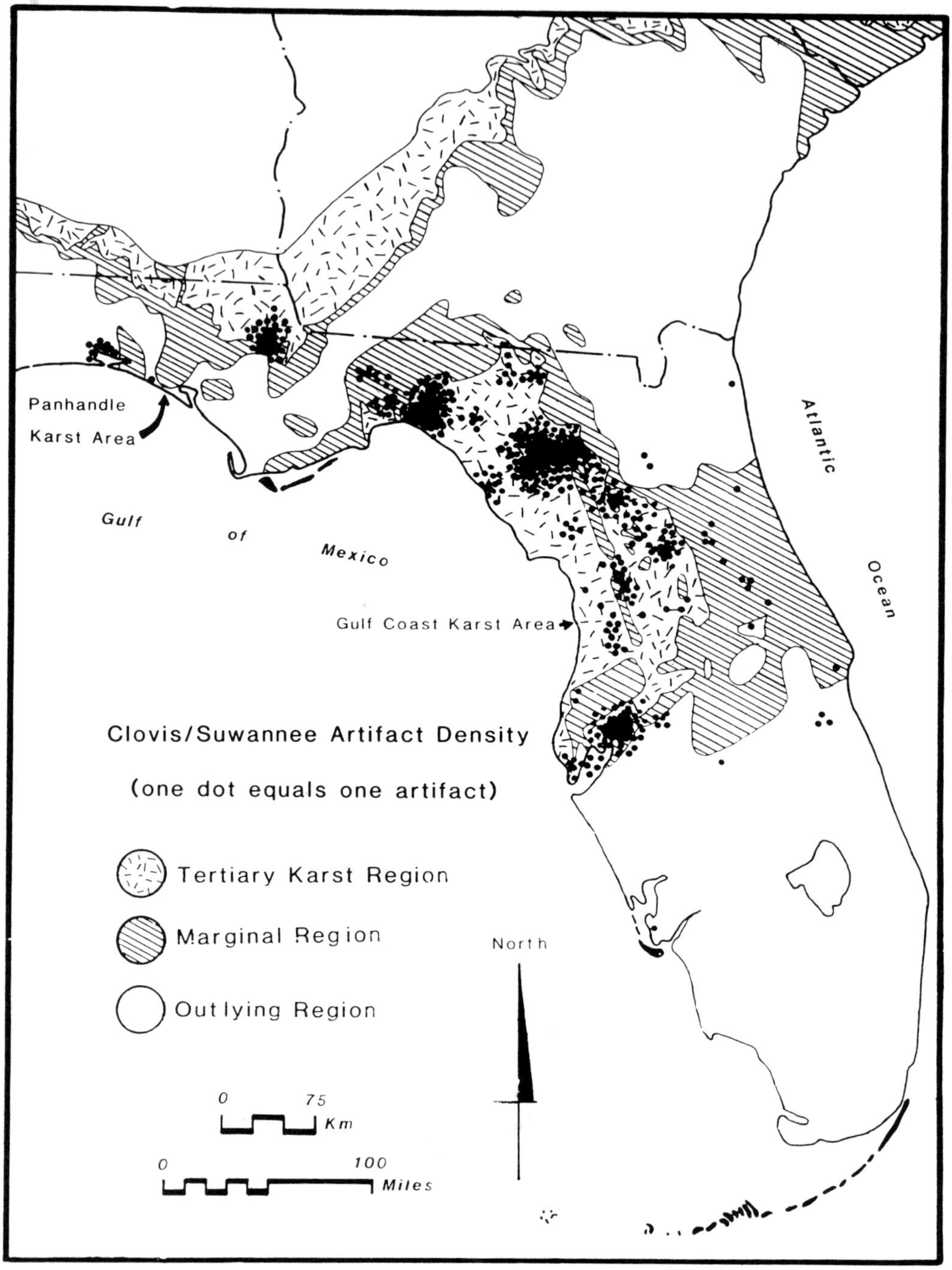

Figure 3. Map showing the concentrations of Paleoindian diagnostic artifacts in relation to the three major geohydraulic regions in Florida. The major concentration of Paleoindian sites is located in the Tertiary karst region where both water from the Floridian Aquifer and chert are available. The Tertiary karst region borders and extends offshore an undetermined distance from the north-central Florida Gulf Coast.

are particularly rich in chert deposits. These observations have led to the following model of Paleoindian settlement (Dunbar, 1991):

> The river basins in the (two) Tertiary Karst Regions of Florida have the greatest concentration of Clovis/Suwannee Paleoindian sites because unique environmental conditions created natural resource accumulations that complemented technology and subsistence behavior. Stable habitats in the karst regions supported grazing animals but dry intervals confined game herds to oasis locales. During droughts, oases in the karst river and lake bottoms offered water, food, bone and lithic resources for Paleoindian exploitation. As a result, major site clusters in Florida became centered around rivers like the Santa Fe and Aucilla because multiple resources were available and repeated exploitation could be supported. During wet periods when intermittent water sources were both available and widespread, game herds migrated away from oases locales, and mobile bands of Paleoindian hunters and gathers followed them. A semi-sedentary Paleoindian lifeway may have existed in Florida with prolonged occupations around karst rivers and lakes and less frequent periods of hunter/gatherer high mobility.

The coastal area surrounding Apalachee Bay has one of Florida's major Paleoindian site clusters, and the location of the sites conforms to the above model. Therefore, we predict that where the geohydraulic conditions and the abundance of lithic resources were similar on the continental shelf during Paleoindian times, similar site-distribution patterns should occur. However, further out to sea and near the Clovis shoreline, predictive modeling should change because geohydraulic conditions differed. To test this hypothesis, we drew from the data regarding submerged and terrestrial sites around the Aucilla River basin and, in particular, on one of its landlocked segments, Half-Mile Rise, to predict where sites might occur immediately offshore in Apalachee Bay (Figure 1).

The Half-Mile Rise project has involved several phases, including site surveys on land and underwater, random and controlled underwater surface collections from eroded sites, and one controlled excavation at the Page/Ladson site (Figure 4; Dunbar et al., 1988; Willis, 1988; Richardson, 1988; Alexon, 1988). The majority of inundated sites in the Aucilla basin vary from about 2 to 10 m below sea level, even though they are located several kilometers inland from the Gulf of Mexico. Knowing the geographic distribution of both land and underwater sites in the Aucilla River area has allowed a degree of accuracy regarding regional predictability of site locations.

A steadily increasing body of evidence from Half-Mile Rise supports the oasis hypothesis and provides an intriguing picture of Pleistocene/Holocene environments for the Apalachee Bay area. Deep sinkholes in the Aucilla River were shallow ponds and swamp forests about 18,000 years ago, and most other segments of the river appear to have been dry. Naturally modified bone of late-Pleistocene animals from the Page/Ladson site indicates that animals entered the sink, probably for water, from about 15,000 to 10,500 years ago. Artifacts of ivory and stone suggest that Paleoindians hunted around, and in, the sinkhole. The sinkhole also could have provided a stable source of drinking water for Paleoindian inhabitants. Human activity peaked about 10,000 years ago. Stone tools are the most common type of artifact recovered, but bone tools and carved wood also have been recovered in the thick, highly stratified deposits in the sinkhole.

Lithic resources were essential to early-Indian populations, and chert for tool making was not always present at sinkhole localities. The chert-bearing limestone surrounding the Aucilla River comprises a slightly undulating karst plain sculpted by higher sea level stands. Extensive, but thin deposits of clastic sediments, predominantly sand, bury the karst surface. Erosion-resistant chert outcrops are an exception and protrude above the ground

Figure 4. Diver conducting test excavations in a rocky shallows at the Page/Ladson site. *In situ* components of the Page/Ladson site have been identified in deeply stratified sequences and contain in them, good, sometimes excellently preserved bone and wood artifacts. Although the Page/Ladson site lies several miles inland from the coast, most of the inundated site lies below the present sea level.

surface. The distribution of chert in the karst plain and in the Aucilla river basis is irregular, but frequent.

Chert boulders are exposed when the limestone surrounding them dissolves, leaving behind pinnacles of erosion-resistant chert. At present, about 100 chert outcrops are known in the area and occur in two types of settings. First, and most common are chert outcrops (boulder fields) in the karst plain. The second, and most exposed type of outcrops are located in river channels. Because they are erosion resistant, river channel outcrops form shallower (5 m or less) areas along the channel bottom. All of the known underwater chert outcrops in the Aucilla River and its tributary, the Wacissa River, have been identified as quarry sites and thus provide evidence that large segments of the river channels were once dry. Underwater quarry sites represent the most conspicuous and, therefore, easy to find type of inundated site because artifact frequency (lithic debitage) is high and chert outcrops are visible surface features.

Paleoindian and early Archaic sites are concentrated in and around the Aucilla River and other areas where clusters of sinkholes most likely formed persistent and multiple drinking-water sources. Isolated sinkholes, located in the karst plain where little of the subterranean cavernous limestone is revealed, appear to have been less frequently utilized by the early inhabitants. Therefore, multiple water sources are assumed to have been preferred over areas in the karst plain where such features were less frequent, more isolated, and less reliable. Chert quarry sites are an exception, being found wherever chert crops out, both in river channels and in the surrounding karst plain.

SITES IN APALACHEE BAY—THE NEARSHORE SEARCH

Apalachee Bay is situated in the northern part of the karst shelf. Several spring-fed rivers flow into Apalachee Bay, including the Ochlockonee, St. Marks, Pinhook, Aucilla, Econfina, and Steinhatchee (Figure 1). Only the Ochlockonee River, which displays many nonkarst characteristics, carries a particulate sediment load to the bay. The other karst rivers lack sediment load (Vernon, 1951). The bay has a shallow, sometimes silty, but typically sandy bottom in water depths less than about 3 m which is typically overgrown by various species of grasses. These nearshore grass-bed areas are sometimes interrupted by rock outcrops and oyster bars which alter the ''typical'' shallow-water sedimentary profile. However, bay scallops, which seasonally inhabit the grass beds, add their shells to the otherwise fine-grained sediment in a relatively uniform manner. At depths greater than about 5 m, the coastal grass beds thin out and the bottom terrain changes to barren sand flats and rocky bottom supporting benthic marine organisms such as sponges, corals, and sea fans. In these deeper waters, surface sand generally covers a bioclastic detritus consisting of the broken-up exoskeletons of marine shells and corals.

The lack of massive erosion and redeposition has not only helped to preserve significant sites in rivers like the Aucilla, it has also prevented Apalachee Bay from becoming blanketed by deltaic sediments spread by marine action. The bottom of much of the bay is covered by a thin cover of sediment, containing a biogenic shell hash, over the karstic bedrock. On the other hand, sinkholes in the bay contain sediment accumulations several meters deep which probably contain subaerial and freshwater deposits that, like those of the Page/Ladson site, span the Pleistocene/Holocene boundary. Nearshore rock outcrops, exposed at low tide, are abundant and consist of either erosion-resistant dolomite or chert (Yon, 1966). Further offshore, numerous rock outcrops, springs, and other remnant features of past terrestrial or freshwater environments are covered with marine growth.

Both east and west of Apalachee Bay, the potential for easy site identification does not seem as good. West of the bay, offshore surface karst is buried by sediments deposited by the nonkarst Apalachicola River. Parts of the paleochannel of the Apalachicola River have been traced offshore and, unlike the nearby karst channels, the Apalachicola River was larger in the Pleistocene. During Holocene transgression, the Apalachicola River valley became buried and its offshore extent is now more deeply buried and less accessible than areas in Apalachee Bay (Donoghue, 1990). Buried features such as infilled river channels and sinkholes could not be identified without the aid of sophisticated remote-sensing and coring equipment and would be impossible to test without the use of some type of massive excavation equipment. East of the Steinhatchee River in Apalachee Bay, chert resources and the concentration of Archaic and earlier sites diminish even though the limestone formations of the Ocala group are not deeply buried. Within the Tertiary karst region, chert-bearing limestones are more or less absent along the coast where the Ocala group outcrops, and chert as well as concentrations of early sites pick up again further south, in the New Port Richey to Tampa Bay area where the Suwannee and St. Marks Formation limestones outcrop again.

Because high concentrations of early sites, including Paleoindian, occur in the Apalachee Bay area and the thin sediment cover allows easy access to them, the opportunity to complete a successful offshore survey appeared excellent. Apalachee Bay is also a low-energy section of the shelf compared to the site concentration area near Tampa, leading to a greater expectation of preserved site integrity.

Offshore rock outcrops are one of the most easily identifiable seafloor features on the Tertiary karst shelf. The old sponge-diving grounds in Apalachee Bay between the St.

Marks and Steinhatchee Rivers is such an outcrop area. Inundated chert outcrops on the seafloor were considered to be the easiest potential archaeological sites to identify. The irregular topography associated with protruding outcrops attracts fish and other marine life which have, in turn, attracted fishermen and divers (Serbousek, 1988). Loran-C coordinates for potential site locations were provided by several sport divers (mostly spear fishermen). However, few archaeological site locations were known by these divers since the purpose of their activity was to spear fish around the outcrops. Prior to our survey attempts, only one site in Apalachee Bay was known to have produced a diagnostic artifact, an Archaic stemmed point, picked up by a diver as an item of curiosity (Florida Master Site File information).

Four short field sessions, comprising 17 total days of survey and limited excavation, have been undertaken since the summer of 1986. In addition to rock outcrops, relict river channels, springs, and sinkholes have also been observed in the often monotonous marine terrain. It should be noted that the words "easy to locate" take on new meaning in an underwater environment where search operations take place by swimming, diver towing, and fathometer bathymetry. These methods are more time consuming than are land-based surveys, but are just as productive. The majority of sites were located by towing a diver behind the boat to facilitate inspection of large areas of the seafloor in areas where frequent rock outcrops were known to occur. The diver would scan the bottom and drop off the tow rope when a target seemed worthy of closer inspection. Once a site was located, additional divers would enter the water to assist with the survey and a series of hand-fanning tests of the area would be conducted. The diver tow method is very much dependant on good water visibility.

As chert outcrops were the targets for testing site potential, we first had to determine if any of the rock in the outcrop area was chert. Sometimes artifacts were found on the surface, but most often, a chert outcrop was identified and subsequent hand-fan testing yielded the buried evidence of prehistoric human utilization. Once a site was confirmed, its location in both Loran-C and geographic coordinates (longitude and latitude) were gathered and the site was recorded.

Initial Survey—1986

In 1986, the first offshore survey work was conducted over a 4-day period. The survey began by searching for and inspecting rock outcrops. With the guidance of spear fishermen, features such as springs and sinkholes also were visited. Six sites were located, ranging from 1 to 4.8 km offshore, and at depths ranging from 0.5 to 3 m below mean sea level. Similar to our land-based findings, all chert outcrops discovered offshore proved to be quarry sites. One dolomite outcrop was not a quarry location, but did produce a few flakes of bifacial retouch.

A relict river channel segment was identified about 5 km offshore from the mouth of the Econfina River. This channel was identified by locating its topographic profile with a fathometer. Features in the channel which suggested past terrestrial origins included river-bank areas covered with oyster shell, rock outcrops of chert in and adjacent to the channel, cypress tree roots, and lithic debitage scattered almost everywhere.

This drowned reach of the Econfina River once formed a rocky shallows in a channel segment less entrenched than the channel areas above and below it because the chert formed an erosion-resistant part of the lithology. In testament to its prehistoric utilization, chipped stone artifacts, including quarry debris, cores and reduction flakes, and, to a lesser extent, bifaces, occur in abundance. Testing was limited to hand fanning and surface collection.

However, various samples were recovered, including cypress root samples which yielded a carbon date 5160 ± 100 B.P. (Faught, 1988). Clearly, the Econfina Channel site was the most significant of our initial discoveries.

Ray Hole Spring Expedition—1986

Later, in the fall of 1986, another expedition was undertaken to investigate Ray Hole Spring located approximately 38.5 km offshore. This site lies near the boundary of Apalachee Bay and open gulf waters and is due south from the mouth of the Aucilla River. Ray Hole "Spring" was plugged with sediments and looked more like a partially infilled sinkhole than a surface vent for the Floridian Aquifer. The feature was surrounded by sheer rock walls on three sides and a gentle sediment bank descended into the sink on the other. Several years earlier, Ray Hole Spring was reported to be large enough for cavern diving and was observed on one occasion to have a minor flow (Rosenau et al., 1977).

The accounts of divers indicate that Ray Hole Spring was flowing prior to 1985. The full force of hurricane and tropical storm activity usually spares Apalachee Bay (Jordan, 1984), perhaps because the bay forms a protected pocket (the so-called "Big Bend") where the peninsula meets the panhandle, and it is surrounded by land on three sides. However, in 1985, two large storms passed just south of the bay and then, in November, Hurricane Kate struck hard, running up the bay's west side. We have no accounts of an open sinkhole in the Ray Hole Spring area after 1985, suggesting that it was filled as a result of this storm activity.

The natural bottom surrounding the Ray Hole sink consists of a thin veneer of bioclastic detrius and sand above the flat limestone bedrock of a karst plain. Similar to the pocked surface of the onshore karst plain, small crevices and solution holes are probably abundant, but hidden by the sediment fill and difficult to identify without subbottom-profiling equipment. Crevices surrounding the sinkhole held sediment sequences up to 1 m thick. It is reasonable to expect that some crevices might be even deeper, providing excellent locations for testing.

The sinkhole was the only topographic interruption in the otherwise monotonous marinescape. A rock escarpment descended into the sink and outcrops were common around its margins. The base of a rock cliff adjacent to the deepest part of the sinkhole was undercut, but only large enough to form a refuge for small fish. Terrestrial sediment, potentially held deep within the sinkhole, was not reached by auger testing even though the deepest auger hole reached 4 m into sinkhole sediments.

A small induction dredge was used to place several test pits around the margins of the sinkhole in an effort to locate sediment pockets that might have trapped artifacts. Several chert items were recovered which may be artifacts. These chert items look like debitage, but were too poorly preserved for positive identification. The chert specimens appear to have patination or corrosion, presumably from exposure to the marine chemical environment, and also have been affected by marine organisms capable of dissolving voids in the rock. All of the "flakes" were recovered from marine sediments, which indicates they were displaced from an unknown context. Samples taken from the natural rock outcrops resemble the onshore dolomite of the Oligocene Suwannee Formation Limestone.

Induction dredge excavation of a large crevice revealed typical marine sediments consisting of sand and marine shell detrius resting on a lower unit of oyster shell. A large chunk of "live oak" wood, *Quercus virginiana,* was recovered resting on the limestone bedrock in the basal portion of the oyster shell unit. The live oak wood had not been damaged by marine or brackish-water organisms such as teredo worm and appeared to be identical to the freshwater driftwood common in onshore rivers and springs.

The oak wood from Ray Hole Spring is suggestive of coastal and inland hammocks unaffected by saltwater intrusion. Oyster shell is, of course, indicative of a brackish-water environment where saltwater becomes diluted by the influx of freshwater adjacent to the coast and certain fannal species, notably oysters, flourish. The overlying bioclastic debris (primarily shell fragments), sand, and chunks of rock, including an occasional "flake-like" piece of chert, are characteristic of the marine sediment at Ray Hole Spring. Marine sediment continues to form, although slowly, as the shells of marine animals are broken and added to the rubble lying on the surface of the degrading limestone. The marine sediments are transient with storm surge, unless they settle into traps like Ray Hole Spring (Doyle and Sparks, 1980).

The oak wood was C-14 dated to 8220±80 B.P. and oyster shell yielded an age of 7300±60 B.P. (Anuskiewicz, 1988). Based on these carbon dates, a brackish coastal environment had replaced terrestrial/freshwater habitats by around 7300 years ago, which is consistent with our model of sea level rise for the Gulf shelf. Thus, the marine transgressive interpretation of the sediment sequence adjacent to Ray Hole Spring seems correct, although the potentially much deeper sediment sequence buried in the sinkhole might provide a greater level of confidence.

In general, the marine conditions at Ray Hole Spring reflect a higher-energy environment compared to the protected inner reaches of Apalachee Bay. At Ray Hole Spring, with typically sparse sediment above bedrock, the most promising areas for future tests are the crevices and pockets greater than 1 m deep surrounding the sinkhole. Coring of the thick sediment column in the sinkhole should be conducted to determine if similar, but thicker and better preserved brackish and freshwater sediments occur below the marine deposits. Assuming Ray Hole Spring was open during the late Pleistocene, it should minimally contain an environmental record of its terrestrial-to-marine natural history, as many other sinkholes on the shelf probably do. If the area was occupied prior to sea transgression, it may also provide a record of human prehistory.

Port Paradise Springs—1987

In the late summer of 1987, an exploratory dive to Port Paradise Springs was arranged by Don Serbousek (a long-time Florida avocational archaeologist/paleontologist). This sinkhole is located due west of Crystal River, approximately 40 km offshore and about 70 km south of Apalachee Bay (not on map). Even though Port Paradise Springs is outside the Apalachee Bay study area, it was investigated because it is an example of an open cavernous system. Port Paradise Springs had not been filled by marine sediments subsequent to sea level transgression.

The shelf gently slopes from the Crystal River area to a distance of 40 km offshore, where it reaches an average depth of about 9 m below sea level. The sinkhole is surrounded by a flat limestone bottom with very little sediment. A number of small rock ledges about 20 to 30 cm high are randomly scattered across the bottom and are moderately overgrown by sponges, seafans, and small corals. The local limestone is believed to represent one of the Eocene, Ocala Group formations, but this identification is based on field observations and has not been confirmed.

The Port Paradise "spring" forms a cenote 9 m in diameter which penetrates the limerock 28 m before it breaches a system of lateral caves. A waterline bench or ledge is etched into the limestone walls of the sinkhole at a depth of about 15 m below sea level, indicating that the upper 6 m of the sinkhole above the ledge was exposed during a lower sea level stand. The "spring" was nonflowing, although a thermocline that separated 82°F gulf water from 72°F groundwater was encountered at a depth of about 17 m below sea level or about 2 m below the ledge. Water below the thermocline was cloudy and, according to the cave diver who explored below these depths, smelled and tasted of sulfur. Both the water temperature and a sulfur content are consistent with water originating from the Floridian Aquifer along this section of coast (Rosenau et al., 1977). Concealed by the thermocline and dim haze of cloudy water, a sediment cone ascends from its base at 37 m below sea level to its peak at 24 m. A cave system with an approximate 12-m floor-to-ceiling height extends an unknown distance into the limestone running laterally away from the sinkhole entrance (Serbousek, 1988). All of this evidence suggests that Port Paradise has been and continues to be a conduit connected to the Floridian Aquifer.

Clearly, Port Paradise Springs experienced partial terrestrial conditions during its evolution and promises to hold undisturbed Pleistocene/Holocene sediment in the silt cone. However, Ray Hole Spring promises to have better archaeological potential because it is located in an area of the karstic shelf having chert-bearing limestone. Sinkhole locations like Port Paradise and Ray Hole Spring do exist on the karstic shelf, and probably many more could be identified with the aid of remote-sensing equipment.

Econfina Channel Site Second Expedition—1988

In July of 1988, research resumed in Apalachee Bay with 5 days of excavations at the Econfina Channel site (8Ta139; Faught, 1988) and 1 day searching for new sites further offshore.

Several test pits were placed in the channel and on its margins. Sediment levels in the channel were about 1 m deep and were not clearly defined, having wood (roots?)-occupying sediment that also contained brackish-water and marine shells. The sediment column on the eastern bank was sealed beneath an undisturbed level of articulated oyster shell; therefore, the stratigraphy below the articulated oysters was undisturbed subsequent to their growth over the site. The deepest stratigraphy was revealed in a test pit placed on the eastern margin of the channel where the articulated oyster shell deposit concealed cultural deposits. The artifact-bearing sediment below the articulated oyster shell level contained much debitage and, like the channel deposits, was curious because wood occurred with brackish-water shells. Cypress wood, which has been collected from the site, is indicative of freshwater, not brackish water, environments.

Excavation on the eastern bank revealed three stratigraphic levels reflecting marine, coastal, and possibly terrestrial environments. The uppermost 10 to 15 cm are transient marine deposits containing broken shell and minor amounts of debitage in a sandy matrix. This overlies a layer of oyster shell, 15 to 20 cm thick, which contains articulated shells, often in clumps, but no debitage or other cultural material. Below this, Level 3a consists of a structureless brackish-water shell bed (no articulated shells). This layer is about 1 m thick, and had the highest concentration of chert artifacts encountered during testing. It is tempting to view this level as a midden deposit, but no other evidence was found to strengthen this argument. The contact between Levels 3a and 3b is gradational and lies approximately 1.30 to 1.40 m below the seafloor.

At about 1.40 m, the shell content, although still considerable, diminishes, and well-preserved wood that appears to represent tree roots is first encountered. The occurrence of natural rock boulders and cobbles also becomes much more prevalent at about 1.40 m and increases until, at 1.65 m, the excavation had to be halted because it was too difficult and time consuming to continue removing the rock. Two samples of the freshwater mollusc *Psidium* sp. were located among these boulders. Rocks from this level were smooth and rounded, similar to limestone boulders and cobbles uncovered on land-based excavation. The rounded nature of the rock is presumed to have resulted from chemical solution in the soil. This is in striking contrast to the jagged and sculpted appearance of limerock exposed to flowing-water environments in this region.

Materials collected during the 1988 field session were generally more diagnostic than, but similar to the material gathered in 1986. Bone specimens were absent, with the exception of a few well-mineralized dugong (sea cow) rib fragments, eroded out of the Oligocene Suwannee Formation Limestone.

Debitage collected from the surface was heavily patinated with a black exterior and white, chalky interior. Debitage buried below surficial deposits had a gray, somewhat chalky-feeling exterior and were patinated similar to debitage from terrestrial sites, having a center of unaltered "fresh" chert. Like the rock samples from Ray Hole Spring, flakes and other stone artifacts from Econfina Channel, particularly from the surface, appeared to have been affected by chemical reaction to saltwater.

As a general observation, chert patina on terrestrial sites occurs in soils that are subject to alternating wet-dry conditions (Clark and Purdy, 1979). Terrestrial patination turns the outer surface of chert artifacts white or tan and, as the patina becomes more advanced, the outer rind becomes chalky. Debitage from freshwater wetland and underwater locations shows little if any patina (McNutt, 1978) but, instead, is usually darkly stained. Thus, in the Apalachee Bay area, chert artifacts that have been recovered from marine, freshwater, and terrestrial environments have been affected differently, and their visual appearance seems to be indicative of the environment of origin. Because chert artifacts at sea have been exposed to at least two, and potentially three, different environments (terrestrial and/or freshwater and, finally, marine), the prospect of trying to determine a site history through the study of postdepositional lithic taphonomy promises to be an interesting avenue for research.

Most stone artifacts collected from the Econfina Channel site are quarry and reduction waste, with lesser amounts of refinished materials and tools. Among the few tools recovered, the base of an Archaic projectile point indicates an age range of about 7000 to 3000 B.P. for the Level 3a deposit (Figure 5, top left, is a Marion or Putnam type; Bullen, 1975). The 5160 ± 100 B.P. carbon date obtained on the cypress sample recovered during the 1986 expedition agrees with this time frame.

Prior to its inundation by the sea, the Econfina Channel site offered chert resources as a potential draw to prehistoric peoples. By at least 5000 years ago, the site area was affected by rising groundwater tables and became shallows in the Econfina River channel. The site was also situated less than 1 km from a spring (currently open and flowing) which flowed west to a confluence with the river channel. Although the site might have had potential for human utilization when freshwater sources occurred in close proximity, the evidence for prehistoric utilization does not show up until a brackish-water environment became established. Once the site area became coastal, freshwater and brackish-water resources as well as chert formed a time-limited location ideally suited for prehistoric utilization.

As the sea level rose, the river and coastline migrated toward their present positions. Clearly, the time of significant artifact accumulation coincided with the deposition of disarticulated shells when the site area was coastal. However, this environment ended when continued inundation isolated the site from the shoreline. Locations like the Econfina Channel site, where accumulated sediment consisted of rock and shell, provided the opportunity for oyster larvae attachment and growth. Therefore, inundation promoted oyster bar development as undetermined generations of the animal grew atop their solid foundation and added to it in a way that was self-perpetuating. Eventually, though, transgressing seas

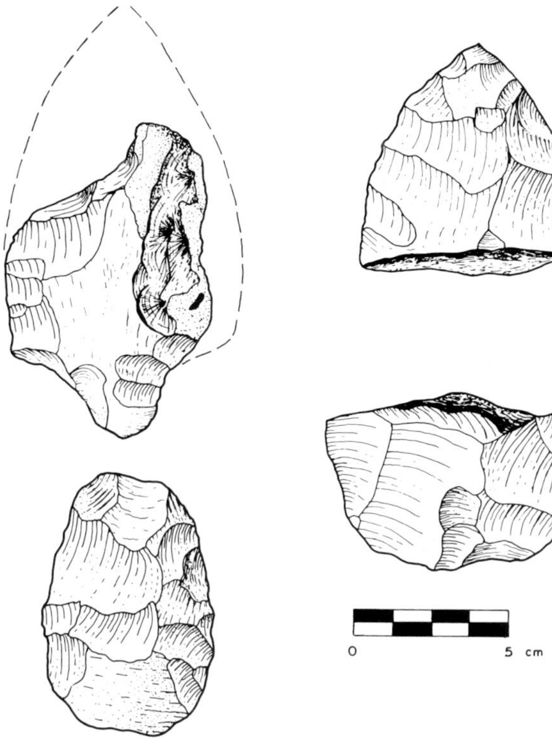

Figure 5. Lithic biface artifacts from the Econfina Channel site. From left to right, top row: mid-Archaic (Putnam type) projectile-point fragment, broken distal end of a preform. Bottom row: unfinished biface, broken preform base.

led to their demise. Once the river and its freshwater diluting influence withdrew, modern marine conditions became established over the site area and oyster colonies died off in water too salty for their continued survival. These features make excellent markers for stillstand appraisal.

During the 1988 field expedition, the one day spent searching for new sites further offshore resulted in the identification of two new sites associated with chert outcrops. Both were about 10 km offshore in areas covered with little sea-grass growth, but substantial sponge and seafan populations (Figure 1). The smaller Stallings site was about 4.5 m below mean sea level, while the larger Fitch site was about 5.5 m below mean sea level.

The Stallings site is located a few kilometers east of the relict Aucilla River on a slightly more elevated inundated karst plain. It was discovered by navigating to a known area of rocky bottom, and then conducting an in-water search of the area. There were no recognizable karst depressions associated with the site; rather, it was an area of relatively flat terrain interrupted by occasional rock outcrops. Chert rock was identified in two areas and protruded as much as 50 cm above the sandy bottom. Sampling of the site was limited to hand-fan excavation and taking rock samples from outcrops. Lithic debitage and a unifacially retouched scraper were recovered, but the frequency of cultural material was low.

The Fitch site (8Ta739) consisted of two adjacent and elongate rock outcrops surrounded by sand and detrius. A casual surface collection produced a number of specimens, including one conical core reminiscent of late-Paleoindian or early-Archaic material. The Fitch site appeared to have good potential and was chosen for testing in 1989.

The Fitch Site Expedition—1989

In July 1989, logistics had improved and testing on the Fitch site was conducted from Florida State University's research vessel, Nectes, while smaller craft assisted with on-site operations or were utilized to conduct auxiliary tow-sled surveys in the surrounding area.

Rock outcrop areas of the Fitch site consisted of fractured rubble ranging in size from large boulders to pieces smaller than 8 cm in diameter. Two outcrops form ridges with little sediment fill. Rock size and topographic irregularity were greatest toward the center of the outcrops and diminished toward their margins. In the adjacent sand flats, rocks displaced from the outcrops seemed to decrease in size as the distance increased from their apparent source.

Considerable amounts of the lithic debris appeared to be heat-exploded spherical spalls. The circumstance that caused the rock to spall is not clear, but its terrestrial origin is certain if heat fracture was, indeed, the cause. By whatever means, the outcrops became broken into smaller components, and marine processes displaced smaller rock onto the adjacent and somewhat lower ground.

Test excavations in the surrounding sandy areas revealed a deeper sediment column than expected. Test pits placed in these areas could not be completed due to the difficult digging conditions at some levels, and, as a result, only part of the stratigraphic sequence can be discussed.

The surface level consists of transient sand containing pebbles and shells. Ten to 15 cm below the seafloor is a layer of cobble-sized rock embedded in a matrix of marine shell. Occasional chert flakes were recovered in this shell matrix. The cobble-sized rock rests on dense, clayey sediment which was difficult to excavate, but formed a convenient boundary for separating upper and lower stratigraphic sequences.

A marine shell bed averaging 60 to 75 cm thick occurs below the clay level. The shell bed is primarily made up of scallop shells, with lesser amounts of other types of marine shells, wood, and artifacts. Slender, root-like wood stretched laterally and slightly downward, branched off in places, and extended fully across the 1-m^2 test pits. The wood was poorly preserved and most of the outer bark was missing. The wood probably represents root growth because its fragile structure remained intact and spread randomly throughout the level. It also suggests that the deposit has not been disturbed since the time the wood became incorporated into the shell bed; otherwise, it would not have survived intact.

A second level of rounded, mostly tabular cobbles and small boulders (up to about 0.25 m in diameter) lay underneath the marine shell bed and formed a barrier to excavation. The rocks appeared to be mostly dolomite, but also included occasional chert nodules.

Below the lower rock level, a thin level of marine shell detrius (about 5 cm) rested above fossiliferous clay. The shell and clay levels did not contain artifacts, but fossil bone, including the remains of extinct Pleistocene fauna, was recovered. The surface of the clay had been eroded and pockets 15 to 20 cm deep also contained fossil bone. The fossiliferious clay contains embedded bone and is the likely origin for all of the vertebrate fossils collected.

The assemblage of vertebrate fossils, both from the clay layer and from the overlying shell (Table I), is generally well preserved, and most of the bones have a similar light brown color, indicating that the assemblage is homogeneous. The exceptions are the *Carcharodon, Diodon,* and *Dugongidae* bones which are glassy, black, and show considerable water wear. They are evidently reworked from one of the older marine formations in the region, possibly the Oligocene Suwannee or Miocene St. Marks Formations. The rest of the fauna are a mixture of freshwater and terrestrial species, all reasonably regarded as late Pleistocene in age. They include four extinct species, *Geochelone crassiscuttata, Eremoth-*

TABLE I. Vertebrate Fossils from Fitch Site

Class Choldrichthyes (sharks and rays)
Carcharhinus sp. (bull shark)
Carcharodon megalodon (extinct great white shark)
Class Osteichthyes (bony fishes)
Amia calva (bowfin fish), freshwater
Diodon sp. (parrot fish), salt water
Class Reptilia
Geochelone crassiscuttata (giant tortoise)
Terrapene carolina (box turtle)
Apalone ferox (softshell turtle)
Macroclemys temminki (alligator snapper)
Chrysemys scripta (red-eared turtle)
Kinosternon sp. (mud turtle)
Alligator mississipiensis (alligator)
Class Mammalia
Eremotherium mirabile (giant ground sloth)
Procyon lotor (raccoon)
Dugongidae, inc. sed. (extinct sea cow)
Equus sp. (Pleistocene horse)
Palaeolama mirifica (short-limbed llama)

erium mirabile, Equus sp., and *Palaeolama mirifica,* most of which are generally thought to have become extinct about 11,000 years ago. *Eremotherium,* however, may be an exception, as it has not been reliably dated with C-14 dates in deposits younger than 12,000 radiocarbon years ago (Grayson, 1987). On the other hand, this genus is generally rare in the late Pleistocene of eastern North America and has still not been reliably dated. A set of C-14 dates for this site would provide not only further information on the Fitch Site, but also on the time range of *Eremotherium* in eastern North America.

The calcitic clay level of the site is clearly nonmarine. The two principal fauna components are freshwater and terrestrial species, and of these, the freshwater elements are more diverse and more abundant. Some species, notable *Carcharhinus,* are probably estuarine, but could range far upstream in a high-ionic stream such as the ancestral Aucilla, which drains predominantly limestone terrain. The aquatic species include a relatively large variety of freshwater inhabitants, and include about an equal balance between lentic (moving water) and lotic (quiet water) species. The fine-grained sediments in which the vertebrates accumulated suggest a possible backwater or an intermittently flowing channel. Evidently, the large terrestrial faunal elements were transported from adjacent savanna or scrub which supported both grazing and browsing species of large herbivores. The most important environmental conclusion regarding the vertebrate sample from the Fitch site is that no elements from the in-place fauna represent exclusively marine environments, and only one (a small, euryhaline bull shark) hints at an estuarine condition.

The Fitch site strata appear to have been deposited over one or two glacial/interglacial episodes. Two levels of clay capped by corresponding levels of rock are similar and are believed to represent cyclical events. These sequences are separated by a level of marine shells with inclusions of terrestrial elements. The lowest clay level represents either a pre-Wisconsin, mid-Wisconsin, or post-Wisconsin event during a time when the sea level was only slightly lower and Pleistocene fauna roamed the area. Although the pre-Wisconsin reconstruction seems most likely, two possibilities are outlined below.

Pre-Wisconsin Reconstruction

If the giant sloth *Eremotherium* was a coastal dweller, as fossil evidence suggests (Webb, 1974), and became extinct prior to 12,000 years ago, a pre-Wisconsin or mid-Wisconsin age (probably Sangamon) is indicated for the fossiliferous, freshwater clay. This low-energy freshwater or wetland environment was replaced by a rising sea that overtook the area. The mid-Wisconsin possibility for this occurrence is eliminated if, as some have suggested, the Wisconsin sea remained below its near-present position during the period of the glacial decline between the Wisconsin I and II advances (Barnola et al., 1987; Genthan et al., 1987). The rock level above this fossil-bearing stratum suggests it was formed in a high-energy environment, possibly when the surrounding rock outcrops were situated along the coastline and subjected to surf action. The subsequent accumulation of seashell above the rock took place during a period of full marine conditions similar to today. Confounding matters, the scallop level also possessed evidence of terrestrial exposure, suggesting another interval of lower sea level. Artifacts which occur in the topmost portions suggest this shell bed was terrestrial during the late-Pleistocene/early-Holocene time frame. The upper clayey silt was deposited during a low-energy interval and, in that respect, resembles other quiet-water deposits. The rock deposit above the clay marks another episode of high energy and erosion that may similarly have resulted from surf action. Seashells are again actively, although slowly, accumulating, as the shell sifts downward through transient sediments during periods of surface-sand mobility.

Post-Wisconsin Reconstruction

The lower clay level of the Fitch site is the only *in situ* late Pleistocene sediment containing evidence of wetland and terrestrial origin. The bones of land and freshwater animals accumulated around this formerly quiet-water location and became buried in erosion-resistant clay. Holocene sea level rise inundated the site and introduced a more dynamic marine environment. The two rock levels were deposited during high-energy events, perhaps hurricanes. Between these erosive events, marine conditions similar to the present existed and the shell-bed accumulation was deposited. Artifacts and the root-like wood were introduced in the shell bed as a result of secondary deposition in a manner not yet understood.

Survey Around the Relict Aucilla

While one crew studied the Fitch site, another formed a survey party to search for new sites. Five additional sites were discovered (Figure 1), and a relict segment of the Aucilla River channel was identified. The most promising site (8Ta740) was located on ''high ground'' adjacent to the relict channel segment. Surface collection of the area produced a variety of lithic tools and bone associated with a possible oyster midden (Figure 6).

Our survey work has helped to demonstrate that within the first few dozen kilometers of land, there seem to be many sites from which much remains to be learned.

Modeling the Clovis Shoreline

Other than being located in the Gulf of Mexico, the sites which have been investigated fit the land-based distribution model. Offshore sites were encountered near the predicted topographical targets, even though the marine environment made things look unfamiliar. Offshore sites were also located at depths shallower than, or equal to, some of the underwater components of the inland Page/Ladson site. The predictability of sites further offshore, in deeper water, is not expected to conform to land-based modeling because geohydraulic conditions and resource potentials differed closer to paleoshorelines.

During the late Pleistocene, water tables in the interior of Florida were lower and sites became focused around deep karst features. Sites like Page/Ladson were then hundreds of kilometers inland and many meters above sea level. Offshore locations like Ray Hole Spring and Port Paradise Springs also fell within the area of inland depressed water tables during the time of the Clovis sea. Evidence of a water table 6 m below the rim of Port Paradise Springs supports this assumption. However, closer to the Clovis shoreline, surface water would have been more abundant because the aquifer would have been near the surface and site distributions would probably have differed.

The top of the Floridian Aquifer would have come to the surface as it met its base level—the sea. Assuming the outer part of the shelf was habitable, coastal areas during the Paleoindian period promise to differ from the interior model in two respects. First, the potential for exploiting different resources from more varied coastal habitats could possibly result in different tool kits. Second, with the aquifer near the surface, potable water may have been widespread and settlement opportunities more dispersed. With dramatically different geohydraulic conditions and little comparative site information to go by, predictive modeling of these sites becomes more difficult. However, certain factors relevant to a modeling attempt are worth consideration.

First, and probably most important, what type of water would the coastal Floridian Aquifer have held near a paleoshoreline? Was the aquifer contaminated by saltwater intrusion into its exposed coastal margins, or like today, did it discharge fresh water along the coast? A contaminated aquifer suggests scarce drinking water, saltwater coastal habitats, and conditions not likely to promote major coastal occupations. Conversely, an aquifer discharging fresh water suggests plentiful drinking water, rich, brackish-water coastal habitats, and widespread settlement options.

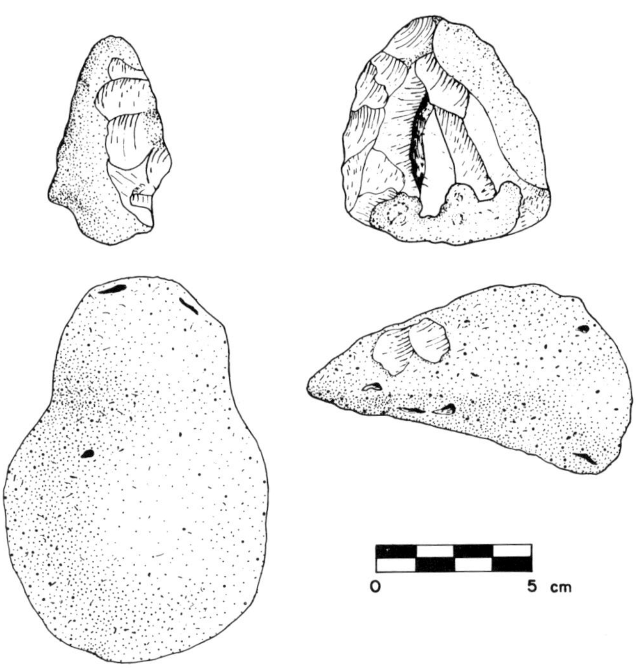

Figure 6. Lithic artifacts from 8Ta740. From left to right, top row: water-abraded early-Archaic kirk serrated projectile point, water-abraded conical chert core. Bottom row: water-abraded chert object, water-abraded early-Archaic Hendrix scraper.

Its seems most likely that the Floridian Aquifer discharged fresh water along the coastal strand, particularly after about 15,000 years ago, when the aridity of the glacial maximum gave way to climatic moderation in north Florida (Watts, 1983). Sinkholes on the continental shelf promise to provide much of the missing paleoenvironmental information needed to develop a better understanding of site potentials on the outer shelf.

Second, in Florida as elsewhere in the Southeast, chert-bearing regions have a disproportionately high number of Paleoindian sites. For example, the Aucilla River is located in an area of chert-bearing limestone and numerous Paleoindian sites are known to exist. Conversely, Crystal River is located in an area where the limestone contains little chert and only a few, isolated Paleoindian artifacts have been recovered.

Assuming sites near the paleoshoreline will be difficult to locate, chert quarry locations, should they exist, might be the most obvious, and therefore the most logical, to target for an initial search. Chert outcrops have proved to be good visual targets in shallow water, but promise to provide even better targets for sidescan sonar in water deeper than 10 m (Garrison, personal communication, 1989). If chert-bearing limestones are not buried and extend to the Clovis (i.e., 12,000 B.P.) shoreline, the chances of finding a site so deep and so far offshore will be greatly enhanced. Assuming a chert quarry can be located, then the search for other types of nearby activity sites should be initiated.

Several Tertiary limestone formations on land contain chert nodules, but only two contain abundant quantities throughout their outcrop areas. The Suwannee Limestone of Oligocene age and St. Marks Formation of early-Miocene age contain more silica impurities than other exposed Tertiary limestones (Bishop and Dee, 1961; Cooke, 1945; Carr and Alverson, 1959). Eocene limestones are known for their almost pure calcium carbonate content and are believed to have been locally silicified in areas where the mid-Miocene-age Hawthorne Formation supplied the dissolved silica gel necessary for silification to take place (Willams et al., 1977). Eocene limestones along the present coastline where the Hawthorne Formation is absent contain few chert nodules. Conversely, the St. Marks and Suwannee limestones contain abundant chert near the coast as well as offshore. The probability of locating chert outcrops seems most likely where Suwannee and St. Marks limestones are exposed on the seafloor.

The first step toward conducting a site survey on the continental shelf would be to determine how far offshore chert sources extend. As both the Suwannee and St. Marks limestones extend into Apalachee Bay, it would make sense to track them farther offshore.

Conclusions

The sediment-starved karst shelf of Apalachee Bay has given researchers a rare opportunity to easily access potential archaeological site locations, visually identify site remains, and test them with a minimal amount of effort. Continued research on the karst shelf promises to yield a model for predicting archaeological site locations in the eastern Gulf. To date, the research objectives and work carried out in Apalachee Bay have been modest, but have demonstrated that archaeological sites exist, are fairly abundant, and can be located by employing land-based predictive modeling.

As the sea level migrated in and out during the Wisconsinan stage, the water level in then-terrestrial sinkholes rose and fell in response to similar adjustments in the surface of the Floridian Aquifer. Sinkholes, like Ray Hole Spring, may have been dry or held water at vertical depths well below the surrounding terrain during the low point of the last Wisconsin advance. When the sea level began to rise, the level of the aquifer rose closer to the ground's surface and brought with it an increase in soil moisture. As during the

Holocene, when vegetation patterns changed to the modern terrestrial assemblage with the establishment of near-modern aquifer levels (Watts, 1983), late-Wisconsin vegetation patterns on the continental shelf most likely changed as the aquifer and sea began rising and vegetation zones were chased to higher elevations. The continued march of transgression eventually led to the inundation of places like Ray Hole Spring. At first, Ray Hole Spring was situated near coastal habitats, but eventually came to a position far out to sea. From the perspective of prehistoric use, that time from the first opportunity for human occupation (here considered to be Clovis) until a site area became too deeply inundated to be accessed in a bipedal fashion is the period encompassing the greatest potential for archaeological site formation. From the perspective of archaeological significance, understanding the formation of a site and the processes that have affected it subsequent to inundation are of the utmost concern. In particular, reconstruction of the nature of the sedimentation and paleoenvironment are most critical if we are truly serious about trying to understand prehistoric occupation of the shelf and its significance compared to other parts of the resource base.

Sediments encountered during the offshore surveys varied in content and ranged from thin deposits covering bedrock to deep stratigraphic sequences of unknown thickness. Despite stratigraphic differences from site to site or in some cases from one test pit to another, the sediments that were encountered can be tentatively separated into five major categories.

All of the sites had several centimeters of transient sediment, which appear to be more stable in areas of marine growth and less stable in areas of barren seabed. Near sea-grass beds, this sediment appears to possess some humic content, but in barren areas it is primarily made up of sand with bioclastic content. These unconsolidated surface deposits are believed to shift during periods of storm-surge activity.

A second, frequently encountered deposit consists of a consolidated layer of compact marine sediments resting below these loose surficial deposits. Although the particle size and makeup varies from location to location, it contains the bioclastic detrius of dead shellfish as a characteristic marker. Scallop shell-dominated levels are common closer to shore near the sea-grass beds, while varied and decreased shell content seems common farther offshore at locations like Ray Hole Spring. This type of deposit is absent in sediment-starved areas where only transient sediment covers the bedrock.

Marine shell beds containing root-like wood and other evidence of terrestrial exposure are a third class of deposit. Strata of marine origin may contain the subsequent intrusive material of prehistoric occupation because lower sea levels exposed former marinescapes as usable land. The presence of terrestrial material in an otherwise marine sediment promises to be significant, particularly on sediment-starved portions of the shelf where archaeological sites are more likely to have become incorporated into, or rest on, marine deposits.

Brackish-water sediment, primarily consisting of oyster shell, is a fourth sediment classification. Oyster bars and disarticulated brackish-water shell concentrations were discovered with surprising frequency. Most oyster deposits consisted of disarticulated jumbles of whole and broken shells, but the Econfina Channel site contained articulated shell indicative of *in situ* growth and deposition. Oyster and other brackish-water shells are characteristic markers of the deposits, but they have also produced wood, including in-place tree roots and secondarily deposited driftwood.

A fifth sediment class includes a range of terrestrial and freshwater deposits such as the fine-grained fossiliferous clay at the Fitch site. Deposits fitting into this category contain evidence of terrestrial origins, but do not have brackish-water or marine components.

Investigation of sites further offshore should include the establishment of a local sea level curve as a necessary step toward predicting site locations. From Apalachee Bay extending south onto the outer shelf, offshore sinkholes may contain *in situ* datable material, protected from transgression, and capable of providing absolute sea level data points for the eastern Gulf of Mexico.

Assuming that the environmental data recovered from sinkholes indicates coastal conditions that would have favored human occupation, then reliable predictive methods need to be developed. At present, the only clear modeling parameter that appears as if it will apply is the occurrence of chert. Site location and predictability on the outer continental shelf promise to be most difficult but not impossible, particularly if chert outcrops can be located. But until an attempt to find sites is made, we will be left to speculate about the Clovis shoreline and Paleoindian coastal adaptation.

REFERENCES

Alexon, R.C. (1988) Gingery cache site (8Ta99): a cache of possible Paleolithic tools found with mastodon bones. *Fla. Anthropol.* 41: 483.

Anderson, D.G. and Hanson, G.T. (1988) Early Archaic settlement in the Southeastern United States: a case study from the Savannah River Valley. *Am. Antiq.* 53: 262.

Anderson, D.G. (1990) The Paleoindian colonization of eastern North America: a view from the Southeastern United States. In: *Early PaleoIndian Economies of Eastern North America.* Tankersley, K. and Isaac, B., Eds. Res. Econ. Anthropol. Suppl. 5, JAI Press, Greenwich.

Anuskiewicz, R.J. (1988) Preliminary archaeological investigation at the Ray Hole Spring in the eastern Gulf of Mexico. *Fla. Anthropol.* 41: 181.

AENA. (1982) A compilation of fluted points of eastern North America by count and distribution: An AENA project. *Archaeol. East. North Am.* 10: 27.

Ballard, R. and Uchupi, (1970) Morphology and Quaternary history of the Continental Shelf of the Gulf Coast of the United States. *Bull. Mar. Sci.* 20: 547.

Barnola, J.M., Raynaud, D., Korotkevich, Y.S., and Lorius, C. (1987) Vostok ice core produces 160,000 year old record. *Nature.* 329: 408.

Bishop, E.W. and Dee, L.L. (1961) Rocks and Minerals of Florida — A Guide to Identification, Occurrences, Production and Use. Spec. Publ. 8, Florida Division of Geology, Tallahassee.

Blackwelder, B.W., Pilkey, O.H., and Howard, J.D. (1979) Late Wisconsinan sea levels on the Southeast U.S. Atlantic Shelf based on in-place shoreline indicators. *Science.* 204: 618.

Bloom, A.L. (1983) Sea level and coastal morphology of the United States through the late Wisconsin glacial maximum. In: *Late Quaternary Environments of the United States,* pp. 215–229. Wright, H.E., Jr., Ed. University of Minnesota Press, Minneapolis.

Brooks, H.K. (1973) Geological oceanography. In: *A Summary of Knowledge of the Eastern Gulf of Mexico,* pp. IIE 1–50. Jones, J.I., Ed. State University System, Florida Institute of Oceanography, St. Petersburg.

Brooks, H.K. (1974) Lake Okeechobee. In: *Environments of South Florida: Present and Past,* pp. 256–286. Gleason, P.J., Ed. Mem. 2, Miami Geological Society, Miami.

Bullen, R. (1975) *A Guide to the Identification of Florida Projectile Points.* Kendall Books, Gainesville.

Bullen, R.P., Webb, S.D., and Waller, B.I. (1970) A worked mammoth bone from Florida. *Am. Antiq.* 35: 203.

Carr, W.J. and Alverson, D.C. (1959) Stratigraphy of Middle Tertiary Rocks in Part of West-central Florida, Bull. 1092, U.S. Geological Survey, Washington, D.C.

Coastal Environments. (1977) Cultural Resources Evaluation of the Northern Gulf of Mexico Continental Shelf. Prepared for U.S. Department of the Interior, Office of Archaeology and Historic Preservation, Interagency Archaeological Services, NPS, Washington, D.C.

Cooper, H.H., Jr., Kenner, W.E., and Brown, E. (1953) Ground Water in Central and Northern Florida. Rep. Invest. 10, Florida Geological Survey, Tallahassee.

Clark, D.E. and Purdy, B.A. (1979) Electron microprobe analysis of weathered Florida chert. *Am. Antiq.* 44: 3.

Clausen, C.J., Brooks, H.K., and Wesolowsky, A.B. (1975) The Early Man site at Warm Mineral Springs, Florida. *J. Field Archaeol.* 2: 191.

Clausen, C.J., Cohen, A.D., Emiliani, C., Holman, J.A., and Stipp, J.J. (1979) Little Salt Spring, Florida: a unique underwater site. *Science.* 203: 609.

Clayton, D.H. (1981) Part III: faunal report for 8Hi393c/uw. In: Report on Phase II Underwater Archaeological Testing at the Fowler Bridge Mastodon Site (8Hi393c/uw) in Hillsborough County, Florida, pp. 107–135. Florida Division of Archives, History and Records Management, Tallahassee.

Clayton, D.H. (1983) Unusual marks found on giant land tortoise remains in Hillsborough River. *Fla. Anthropol.* 36: 101.

Cockrell, W.A. (1980) Drowned sites in North America. In: *Archaeology Under Water: An Atlas of the World's Submerged Sites,* pp. 138–145. Muckelroy, K., Ed. McGraw-Hill, New York.

Conover, C.S., Geraghty, J.J., and Parker, G.G. (1984) Ground water. In: *Water Resources Atlas of Florida,* pp. 36–53. Fernald, E.A. and Patton, D.J., Eds. Florida State University Press, Tallahassee.

Cooke, W.C. (1945) The Geology of Florida. Bull. 29, Florida Geological Survey, Tallahassee.

Curray, J.R. (1965) Late Quaternary history, continental shelves of the United States. In: *The Quaternary of the United States,* pp. 723–735. Wright, H.E., Jr. and Frey, D.C., Eds. Princeton University Press, Princeton, NJ.

Davies, W.E. and LeGrand, H.E. (1972) Karst of the United States. In: *Karst: Important Karst Regions of the Northern Hemisphere,* pp. 467–505. Herak, M. and Springfield, V.T., Eds. Elsevier, New York.

Donoghue, J. (In press) Lake Wisconsin and Holocene depositional history of the northeast Gulf of Mexico. *Mar. Geol.*

Donoghue, J. (1990) Personal communication. Department of Geology, 108 Caraway Blvd., Tallahassee, (904) 644-2703.

Doran, G.H. and Dickel, G.H. (1988) Multidisciplinary investigations at the Windover site. In: *Wet Site Archaeology,* pp. 263–290. Purdy, B.A., Ed. Telford Press, Caldwell, NJ.

Doyle, L.J. and Sparks, T.N. (1980) Sediments of the Mississippi, Alabama, and Florida (MAFLA) Continental Shelf. *J. Sediment. Petrol.* 50: 3.

Dunbar, J.S. (1981) The effect of geohydrology and natural resource availability on site utilization at the Fowler Bridge mastodon site (8Hi393c/uw) in Hillsborough County, Florida. In: Report on Phase II Underwater Archaeological Testing at the Fowler Bridge Mastodon Site (8Hi393c/uw) Hillsborough County, Florida, pp. 63–106. Florida Division of Archives, History and Records Management, Tallahassee.

Dunbar, J.S. (1991) Resource orientation of Clovis and Suwannee age Paleoindian sites in Florida. In: *Clovis Origins.* Bonnichsen, R. and Turnmire, K., Eds. Center for the First Americans, Oregon State University, Corvallis, Oregon.

Dunbar, J.S., Webb, S.D., and Cring, D. (1990) Culturally and naturally modified bones from a Paleo-Indian site in the Aucilla River, north Florida. In: *Bone Modification,* pp. 473–497. Bonnichsen, R. and Sorge, M., Eds. Center for the First Americans, University of Maine, Orono.

Dunbar, J.S., Webb, S.D., and Faught, M.K. (1988) Page/Ladson (8Je591): an underwater Paleoindian site in northeastern Florida. *Fla. Anthropol.* 41: 442.

Dunbar, J.S. and Waller, B.I. (1983) A distribution analysis of the Clovis/Suwannee Paleoindian sites of Florida—a geographic approach. *Fla. Anthropol.* 41: 18.

Edwards, R.A. (1948) An abandoned valley near High Springs, Florida. *Q. J. Fl. Acad. Sci.* 2: 125.

Faught, M.K. (1988) Inundated sites in the Apalachee Bay area of the eastern Gulf of Mexico. *Fla. Anthropol.* 41: 185.

Faught, M.K. (1988) Preliminary Report of Excavations at the Econfina Channel Site (8Ta139). Unpublished file report, Florida Bureau of Archaeological Research, Tallahassee.

Flemming, N. (1983) Survival of submerged Lithic and Bronze Age artifact sites: a review of case histories. In: *Quaternary Coastlines and Marine Archaeology,* pp. 135–173. Masters, P.M. and Flemming, N.C. Eds. Academic Press, London.

Garrison, E. (1989) Personal communication. NOAA Marine and Estuary Division, Office of Ocean and Coastal Resource Management, 1825 Connecticut Ave., Washington, D.C., (202) 673-5126.

Garrett, S.E. (1983) Coastal Erosion and Archaeological Resources on National Wildlife Refuges in the Southeast. Archaeological Services Branch, NPS Southeastern Region, Atlanta.

Genthan, C., Barnala, J.M., Raynaud, D., Lorious, C., Jouzel, J., Barkov, N.I., Korotkevich, Y.S., and Kotlyakov, V.M. (1987) Vostok Ice Core: Climatic response to CO_2 and orbital forcing changes over the last climatic cycle. *Nature*. 329: 414.

Gerrell, P.R., Scarry, J.F., and Dunbar, J.S. (In press) Analyses of early Archaic unifacial adzes from north Florida. *Fla. Anthropol.*

Gilliland, M.S. (1988) Marco's buried treasure: wetlands archaeology and adventure in nineteenth century Florida. In: *Wet Site Archaeology*, pp. 255–261. Purdy, B.A., Ed. Telford Press, Caldwell, NJ.

Gluckman, S.J. (1967) Underwater Archaeology: Theory and Method. M.S. thesis, Department of Anthropology, University of Florida, Gainesville.

Goggin, J.M. (1962) Recent developments in underwater archaeology. In: Proc. 16th Southeastern Archaeological Conf. Newslett. Vol. 8.

Goodyear, A.C. (1988) A hypothesis for the use of cryptocrystalline raw materials among Paleoindian groups of North America. In: *Eastern Paleoindian Lithic Resource Use*, pp. 1–9. Ellis, C.J. and Lothrop, J.C., Eds. Westview Press, Boulder.

Goodyear, A.C. and Warren, L.O. (1972) Further observations on the submarine oyster shell deposits of Tampa Bay. *Fla. Anthropol.* 25: 52.

Grayson, D.K. (1987) An analysis of the chronology of late Pleistocene mammalian extinctions in North America. *Quat. Res.* 28: 281.

Haynes, V.C. (1968) Geochronology of late-Quaternary alluvium. In: *Means of Correlation of Quaternary Successions*. Vol. 8. *Proceedings of the VII Congress of the International Association for Quaternary Research*. University of Utah Press, Salt Lake City.

Haynes, V.C. (1980) The Clovis culture. *Can. J. Anthropol.* 1: 115.

Haynes, V.C. (1984) Stratigraphy and late Pleistocene extinction in the United States. In: *Quaternary Extinctions*, pp. 345–353. Martin, P.S. and Klein, R.G., Eds.

Hutton, J.G., Hine, A.C., Evans, M.W., Osking, E.B., and Belknap, D.F. (1984) Influence of a karstified limestone surface on an open-marine, marsh dominated coastline: west central Florida. In: *Sinkholes: Their Geology, Engineering and Environmental Impact*, pp. 35–42. Beck, B.F., Ed. Florida Research Institute, University of Central Florida, Orlando.

Jordan, C.L. (1984) Florida's weather and climate: implications for water. In: *Water Resources Atlas of Florida*, pp. 18–35. Fernald, A.A. and Patton, D.J., Eds. Institute of Science and Public Affairs, Florida State University, Tallahassee.

Jenks, A.E. and Simpson, H.H. (1941) Beveled artifacts in Florida of the same type as artifacts found near Clovis, New Mexico. *Am. Antiq.* 6: 314.

Jennings, J.N. (1985) *Karst Geomorphology*. Blackwell, New York.

Kraft, J.C. (1978) Coastal stratigraphic sequences. In: *Coastal Sedimentary Environments*, pp. 361–383. Davis, R.A., Jr., Ed. Springer-Verlag, New York.

Kraft, J.C., Belknap, D.F., and Kayan, I. (1983) Potentials of discovery of human occupation sites on the continental shelves and nearshore coastal zones. In: *Quaternary Coastlines and Marine Archaeology*, pp. 87–120. Masters, P.M. and Fleming, N.C., Eds. Academic Press, London.

Lazarus, W.C. (1965) Effects of land subsidence and sea level changes on elevation of archaeological sites on the Florida gulf coast. *Fla. Anthropol.* 13: 49.

LeGrand, H.E. and Stringfield, V.T. (1973) Karst hydrology—a review. *J. Hydrol.* 20: 97.

Mayer-Oaks, W.J. (1984) Fluted projectile points: a North American shibboleth viewed in South American perspective. *Archaeol. East. North Am.* 12: 231.

Milanich, J.T. and Fairbanks, C.H. (1980) *Florida Archaeology*. Academic Press, New York.

Milliman, J.D. and Emery, K.O. (1968) Sea levels during the past 35,000 years. *Science*. 162: 1121.

Morner, N.A. (1984) Eustasy, geoid changes, and multiple geophysical interaction. In: *Castastrophes and Earth History*, pp. 395–415. Berggren, W.A. and Van Couvering, J.A., Eds. Princeton University Press, Princeton, NJ.

McNutt, R.S. (1978) Comparison of Patination of Underwater Lithics vs. Terrestrial Lithics: An Archaeological Approach. Unpublished report, Florida State University, Tallahassee.

Neill, W.T. (1964) Points and extinct animals in Florida. *Fla. Anthropol.* 17: 17.

Opdyke, N.D., Spangler, D.P., Smith, D.L., Jones, D.S., and Lindquist, R.C. (1984) Origin of the epeirogenic uplift of Pliocene-Pleistocene beach ridges in Florida and development of the Florida karst. *Geology.* 12: 226.

Palmer, A.N., Palmer, M.V., and Queen, J.M. (1977) Geology and origin of caves of Bermuda. In: Proc. 7th Int. Congr. Speleology, pp. 366–369.

Perlman, S.M. (1980) An optimum diet model, coastal variability, and hunter-gatherer behavior. In: *Advances in Archaeological Method and Theory,* pp. 257–310. Vol. 3 Schiffer, M., Ed. Academic Press, New York.

Pirazzolli, P.A., Grant, D.R., and Woodworth, P. (1989) Trends of relative sea level change: past, present and future. *Quat. Int.* 2: 63.

Purdy, B.A. (1988) Archaeological wet sites: untapped archives of prehistoric documents. In: *Wet Site Archaeology,* pp. 325–336. Purdy, B.A., Ed. Telford Press, Caldwell, NJ.

Richardson, S. (1988) Survey of the Aucilla River from Ward Island on the Jefferson-Taylor county line, Florida. *Fla. Anthropol.* 41: 471.

Rosenau, J.C., Faulkner, G.L., Hendry, C.W., and Hull, R.W. (1977) Springs of Florida. Bull. 31 (Revised), Florida Department of Natural Resources, Bureau of Geology, and Florida Department of Environmental Regulation, Bureau of Water Resources Management, Tallahassee.

Ruddiman, W.F. and Duplessy, J.C. (1985) Conference on the last deglaciation: timing and mechanism. *Quat. Res.* 23: 1.

Schmidt, W. and Scott, T.M. (1984) Florida karst—its relationship to geologic structure and stratigraphy. In: *Sinkholes: Their Geology, Engineering and Environmental Impact,* pp. 11–16. Beck, B.F., Ed. Florida Sinkhole Research Institute, University of Central Florida, Orlando.

Scholl, D.W., Craighead, F.C., Sr., and Stuiver, M. (1969) Florida submergence curve revised: its relation to coastal sedimentation rates. *Science.* 163: 562.

Science Applications, Inc. (1979) Cultural Resource Survey of the Continental Shelf from Cape Hatteras to Key West. In: Prehistoric Archaeology. Vol. 2. Bureau of Land Management, New Orleans, La.

Serbousek, D. (1983) Exploration of a Paleoindian site on the Aucilla River. *Fla. Anthropol.* 36: 89.

Serbousek, D. (1988) An example of an offshore sinkhole in the Gulf of Mexico with good archaeological potential. *Fla. Anthropol.* 41: 190.

Simpson, C.J. (1948) Folsom-Like points from Florida. *Fla. Anthropol.* 1: 11.

Stapor, F.W., Mathews, T.D., and Lindfors-Kearns, F.E. (1988) Episodic barrier island growth in southwest Florida: a response to fluctuating Holocene sea level. *Miami Geol. Soc. Mem.* 3: 149.

Stright, M.J. (1986) Human occupation of the Continental Shelf during the late Pleistocene/early Holocene: methods for site location. *Geoarchaeology.* 1: 347.

Stubbs, S.A. (1940) Solution A dominant factor in the geomorphology of peninsular Florida. *Proc. Fla. Acad. Sci.* 5: 149.

Sweeting, M.M. (1981) Karst geomorphology. In: *Benchmark Papers in Geology 59.* Hutchinson Ross, Stroudsburg, IL.

Vernon, R.O. (1951) Geology of Citrus and Levy Counties, Florida. Bull. 33, Florida Geological Survey, Tallahassee.

Waller, B.I. (1969) Paleo-Indian and other artifacts from a Florida stream bed. *Fla. Anthropol.* 22: 37.

Waller, B.I. (1970) Some occurrences of Paleoindian projectiles in Florida. *Fla. Anthropol.* 30: 79.

Warren, L.O. (1970) The Kellogg fill from Boca Ciega Bay, Pinellas County, Florida. *Fla. Anthropol.* 23: 163.

Watts, W.A. (1983) Vegetational history of the Eastern United States 25,000 to 10,000 years ago. In: *Late Quaternary Environments of the United States.* Vol. 1. *The Late Pleistocene.* Porter, S.C., Ed. University of Minnesota Press, Minneapolis.

Watts, W.A. and Hansen, B.C.S. (1988) *Environments of Florida in the Late Wisconsin and Holocene.* In: *Wet Site Archaeology,* pp. 307–323. Purdy, B., Ed. Telford Press, Caldwell, New Jersey.

Webb, S.D. (1974) *Pleistocene Mammals of Florida*. University Presses of Florida, Gainesville.

Webb, S.D. (1981) A Cultural Resource Survey of the Continental Shelf from Cape Hatteras to Key West. Vol. 1. Introduction and Physical Environment, pp. 73–112. File Rep., U.S. Bureau of Land Management, McLean, VA.

Webb, S.D., Milanich, J.T., Alexon, R., and Dunbar, J.S. (1984) A Bison antiquus kill site, Wacissa River, Jefferson County, Florida. *Am. Antiq.* 49: 384.

Williams, K.E., Nicol, D., and Randazzo, A.F. (1977) The Geology of the Western Part of Alachua County, Florida. Rep. Invest. 85. Florida Geological Survey, Tallahassee.

Willis, C. (1988) Controlled surface collection of the Little River Rapids site (8Je603): a stratigraphically deflated site in the Aucilla River, north Florida. *Fla. Anthropol.* 41: 453.

Yon, J.W. (1966) Geology of Jefferson County, Florida. Bull. 48, Florida Geological Survey, Tallahassee.

Part II

Unstable Land

The chapters in this section describe areas where coastal change is complex, defying simple modeling. As in the previous section, there are contrasts between islands and the mainland, coastal peoples, inland peoples, and those who seasonally move from one focus to another. These different settlement patterns led to different prehistoric responses to coastal change.

Islands

Lucille Lewis Johnson
Department of Anthropology
Vassar College
Poughkeepsie, New York

INTRODUCTION

Winslow and Johnson are preoccupied with establishing paleoshorelines as they are studying coastal cultures in a seismically active area. Winslow's model of coastal change takes into account eustatic, isostatic, and tectonic effects. She shows that there are radical differences in contours when tectonism is not considered. These are particularly acute in the inner Shumagin Islands where uplift has been very rapid, but Johnson and Winslow show that even in the less active outer Shumagins, prehistoric settlement patterns cannot be understood if only eustatic and isostatic changes are studied.

Chapter

7

Modeling Paleoshorelines in Geologically Active Regions: Applications to the Shumagin Islands, Southwest Alaska

Margaret Anne Winslow
Department of Earth and Planetary Sciences
City College of New York
New York, New York

INTRODUCTION

Interpretation of prehistoric coastal settlement patterns in formerly glaciated, geologically active regions requires reconstruction of the ancient shorelines. The geographic and vertical positions of late Quaternary coastlines depend on sea level changes, postglacial isostatic response, and tectonic uplift or subsidence. These factors interact to create a complicated history of local coastal emergence or submergence, resulting in the creation or destruction of potential harbors. An empirical method is developed in this chapter for reconstructing former sea levels and for creating paleoshoreline maps for an archaeologically important sector of southwest Alaska.

Recognition of a major rise in sea level during the late Pleistocene-early Holocene has led to the belief that early Holocene coastal archaeological sites must lie offshore in 20 to 100 m of water (or more), thus placing their study in the special field of marine archaeology. Postulated encroachment of the surf zone over these prehistoric sites during the Holocene transgression leads to some pessimism about the completeness and even the existence of older sites, especially along the highly erosive Alaska coast (Figures 1 and 2; see also Hopkins, 1967; Black, 1977).

Paleoshorelines along the Alaska coast were modeled by Black (1977) and more generally as part of a global survey by Bloom (1983). Their reconstructions for the Alaska margin suggest that early Holocene sites will lie far offshore of present coastlines, and thus might be difficult to study or document. The second implication of their reconstructions is that the present coastlines should not have coastal sites that are more than about 4000 to 5000 years old, although a few coastal sites in southwest Alaska date to 7000 to 8900 B.P. (e.g., Laughlin, 1963; Laughlin and Laughlin, 1981; Dumond, 1987).

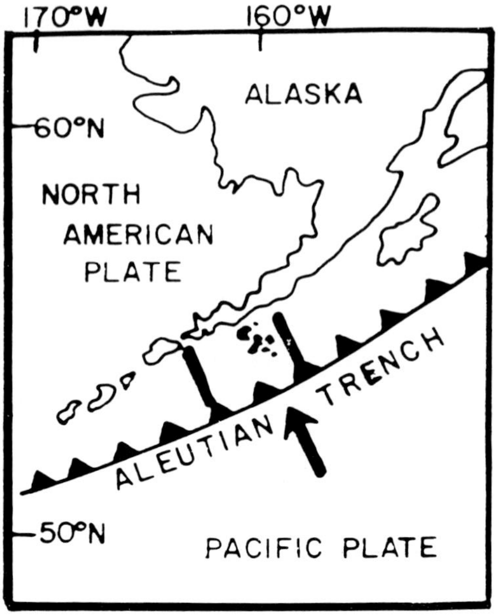

Figure 1. Geological setting of Alaska. Heavy lines mark areas of study. Black islands are the Shumagin Islands. Arrow is direction of convergence of the Pacific plate.

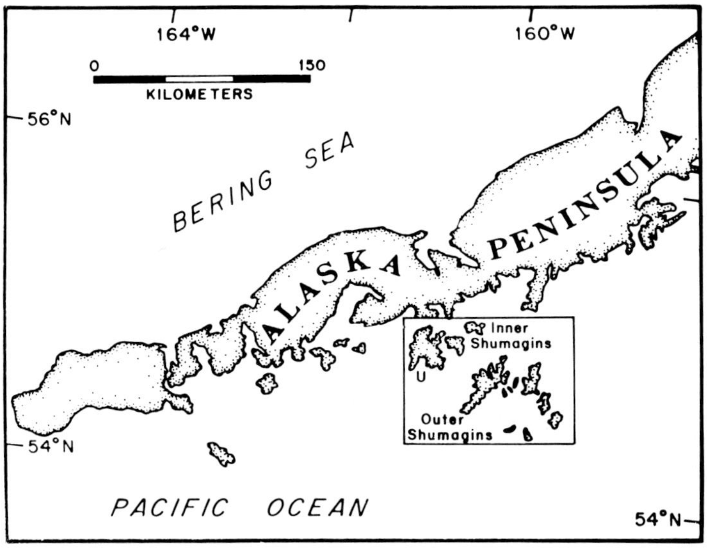

Figure 2. Study area: Alaska Peninsula and (inset) inner Shumagin and outer Shumagin Islands.

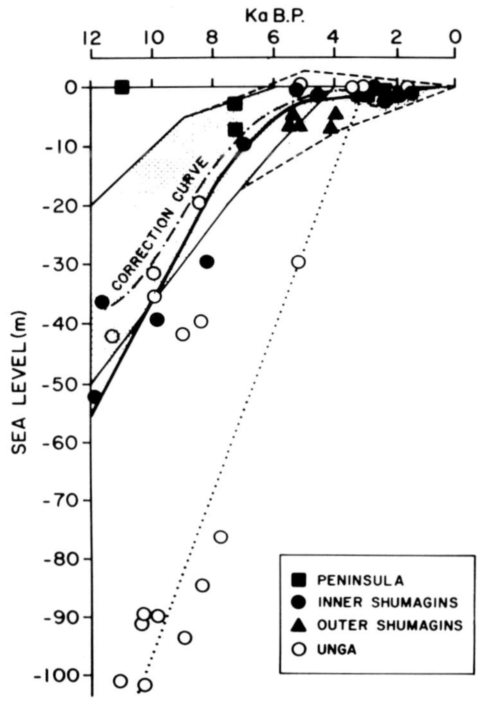

Figure 3. Empirical eustatic-isostatic correction curve derived from field data (heaviest line) compared to the range of published eustatic curves (shaded). Correction curve is calculated eustatic-isostatic curve, as developed from theoretical estimates (Winslow, in revision). Data points are vertical corrections required to reach lines of constant uplift rate as demonstrated in text. Empirical curve (heavy line) is least-squares fit to the correction data points.

The main weakness in earlier paleoshoreline reconstructions for the Alaska coastline was that the sea level curves were borrowed from other regions, especially those developed for the tropics (Figure 3; collections in Scholl et al., 1970; Bloom, 1983; Lajoie, 1986). In addition, instead of establishing a "global" sea level curve during the last decade, a new eustatic curve is published every few months. The collection of these curves creates a wide envelope of possible values, especially for the early Holocene (Figure 3). Moreover, geoidal warping of the lithosphere during the glacial and postglacial epochs is now recognized to be a global-scale phenomenon (Peltier, 1980) and thus is not measurable from local glacial sedimentology and geomorphology alone. This means that one must either calculate eustatic and isostatic correction curves from theoretical estimates or develop one's own empirical curve from the local data, as was done for the original tropical curves (method in Clark et al., 1978; Lajoie, 1986; Scholl, 1970; Winslow, 1982, 1988, 1989, in revision).

In addition to the eustatic and isostatic unknowns for a region, many areas of the world are undergoing either tectonic uplift or subsidence due to the collision or separation of lithospheric plates. Even in regions that have undergone mainly tectonic uplift, such as the Pacific margin of Alaska, local areas of downwarping also occur due to the accordian-like folding of the crust over time. For the Holocene alone, the vertical component of crustal warping may be as much as 80 m, either up on anticlinal hinges or down in synclinal troughs! The tectonic factor must be recognized, therefore, in paleoshoreline reconstructions. In the author's opinion, many of the differences among sea level curves around the world derive from an unrecognized tectonic and isostatic component (Bloom, 1983; Peltier, 1980). Thus, data from early-Holocene paleoshorelines and elevations of coastal archaeological sites require eustatic, isostatic, and tectonic corrections in order to identify patterns of uplift, tilting, or subsidence and, thus, to reconstruct the site's original position with respect to an ancient shoreline.

Derivation of the three components that comprise the eustatic-isostatic-tectonic correction curve for the Alaska margin (Figures 1 and 2) using geologic data has been one of the author's objectives for field research during the last 12 years (Winslow and Hickman, 1978; Winslow et al., 1980; Winslow, 1982). The second goal is to relate the ancient sea levels and patterns of relative emergence/submergence of shorelines to coastal archaeological sites in the Shumagin Islands (Figures 1 and 2; Winslow, 1988, 1989; Johnson, 1989; Winslow and Johnson, 1988; Chapter 8, this volume). Recreation of ancient shorelines and coastal elevations at the time of human arrivals places tight constraints on archaeological models and helps explain the complex nature of human settlement distribution in the study area (Winslow and Johnson, 1988, 1989; Chapter 8, this volume).

In this chapter, the author presents an empirical method for reconstructing paleoshorelines in a tectonically active, formerly glaciated area: the Shumagin Islands and nearby Alaska Peninsula. Two reconstructions are compared that include (1) sea level and isostatic changes, but no tectonic component, and (2) the addition of a steady-state tectonic uplift component for individual blocks. The implications of these two models for predicting the locations and ages of coastal archaeological sites are then presented. In Chapter 8 of this volume, Johnson and Winslow apply the same correction method to detailed data from the outermost Shumagin Islands, which are of major archaeological importance (Johnson, 1988).

GEOLOGICAL SETTING OF THE SHUMAGIN ISLANDS

Earthquakes, vertical changes, and tsunami activity along the seismically active Pacific margin of Alaska have been documented during historical times (Veniaminov, 1839 [1984]; Dall and Harris, 1892; Plafker, 1965, 1967, 1969, 1972; Plafker and Rubin, 1978; Minster and Jordan, 1978; Stauder, 1968). These events relate directly to convergence between the Pacific and North American plates (Figure 1). The Pacific plate plunges beneath the North American plate in southwest Alaska at about 70 m/ka toward the northwest (Minster and Jordan, 1978; DeMets et al., 1988). Both historically observed and geologically inferred great earthquakes near the plate boundary have resulted in the uplift of hundreds of thousands of square kilometers of coast and shelf area (Plafker, 1965, 1967, 1969; Brown et al., 1977; Spence, 1977). Eyewitness observations of terrace uplift during earthquakes lead to the assumption that prehistoric Holocene terraces were raised only during episodic, co-seismic uplift (Darwin, 1845 [1962]; Inman and Nordstrom, 1971; Plafker and Rubin, 1978; Lajoie, 1986).

Early historical reports from the eastern Aleutian Islands, Shumagin Islands, and Alaska Peninsula suggested that "the shoreline has increased in the course of time" (Veniaminov, 1839 [1984]). Veniaminov catalogued many raised beaches in the region. From these he surmised that "either the sea once was higher than it is now or there was an extraordinary inundation, or both" (Veniaminov, 1839 [1984]). Interestingly, Veniaminov rejects the possibility of uplift and of a receding sea because of the similar elevation of the raised beaches.

The Shumagin Islands comprise the eastern limit of Aleut occupation (Dumond, 1987;

Johnson, 1988). These islands and the nearby Alaska Peninsula have experienced significant cumulative uplift and deformation during the last 40 million years (Grantz, 1963; Burk, 1965; Moore, 1972, 1973; Morris and Bucknam, 1972; Winslow, 1988, 1989). Given the earthquake patterns to the east along the same plate boundary, most of the uplift should occur co-seismically (Plafker and Rubin, 1978; Beavan et al., 1983; Davies et al., 1981; Boyd et al., 1988).

The most recent uplift-producing earthquake struck the Shumagin Islands in 1788 A.D. (Veniaminov, 1839 [1984]). This event raised several harbors and produced a great tsunami. Large earthquakes since then, in 1857, 1917, and 1946, have produced less than 1 m of uplift per event (Winslow, in revision). Major changes in coastal morphology and shelf depths in the study area, therefore, appear to occur only during very large earthquakes. The relationship between earthquake-produced uplift and prehistoric settlement patterns in the study area is crucial to the understanding of the region's prehistory (Winslow and Johnson, 1989). Indeed, the distribution of coastal archaeological site elevations and ages in the Shumagin Islands presents a complex puzzle if sea level rise is the only contributor to prehistoric paleoshoreline positions (Chapter 8, this volume).

DATA COLLECTION AND METHODS

In order to recreate late Quaternary paleoshorelines in the Shumagin Islands, raised strandlines and marine terraces were excavated by the author and field assistants between 1978 and 1989 (Figure 4). During the last 5 years, the survey and excavations have been done in the context of a geoarchaeological study of prehistoric Aleut settlement and migration patterns with L. L. Johnson (see Chapter 8, this volume; Winslow and Johnson, 1989; Johnson, 1988).

Over 120 localities of elevated marine strandlines and marine terraces have been identified and excavated during this study. Excavation of the marine surface reveals 40 to 470 cm of peat. Beneath the peat is the former beach or intertidal zone, which is a planar,

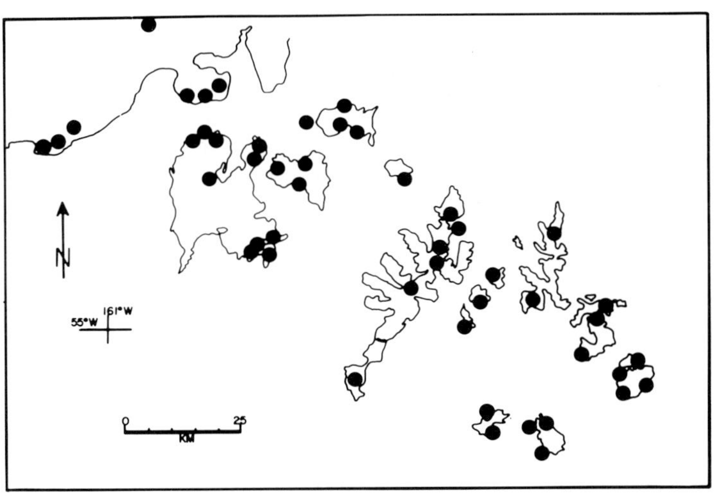

Figure 4. Localities of raised coastal features of the Shumagin Islands and nearby Alaska Peninsula shown as black dots.

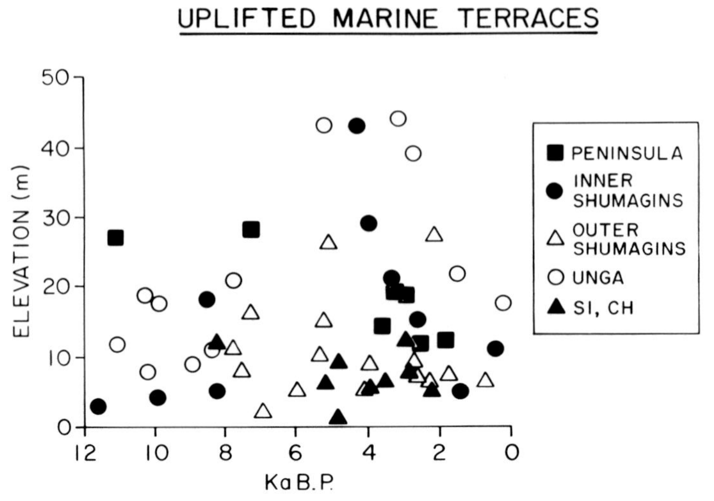

Figure 5. Uncorrected raised shoreline data for the Shumagin Islands and adjacent Alaska Peninsula. Height and age data from Table I. Present elevation corrected for mean sea level ± 1 m. Age is in radiocarbon years B.P. (before 1950 A.D.). Inner Shumagin Islands data are circles; open circle is for Unga Island. Outer Shumagin Islands data are triangles; Simeonof and Chernabura Islands are black triangles. Symbology reflects average age and height error for data.

subhorizontal bedrock surface or 1 to 2 m of beach sand, gravel and driftwood on abraded bedrock. Field tasks included identification of possible raised marine surfaces and verification of their postglacial development and marine position (beach, intertidal, etc.). Direct measurements in the field included present height above sea level (usually measured as height above highest high water and corrected to a calculated mean sea level from tide-gauge and tide-table data; Winslow, in revision) and local tidal range. The basal peat on the marine surface was dated to obtain its radiocarbon age.

Fifty-four samples of peat obtained at the basal nonmarine-marine interface were dated by radiocarbon methods (Figure 5). Raised strandlines and terraces from the study area vary in height from 1 to 44 m above mean sea level, and range in age from approximately 150 ± 50 to 12,000 ± 400 radiocarbon years B.P. (before 1950 A.D.). Numbers of terraces at each site vary from one to eight; typically, one to two levels are recognized in the outer Shumagin Islands, one to eight in the inner Shumagins, and up to seven levels on the south side of the Alaska Peninsula (Winslow, in revision). Each terrace surface represents a minimum of 1 to 3 m of vertical displacement, probably associated with a single earthquake. Because of a nearly 2-m mean tidal range, recognition of smaller vertical displacements is extremely difficult.

MODELING RAISED TERRACE DATA TO RECREATE PALEOSHORELINES

The uncorrected early Holocene terrace data from the study area show a complex age and height distribution (Figure 5). Depending on location, variable heights, numbers, and ages of terraces are found (Winslow and Johnson, 1989; Winslow, 1982, 1988, 1989, in revision). Variations in age and height are evident not only between islands, but across individual islands, as well as in different locations on the Alaska Peninsula. In addition, some localities seem shuffled, such that some older terraces are closer to the present sea level than younger ones (Winslow, 1982, 1988, 1989, in revision).

Raised marine strandlines and terraces are valuable, if imprecise, indicators of former sea levels. Holocene strandlines, unlike Pleistocene and earlier surfaces, do not represent former higher sea levels, but, rather, episodic crustal uplift (discussion in Lajoie, 1986). The development of wide, wave-cut, or wave-built terraces requires that the surface be uplifted quickly and permanently following a long period of stability (Darwin, 1845 [1962]; Inman and Nordstrom, 1971; Matsuda et al., 1978; Clark et al., 1978; Clark and Lingle, 1979; Yonekura and Shimazaki, 1980; Peltier, 1980; Bloom, 1983; Hanks et al., 1984; Lajoie, 1986). Given that wave-cut terraces of the Shumagins abrade granodiorite, volcanic rocks, and metamorphic rocks, uplift must follow a long period (hundreds of years) of stability or must have uplifted at the same rate as the sea level rise (Winslow, in revision).

Former sea levels can be reconstructed, however, only when modeled for tidal range, surf base, sediment thickness, eustatic sea level change, isostatic adjustments due to postglacial rebound/subsidence, and shelf loading caused by the Holocene sea level rise (Inman and Nordstrom, 1971; Black, 1977; Clark et al., 1978; Clark and Lingle, 1979; Clark, 1980; Peltier, 1980; Hanks et al., 1984; Lajoie, 1986; Winslow, 1989; Winslow, in revision). The difficulty in correcting age and elevation data for terraces older than 4000 years in any region is twofold: (1) the wide variability of published eustatic curves and (2) the difficulty of estimating the glacio-isostatic correction for new localities. Creation of an empirical eustatic-isostatic correction curve from the local data is possible only if one has sufficient data from both early and late Holocene raised coastal surfaces, as is the case for this study (Winslow, 1988, 1989, in revision).

Eustatic Correction

The dramatic, and poorly constrained, sea level rise of 100 m $\pm$ 50 during the late Pleistocene-early Holocene means that early Holocene strandline and marine terrace elevations must be corrected by a large factor. The exact shape of the "global" sea level curve for the early Holocene is still under revision and needs to be determined for individual regions because of local geoidal warping (Bloom, 1983; Peltier, 1980; Clark et al., 1978). All eustatic curves are similar in shape, however (Figure 3). There was a rapid, linear rise in sea level from 100 $\pm$ 50 m below the present sea level beginning during the late Pleistocene (Figure 3). The rate of sea level rise decreased exponentially beginning sometime between 8000 and 4000 years ago. From about 4000 years ago and continuing to the present, the rate of sea level rise has been slow and linear (1 to 2 m/ka). The sea level correction problem is especially critical in the transition period occurring somewhere between 8000 and 4000 B.P., when the rapid decrease in the rate of sea level rise, coinciding with the tapering off of isostatic adjustments, produces apparent "uplift" or "a marine regression", and also results in terraces of equal height, but with different ages (Lajoie, 1986; Winslow, in revision).

Isostatic Component

Isostatic adjustments due to postglacial rebound or subsidence and shelf loading caused by the sea level rise must be estimated. For the Shumagin-Alaska Peninsula sector, calculation of a glacio-isostatic correction curve involves estimates of glacial ice thickness, duration of loads, and timing of retreat, the proximity of the locality to the Laurentide, Cordilleran, and Peninsula ice caps, and the thickness and duration of loading by the

Shumagin ice sheet. The correction can be negative, or zero if the near-field and far-field effects cancel out (Clark et al., 1978; Peltier, 1980; Black, 1977; Winslow, 1988, 1989, in revision).

Empirical Method

Given the widespread disagreement about "global" sea level and the difficulty of estimating the local isostatic effects, there is another method to arrive at a correction curve: an empirical model (Clark et al., 1978; Clark and Lingle, 1979; Clark, 1980; Peltier, 1980). The method presented here is similar to that of others, except that the corrections from local data are inverted to obtain straight lines identifying equal uplift rate, rather than the complex nested curves of the other methods. The assumption in this and all empirical methods is that the strandline data from any one region have all been affected by the same sea level and local isostatic effect, so that data from different localities differ only in uplift rate.

Correction Curve

Studies from other regions suggest relatively constant, long-term uplift rates for individual localities for the last 4000 years, during which time little or no eustatic or glacio-isostatic correction is needed (Figure 3). One can draw straight lines from present sea level through late Holocene data from each locality to establish its uplift rate. Once the late Holocene rate for a locality is established, the line is extended back to 12,000 years. All data older than 4000 years will need an elevation correction, up or down, to the constant uplift rate line for their localities. Data of the same locality that diverge from the constant uplift line do so because of a common correction. The set of vertical corrections plotted on Figure 3 results in an empirical curve that omits the locally varying tectonic uplift rates, but which represents a combined isostatic-eustatic correction for the region. With this method, it is impossible to separate the eustatic from the isostatic components.

About 78% of the corrections define a least-squares fit curve that falls well within the envelope of published eustatic curves (Figure 3). The correction curve obtained by this method coincides well with eustatic curves derived elsewhere from uplifted marine strandlines and from submerged shelf areas (Figure 3).

Differential Uplift Rates

To identify cumulative uplift and uplift rates for the early Holocene data, one must apply the inverse of the combined eustatic-isostatic curve to the strandline heights (Figure 6; see also Winslow, in revision). The corrected data represent true tectonic uplift and reveal differential uplift occurring in separate blocks across the Shumagin Islands and nearby Alaska Peninsula. Corrected age and height data from 54 uplifted marine strandlines in the study area (Figure 6) indicate 1 to 65 m of cumulative uplift during the last 12,000 years. The average uplift rate is 5 to 6 m/ka, with a range of 1 to 10 m/ka. Systematic differences in heights, numbers, and ages of raised coastal features define northeast-striking domains representing the Alaska Peninsula, inner Shumagin Islands, and outer Shumagin Islands (Figures 2 and 6). During the last 12,000 years, the Alaska Peninsula in the study area has uplifted at approximately 5 to 6 m/ka, the inner Shumagins at 8 to 10 m/ka, and the outer Shumagins at 1 to 3 m/ka. Uplift rates are lowest in the outer Shumagin Islands,

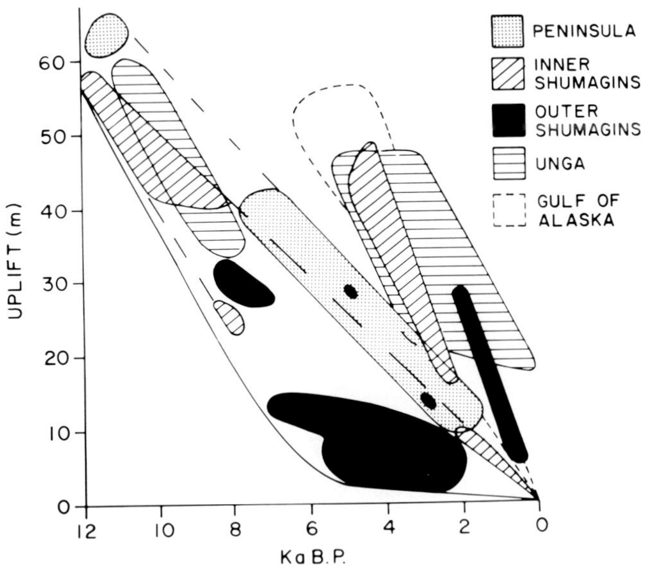

Figure 6. Corrected marine terrace data. Shaded fields represent the data for separate fault blocks. Symbology reflects age and height errors. Vertical scale shows cumulative vertical displacement or cumulative tectonic uplift. Correction curve is the inverse of empirical sea level curve from Figure 3.

especially the two islands near the Aleutian trench (Chernabura and Simeonof Islands). The Peninsula data define a regional average uplift rate of 5 to 6 m/ka. Similar rates (1 to 10 m/ka) have been reported elsewhere around the Pacific margin (compilations in Lajoie, 1986). Further support for semi-independent block uplifts comes from variations in terrace and archaeological site elevations from the outer Shumagins (Winslow and Johnson, 1989). This varying tectonic uplift rate must be applied to each block in the sea level reconstructions to obtain the expected shoreline position (Figure 6).

PALEOSHORELINES

Projected paleoshorelines with no tectonic correction are shown on Figure 7. The empirical correction curve was applied to a bathymetric base map (USGS 16540, scale 1:300,000). Data were averaged in 2 to 3 km^2 grids, depending on the availability of soundings. Much more detailed reconstructions of two of the islands using the same method are in Chapter 8 of this volume. According to the correction curve, the relative sea level rose locally at about 5 m/ka from 12,000 to 8000 B.P., and tapered off rapidly to 1 to 2 m/ka by 5000 B.P. (Figures 3 and 6). This explains why there are no raised marine surfaces older than about 5000 B.P. in the slowly uplifting outer Shumagins, except in a few areas of rapid uplift (Figure 6; Winslow, in revision).

The following paleoshoreline levels were established for islands, assuming no tectonic uplifts:

Figure 7. Paleoshorelines for the Shumagin Islands and nearby Alaska Peninsula without tectonic effect. The corrections use bathymetric base map 16540 at a scale of 1:300,000. Data are averaged in 2 to 3-km^2 grids, depending on available soundings. Soundings were corrected from mean lower low water to mean sea level. Contours represent the position of sea level in the past and are labeled as ka B.P. (thousands of radiocarbon years before present). Contours are for 12, 10, 8, and 6 ka B.P. where bathymetry is available. Z=Zachary Bay, D=Delarof Bay, B=Baralof Bay, S=Simeonof Harbor.

TABLE I.

Age (ka B.P.)	Former Sea Level (Meters Below Mean Sea Level)
12	−56
10	−37
8	−18
6	−6
4	−2.5
2	−1.5
0	0

From the regional-scale paleoshoreline map (Figure 7) it is clear that, in the absence of tectonic uplift, relative sea level rise would have resulted in progressive inundation of the Shumagin shelf, causing increased isolation of islands and also a decrease in their shoreline lengths. All potential early Holocene sites would be located far offshore of present-day coasts, in deeper water with increasing age. At 12,000 B.P., the shoreline would have lain at least 30 km from the Shumagin Islands' present coasts (Figure 7). In the absence of uplift, coastal sites, if established, would lie under 56 m of water. From at least 12,000 B.P. to at least 9000 B.P., the Shumagin Islands would have formed a single, circular peninsula on the Shumagin shelf extending southeastward from the Alaska Peninsula (Figure 7).

By 10,000 B.P., the shoreline would have come within 10 km of several islands' present coasts, defining a west-striking submarine trough that reached Balboa Bay (Figure 7) and a narrow fiord that lay between the future inner and outer Shumagin blocks. At this time, early settlers could have entered some harbors along the south coast of the Peninsula east of Cape Aliaksin and also near Korovin and eastern Popof. Most early Holocene sites on the Shumagin Islands, however, would have been far from today's coast.

At 8000 B.P., the relative sea level would have risen so that all of the outer Shumagins were isolated to create one island, while the present-day inner Shumagins defined two blocks: Unga-Popof and Korovin-Andronica (Figure 7). The Unga-Popof block might have remained attached to the Alaska Peninsula until as late as 7000 B.P. The 8000-year contour would have reached into several Peninsula bays (Figure 7). The shorelines of the inner Shumagin block would have lain only 1 to 2 km from the present coast, but the archaeologically important bays (D and B, Figure 7) would have remained dry. On the outer Shumagin block, several bays would have been open on Nagai Island, but all other islands' shorelines would have lain far from their present coasts, except Chernabura, which would have formed a narrow peninsula closest to the Aleutian trench (Figure 7; see also detailed study in Chapter 8, this volume).

Many of the present islands would have been landlocked until as recently as 7000 B.P. (Figure 7). By 6000 B.P., the shoreline would have moved closer to the present coasts, with a few exceptions. Most critical for archaeological sites in the inner Shumagins, the important bays on Unga (Z, B, and D on Figure 7) would have remained dry, the Popof Strait would not yet have been open, and Korovin's important southern bay would have lain far from the shoreline. All of the outer Shumagin Islands' coasts would have been 1 to 2 km from the present shore, except for Chernabura Island, which appears to have been accessible for the longest period in the study area (see also Chapter 8, this volume). The strait between Simeonof Island and the Koniujis would have been opened by this time. Even at 6000 B.P., many present-day bays would have remained far from the coast.

By 4000 B.P., the shore would have lain close to the present location and has not been drawn because of the scale of the map. In general, it would have lain about 500 m offshore of the present coasts, depending on the local bathymetry. Important exceptions are that Popof Island would have remained attached to Unga Island until quite recently, and some of the bays on Unga would have remained dry until 1000 to 3000 B.P.

Most significant, at 4000 years ago, Simeonof's present coast would still have remained too far inland to corroborate the archaeological data (Chapter 8, this volume). According to a no-uplift model, there should be no coastal archaeological sites older than 1000 to 3000 B.P. in Delarof, Baralof, or Zachary Bay on Unga Island, along upper Popof Strait, or in any locality on Simeonof Island. This contradicts known prehistoric sites in those localities, such as in Simeonof Harbor (Chapter 8, this volume) as well as the sedimentology of these coasts.

In Chapter 8 of this volume, Johnson and Winslow apply detailed corrections to archaeologically important areas of Chernabura and Simeonof Islands. It is clear from their research that, even on islands with slow uplift rates, the tectonic factor is very important for understanding the human settlement distribution through time. Their chapter makes clear why there can be no prehistoric coastal sites on Simeonof Island if there was no land uplift. Therefore, extant archaeological sites and nonmarine stratigraphy at the present coasts of these islands argue that the tectonic effect must be considered.

In this paper, the tectonic correction will be applied to Unga Island only, which is the most rapidly uplifting island (8 m/ka) in the study area (Figure 8). Interpretations from this island can be contrasted to Johnson and Winslow's examples (this volume) for the two most slowly uplifting islands in the Shumagins: Chernabura and Simeonof (1 to 3 m/ka). Tectonic reconstructions of the other islands await further study. In addition to its geological significance, Unga Island appears to have been the most richly populated island in the inner Shumagins during the late prehistoric period and in early historical times until its harbors became too shallow to navigate (Veniaminov, 1839 [1984]; A. Gronholdt, personal communication). Thus, its analysis is most important to future work.

For Unga Island, the paleoshoreline corrections are

TABLE II.

Date (ka B.P.)	Sea Level (wrt mean)	Land Uplift (m)	Current Height on Map (meters above mean sea level)
12	−56	+96	+40
10	−37	+80	+43
08	−18	+64	+46
06	−6	+48	+42
04	−2.5	+32	+29.5
02	−1.5	+16	+14.5
00	0	0	0

Between 12,000 and 6000 B.P., the rate of sea level rise was about equal to the tectonic uplift rate on Unga Island, so that the position of surfaces remained almost equal in height (40 to 46 m above present sea level). This means that raised strandlines equal to or older than 6000 B.P. overlapped stratigraphically. The 8000-B.P. contour is plotted on Figure 8.

During the early Holocene, deep fiords indented Unga's coastline (Figure 8), most of which form wide river valleys, swamps, or lakes at present. Harbors of archaeological importance were deeper and extended far inland to triple their current length. Probably most dramatic is the connection from the west coast to Zachary Bay which existed from at

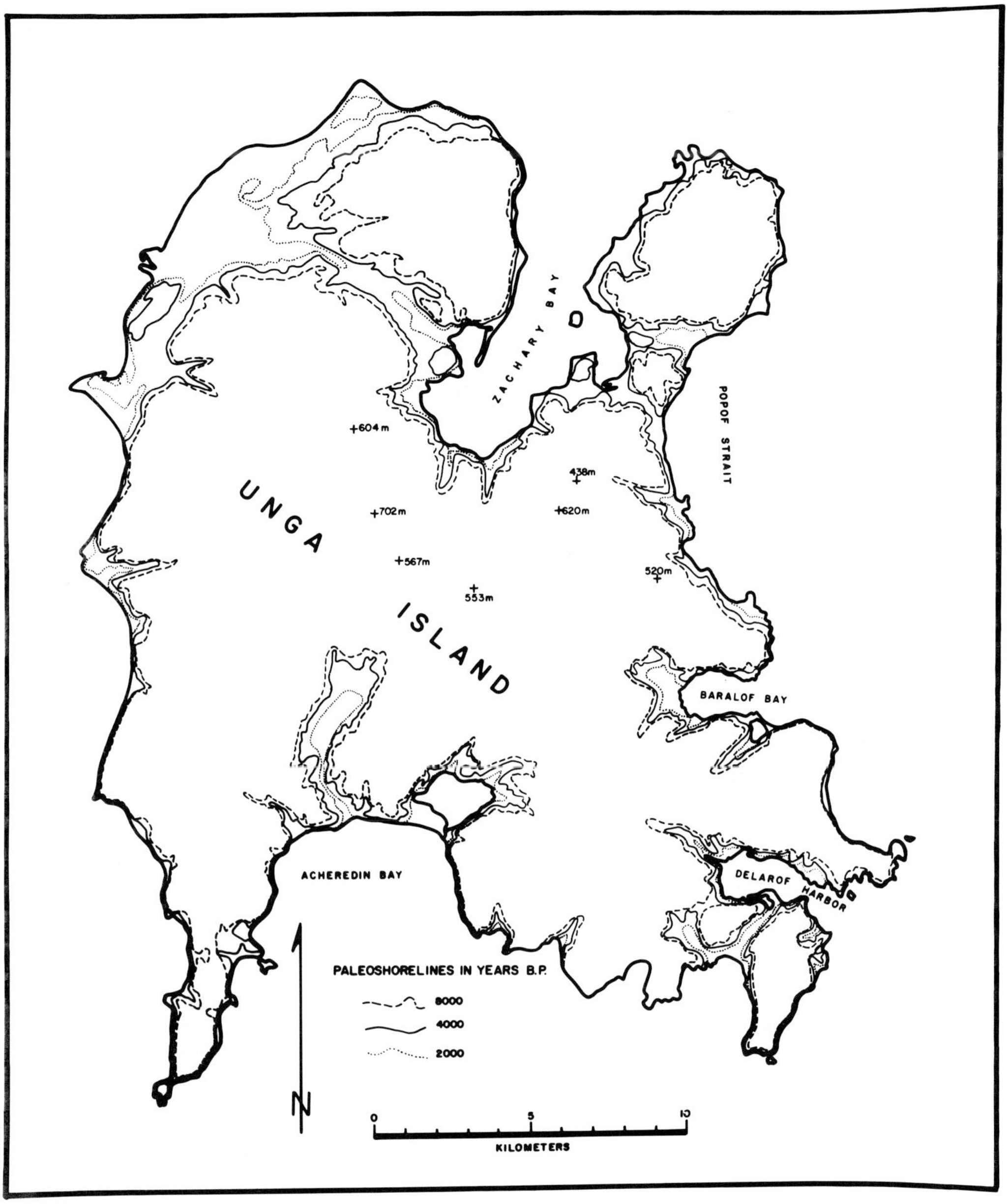

Figure 8. Paleoshoreline reconstruction for Unga Island, including the tectonic effect as well as sea level change. Contours are at 14.5 m (about 47 ft), representing the shoreline at 2000 B.P., 29.5 m (96 ft) for 4000 B.P., and 46 m (150 ft) for 8000 B.P. See text for explanation of ages and heights.

least 12,000 years ago until about 3000 B.P. Its deeply indented coasts should be examined for prehistoric sites, a task begun in 1989 by Johnson and Winslow.

In contrast, due to the slow uplift rate of Chernabura and Simeonof Islands, there is no dramatic areal change observed on either island, but more important changes are seen on Simeonof. Its main harbor either opened recently, perhaps 4000 B.P. if no uplift occurred, or as early as 8000 years B.P. with uplift. Johnson and Winslow (this volume) point out, however, the importance of detailed local topographic and bathymetric measurements, as well as careful assessment of the stratigraphy, before attempting to reconstruct the paleoshorelines of a specific site area.

MODEL IMPLICATIONS: RATES AND FATES OF SITES

"Drowned" Coastlines

The drowned appearance characteristic of some outer Shumagin Islands results from the late Pleistocene-early Holocene inundation of glacially carved valleys by rising seas. Since the sea level rose much more rapidly than any documented global tectonic uplift rate during the early Holocene, glacially carved valleys near stable or even on uplifting coasts would have been inundated. In areas with uplift rates of 8 to 10 m/ka, however, the bottom of an average Alaska margin fiord would have reemerged during the early Holocene (as in the inner Shumagins). In contrast, very slowly uplifting coasts (1 to 3 m/ka) will have only one or two low terraces of late Holocene age, as in the outer Shumagin Islands. Therefore, the characteristic "drowned" appearance of some outer Shumagin Islands does not signify long-term land submergence (cf., Morris and Bucknam, 1972; Grantz, 1963).

Episodic Uplift

Historical data from earthquakes along the Alaska margin, including the Shumagin Islands, indicate that vertical changes occur episodically and not as a continuous uplift. Over long periods of time, however, the slope of the step-like uplifts remains constant (Lajoie, 1986; Plafker and Rubin, 1978). The interplay of nonlinear eustatic and glacio-isostatic effects with episodic uplift events determines the maximum ages of coastal sites and their fates (Figures 7, 8, and 9). Such steps in a tectonic uplift curve result in submergence between seismic events, not because of a tectonic effect, but because of a sea level rise between events (Figure 10). As seismic events occur about every 200 to 800 years in the study area, interseismic submergence during the late Holocene may be up to 20 to 80 cm. During the much more rapid sea level rise of the early Holocene, however, interseismic submergence may be as much as 3 m. Such a change would affect preservation of low coastal sites and change the position of paleoshorelines significantly, especially in very shallow bays.

During periods of rapid sea level rise, uplifted surfaces will reenter the surf zone, experience erosion or burial on the shelf, and then reemerge during subsequent uplift episodes (Figure 10). Submergence and erosion between uplift episodes during sea level rise may result in a temporal gap in the shoreline or settlement record. Interactions between uplift rate and eustatic rate also predict the reemergence date of a marine surface and predict which terrace heights at a particular site may have repeated ages (Figures 9 and 10). Unraveling these factors explains scrambled age and height data in individual sections caused by changes in the slope of the sea level curve and by the episodic nature of the uplift.

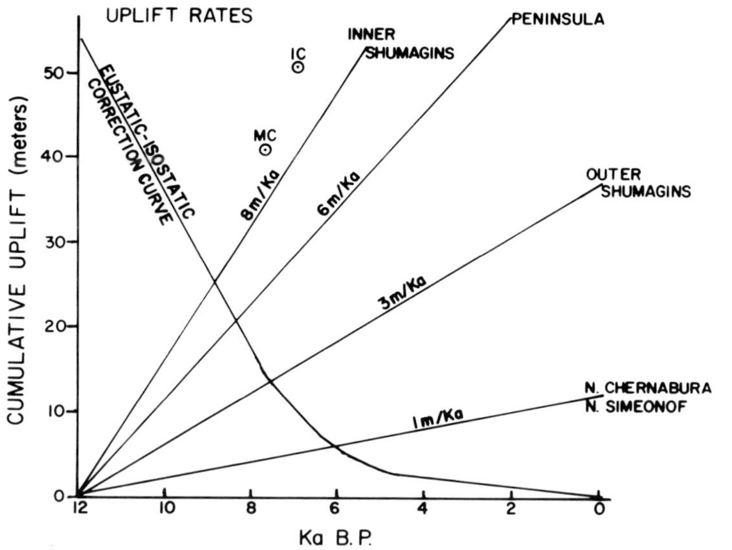

Figure 9. Maximum age of a coastal site in an uplifting region. Straight lines give average uplift rates for largest fault blocks in the study area. Where the inverse of the local eustatic-isostatic curve crosses a line of constant uplift rate, the age of the oldest undunked site may be predicted. Shoreline sites established earlier will experience submergence. See also Figure 10 for an example. MC = Middleton Island, IC = Icy Cape (Plafker references in text).

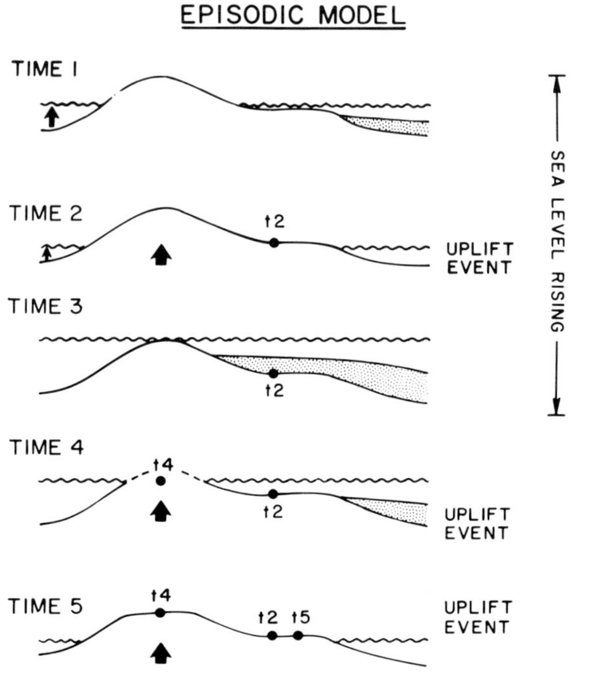

Figure 10. Hypothetical island profile during period of sea level rise Time 1 to Time 3. Episodic model consists of periods of stability during which a wave-cut terrace is abraded and the wave-built terrace is deposited. The uplift rate is 6 m/ka. The model explains how ages of terraces appear scrambled, how a single terrace may yield two different dates, and how terraces formed during periods of sea level rise can experience submergence and later reemergence.

IMPLICATIONS FOR ARCHAEOLOGICAL DATA

Recognition of temporal and spatial patterns of crustal movements is of great importance to the evaluation of the prehistoric archaeological record. Earthquake activity at active plate boundaries, such as the Pacific-North American interface, affects human settlement patterns by generating hundreds of square kilometers of new land surface. Positive effects are offset at least temporarily by tsunamis and destruction of food sources by changes in shelf depth and landslide-produced turbidity. During the Holocene, people arrived in the Shumagin Islands soon after uplift exposed additional coastal surfaces (Winslow and Johnson, 1989). Evidence from the 1788 earthquake in the study area suggests that abandonment of settlements is largely due to destruction of harbors by shallowing and by tsunami activity rather than from co-seismic shaking (Winslow and Johnson, 1989).

Intersections of the tectonic uplift and eustatic-isostatic curves predict the general location (height, age, and the blocks with the highest uplift rate) where the oldest archaeological sites might be found (Figures 6 and 9; Black, 1977; Winslow and Johnson, 1989). Shoreline reconstructions can be used to recreate former harbor geometries, and to predict the areal position of the oldest settlements and of sites most likely to have survived a sea level rise. Paleoshoreline reconstructions must not contradict the requirements of coastal inhabitants to be at the coast and neither too high nor low for the sea level position at the time. Thus, archaeological data place tight constraints upon the sea level reconstructions and provide an independent source of testing of the geological models (Johnson and Winslow, this volume; Winslow and Johnson, 1989).

CONCLUSIONS

Reconstruction of late Pleistocene-early Holocene shorelines in formerly glaciated, geologically active regions, such as the Shumagin Islands, requires determination of local sea level rise, its rate of change, and estimation of isostatic effects. In addition, the total land uplift during the same period and the uplift rate must be measured. When attempting to find one eustatic-isostatic correction curve to apply to the local geological and archaeological data, this study revealed that there were several correction curves, each one representing a fault block with a different uplift rate (Figures 6 and 9). All blocks have the same correction curve in common, however, once the tectonic uplift is removed (Figures 3, 6, and 9).

The relative sea level rise and tectonic uplift curves interact to create a complicated history of local coastal emergence, temporary submergence, and apparent stability. Paleoshoreline maps that include a tectonic correction are complex, but explain the local geological and archaeological data much better than a sea level correction alone. Because of the interaction between land uplift and the rising sea, the creation and destruction of harbors will be highly sensitive to the local bathymetry and topography, thus creating large changes through time among the islands in this study.

When applied to archaeological site locations, elevations, and ages, the implications of choosing the proper shoreline reconstruction becomes clear. In the absence of significant tectonic uplift, there can be no coastal sites above present sea level older than about 4000 to 5000 years. With tectonic uplift, however, the shorelines and harbors most suitable for habitation change dramatically.

Results from this geoarchaeological study of the Shumagin Islands and nearby Alaska Peninsula suggest that temporal patterns in prehistoric human occupation relate directly to former shoreline geometries. Selection of prehistoric coastal settlement locations in the study area was greatly influenced by relative sea level change. If one assumes a sea level model without uplift, the broad Shumagin peninsula was rapidly inundated to form the many small islands of today (Figure 7). In that model, only late Holocene strandlines and coastal archaeological sites should exist above the present sea level. Using a tectonic model, however, there is also major flooding during the early Holocene, but it is offset by rapid uplift in the inner Shumagins (Figure 8). In a tectonic model, the slowly uplifting outer Shumagin Islands, however, did not keep up with rising sea level until the late Holocene.

Modeling of coastal uplift patterns in tectonically active areas can establish the most likely sites for settlements and can explain temporal gaps and the reoccupation of sites in response to prehistoric earthquakes (Winslow and Johnson, 1989; Johnson and Winslow, this volume). The empirical method for reconstructing former sea levels and for creating prehistoric shoreline maps is preferred over applying correction curves from other regions. Application of other methods can produce erroneous interpretations, especially if the local land uplift patterns are ignored.

ACKNOWLEDGMENTS

This project was inspired by the work of R. Black, K. Jacob, J. Davies, and J. Beavan of Lamont-Doherty Geological Observatory and by C. Heusser of New York University. The initial phase was supported by EAR-79-10608. The second phase of geoarchaeology was started with L.L. Johnson and was supported by NGS Grants 3324–86, 3595–87, and 3865–88. Field assistance for the geological part of this work was provided by T. Ray, J. Beavan, G. Guglielmo, Jr., S. Hickman, and J. Stennett. Drafting was done by N. Katz and R. Baroody. Special thanks go to A. Gronholdt, Capt. R. Adams, and to many students at Lamont-Doherty, Vassar, and City College who assisted in the field or in the lab. This manuscript was reviewed in its earlier form by J. Beavan, L. Johnson, T. Johnson, C. Heusser, G. Plafker, and A. Bloom. I would like to thank the current reviewers, L. Johnson and M. Stright, for their helpful suggestions to improve this chapter.

REFERENCES

Beavan, R.J., Bilham, R.G., Mori, J., Wesnousky, S., and Winslow, M.A. (1979) Tree rings reveal Gulf of Alaska earthquakes in 1300, 1420, 1580, 1899. *EOS, Trans. Am. Geophys. Union.* 60: 884.

Black, R.F. (1974) Late-Quaternary sea level changes, Umnak Island, Aleutians: their effects on ancient Aleuts and their causes. *Quat. Res.* 4: 264.

Black, R.F. (1977) Isostatic, tectonic, and eustatic movements of sea level in the Aleutian Islands, Alaska. In: *Geodynamics Symp. Earth Rheology and Late Cenozoic Isostatic Movements.* Stockholm.

Bloom, A.L. (1971) Glacio-eustatic and isostatic controls of sea level since the last glaciation. In: *The Late Cenozoic Glacial Ages,* pp. 355–379. Turekian, K., Ed. Yale University Press, New Haven.

Bloom, A.L. (1977) *Atlas of Sea-Level Curves.* IGCP Proj. 61, UNESCO, Paris.

Bloom, A.L., Broecker, W.S., Chappell, J., Matthews, R.K., and Mesolella, (1974) Quaternary sea level fluctuations on a tectonic coast: new Th230/Th234 dates from the Huon Peninsula, New Guinea. *Quat. Res.* 4: 185.

Bloom, A.L. (1983) Sea level and coastal changes. In: *Late-Quaternary Environments of the United States.* Vol. 2. *The Holocene,* pp. 42–51. Wright, H.E., Jr., Ed. University of Minnesota Press, Minneapolis.

Boyd, T.M., Taber, J.J., Lerner-Lam, A.L., and Beavan, J. (1988) Seismic rupture and arc segmentation within the Shumagin Islands seismic gap, Alaska. *Geophys. Res. Lett.* 15(3): 201–204.

Brown, L.D., Reilinger, R.E., Holdahl, S.R., and Balazs, E.I. (1977) Post-seismic crustal uplift near Anchorage, Alaska. *J. Geophys. Res.* 82: 2981.

Burk, D.A. (1965) *Geology of the Alaska Peninsula-Island Arc and Continental Margin.* Geol. Soc. Am. Mem. 99, Part 1.

Clark, J.A., Farrell, W.E., and Peltier, W.R. (1978) Global changes in post-glacial sea level: a numerical calculation. *Quat. Res.* 9: 265.

Clark, J.A. and Lingle, C.S. (1979) Predicted sea level changes (18,000 years B.P. to present) caused by late-glacial retreat of the Antarctic ice sheet. *Quat. Res.* 11: 279.

Clark, J.A. (1980) The reconstruction of the Laurentide ice sheet of North America from sea level data: method and preliminary results. *J. Geophys. Res.* 85(B8): 4307.

Dall, W.H. and Harris, G.D. (1892) Correlation papers: Neogene. *U.S. Geol. Surv. Bull.* 84: 349.

Darwin, C. (1845) (in paperback, 1962) *The Voyage of the Beagle,* pp. 293–314. Doubleday, New York.

Davies, J.N., Sykes, L., House, L., and Jacob, K. (1981) Shumagin seismic gap, Alaska Peninsula: history of great earthquakes, tectonic setting, and evidence for high seismic potential. *J. Geophys. Res.* 86: 3821.

DeMets et al. (1988) *Geophys. Res. Lett.* 14: 911.

Dumond, D.E. (1987) *Prehistoric Human Occupation in Southwestern Alaska.* Univ. of Oregon Anthropol. Pap. 36, Eugene.

Grantz, A. (1963) Aerial reconnaissance of the outer Shumagin Islands, Alaska. In: *U.S. Geol. Surv. Prof. Pap.* 475–B, pp. B106–B109.

Hanks, T.C., Bucknam, R.C., Lajoie, K.R., and Wallace, R.E. (1984) Modification of wave-cut and faulting controlled landforms. *J. Geophys. Res.* 89: 5771.

Hopkins, D.M. (1967) *The Bering Land Bridge,* Stanford University Press, Palo Alto.

Hopkins, D.M. et al. (1969) Cretaceaous, Tertiary, and early Pleistocene rock from the continental margin in the Bering Sea. *Bull. Geol. Soc. Am.* 80: 1471.

Inman, D.L. and Nordstrom, C.E. (1971) The tectonic and morphologic classification of coasts. *J. Geol.* 79: 1.

Jacob, K.H. (1984) Estimates of long-term probabilities for future great earthquakes in the Aleutians. *Geophys. Res. Lett.* 11: 295.

Johnson, L.L. (1987) Shumagin Island research: results from the fringe. In: *Abstr. 52nd Annu. Meet. Soc. American Archaeology,* Toronto.

Johnson, L.L. and Winslow, M.A. (1991) Paleoshorelines and prehistoric settlement on Simeonof and Chernabura Islands, outer Shumagin Islands, Alaska. In: *Paleoshorelines and Prehistory: An Investigation of Method,* pp. 171–186, Johnson, L.L. and Stright, M., Eds. CRC Press, Boca Raton, FL.

Lajoie, K.R. (1986) Coastal tectonics. In: *Active Tectonics,* pp. 95–124. Wallace, R.E., Ed. National Academy Press, Washington, D.C.

Laughlin, W.S. (1963) The earliest Aleuts. *Alaska Univ. Anthropol. Pap.* 10(2): 633.

Laughlin, W.S. (1980) *Aleuts: Survivors of the Bering Land Bridge.* Holt, Reinhart & Winston, New York.

Laughlin, W.S. and Laughlin, S.B. (1981) Anangula and Chaluka investigations of 1972. *Natl. Geogr. Soc. Res. Rep.* 13: 365.

Matsuda, T., Ota, Y., Ando, M., and Yonekura, N. (1978) Fault mechanism and recurrence time of major earthquakes in southern Kanto district, Japan, as deduced from coastal terrace data. *Geol. Soc. Am. Bull.* 89: 1610.

Minster, J.B. and Jordan, T.H. (1978). Present-day plate motions. *J. Geophys. Res.* 83: 5331.

Moore, J.C. (1972) Uplifted trench sediments: southwest Alaska-Bering shelf edge. *Science.* 166: 1103.

Moore, J.C. (1973) Complex deformation of Cretaceous trench deposits, southwest Alaska. *Geol. Soc. Am. Bull.* 84: 2005.

Morner, N.-A. (1978) Faulting, fracturing, and seismic activity as a function of glacial-isostasy in Fennoscandia. *Geology.* 6: 41.

Morris, R.H. and Buckman, R.C. (1972) Geomorphic evidence of late Holocene vertical stability in the Aleutian Islands, Alaska. *Bull. Seism. Soc. Am.* 62(6): 1365.

Peltier, W.R. (1980) Ice sheets, oceans, and the earth's shape. In: *Earth Rheology, Isostasy, and Eustasy,* pp. 45–63. Morner, N.-A., Ed. John Wiley & Sons, New York.

Plafker, G. (1965) Tectonic deformation associated with the 1964 Alaska earthquake. *Science.* 148: 1675.

Plafker, G. (1967) Surface Faults on Montague Island Associated with the 1964 Alaska Earthquake, pp. G1–G42. *U.S. Geol. Surv. Prof. Pap.* 543–G.

Plafker, G. (1969) Tectonics of the March 27, 1964, Alaska Earthquake. *U.S. Geol. Surv. Prof. Pap.* 543–I.

Plafker, G. (1972) Alaskan earthquake of 1964 and Chilean earthquake of 1960: implications for arc tectonics. *J. Geophys. Res.* 77: 901.

Plafker, G. and Rubin, M. (1978) Uplift History and Earthquake Recurrence as Deduced from Marine Terraces on Middleton Island, Alaska, pp. 687–721. *U.S. Geol. Surv. Open-File Rep.* 78–943.

Scholl, D.W., Craighead, F.C., and Stuiver, M. (1970) Florida curve revised: its relation to coastal sedimentation rates. *Science.* 163: 562.

Spence, W. (1977) The Aleutian arc: tectonic blocks, episodic subduction, strain diffusion, and magma generation. *J. Geophys. Res.* 82: 213.

Stauder, W.S. (1968) Tensional character of earthquake foci beneath the Aleutian trench with relation to seafloor spreading. *J. Geophys. Res.* 73: 7693.

Veniaminov, I. (1839) *Notes on the Islands of the Unalaska District,* pp. 13–16. Black, L.T. and Geoghegan, R.H., Transl. (1984). Alaska Hist. 27, Limestone Press, Kingston, Ontario.

Winslow, M.A. and Hickman, S.H. (1978) Evidence for Quaternary uplifted terraces and tilting in a major seismic gap: the Shumagin Islands, southwest Alaska. *EOS Trans. Am. Geophys. Union.* 59: 1125.

Winslow, M.A., Ray, T.C., and Beavan, R.J. (1980) Geologic and geodetic studies in the Shumagin seismic gap. *EOS Trans. Am. Geophys. Union.* 61: 1110.

Winslow, M.A. (1982) *Uplift and Tilting in the Shumagin Seismic Gap, Southwestern Alaska,* pp. 1–30. Final Rep. EAR-79-10608, National Science Foundation, Washington, D.C.

Winslow, M.A. (1988) Holocene deformation patterns and paleoseismicity in the eastern Shumagin seismic gap: SW Alaska. In: *Abs. with Programs. Geol. Soc. Am. Annual Meeting 20,* No. 7, A382.

Winslow, M.A. and Johnson, L.L. (1988) Paleoseismicity and Aleut migration patterns in a tectonically unstable environment: Shumagin Islands, Alaska. In: *Abstr. Geological Society of America Annu. Meet.* A168.

Winslow, M.A. and Johnson, L.L. (1989) Prehistoric human settlement patterns in a tectonically unstable environment: outer Shumagin Islands, southwestern Alaska. *Geoarchaeology.* 4(4): 297–318.

Winslow, M.A. (1989) Tectonic, eustatic, and isostatic controls of prehistoric coastal settlements in the Shumagin Islands, SW Alaska. In: *Abstr. Geological Society of America Annu. Meet.* A213.

Winslow, M.A. (1989) Late-Quaternary uplift and tilting of marine strandlines in the Shumagin seismic gap, SW Alaska, and implications for prehistoric earthquakes. *EOS Trans. Am. Geophys. Union.* 70(43): 1063.

Winslow, M.A. (in rev.) Late Quaternary uplift patterns of the Shumagin Islands and adjacent Alaska Peninsula, southwestern Alaska. In: *Geological Controls on the Regional Distribution of Archaeological Sites.* Thorson, R., Ed. Geological Society of America.

Yonekura, N. and Shimazaki, N. (1980) Uplifted marine terraces and seismic crustal deformation in arc-trench systems: a role of imbricated thrust faulting. *EOS. Trans Am. Geophys. Union.* 61: 1111.

Chapter

8

Paleoshorelines and Prehistoric Settlement on Simeonof and Chernabura Islands, Outer Shumagin Islands, Alaska

Lucille Lewis Johnson
Department of Anthropology
Vassar College
Poughkeepsie, New York

Margaret Anne Winslow
Department of Earth and Planetary Sciences
City College of New York
New York, New York

INTRODUCTION

In coastal areas in which seismic uplift and tilting combine in complex ways with sea level change, it is necessary to reconstruct the shoreline for various prehistoric periods in order to understand prehistoric settlement patterns. On Chernabura and Simeonof Islands, the southernmost of the Shumagin Island group (see Chapter 7, Figure 2), these reconstructions have provided critical information which indicates that:

1. At 8000 B.P., the islands had not yet been inhabited.
2. The location of the earliest sites between 4000 and 3000 B.P. was determined by both geological and cultural factors.
3. After 3000 B.P., cultural factors were most important in determining settlement. An exception to this is the Chernabura Spit, where geological factors remained primary until 1000 B.P.

CULTURE HISTORY

The outer Shumagin Islands (Chapter 7, Figure 2) are located south of the Alaska Peninsula and west of Kodiak Island. Culturally, the islands formed the eastern terminus of the Aleut culture zone at the beginning of the historic period (Dumond, 1974; Veniaminov, 1984), and the artifactual material gathered to date indicates that they were never part of the Peninsular and Pacific Eskimo zone to the east (Johnson, 1988).

The Shumagin Islanders pursued an effective and obligatory marine adaptation. Resources on land were minor and seasonal: spring greens, a few roots and bulbs, and summer berries. Oceanic resources were rich and fairly stable: the "bay of islands" (Golder, 1922) between Nagai and the other outer Shumagin Islands is less than 30 fathoms throughout. These shallows provided an ideal habitat for seals, sea lions, sea otters, and the fish and shellfish resources the mammals and the human inhabitants depended upon. South of Chernabura and Simeonof Islands, the Shumagin Banks supported huge schools of codfish (Dall, 1870; Elliott, 1886) and the upwelling waters of the Aleutian Trench provided nutrients which attracted migrating whales: herds of humpbacks off of the islands have been reported to be as large and prevalent as those in Prince William Sound (John Martin, personal communication, 1988).

Given the strong focus on marine resources, settlement would be expected to concentrate along the coastlines, as it did. Analysis of a number of Aleutian Islands (McCartney, 1977; Frohlich and Kopjanski, 1981) and of the outer Shumagins (Johnson, 1988, in press) has shown that Aleuts settled in places that had terraces into which their semisubterranean houses could be dug, beaches upon which their boats could be landed, and which trapped driftwood and whales necessary for construction and subsistence, fresh water, and nearby lookout points from which whales and enemies could be spied. In many areas, these conditions can be recognized from offshore, and land checks show the locations to have been occupied.

SEISMIC HISTORY

Geologically, Simeonof and Chernabura Islands are located in the Shumagin seismic gap, which has been very active during the Holocene (see Chapter 7). Therefore, these two islands have been subject not only to a steady sea level rise, but also to episodic seismic uplift during the Holocene. Winslow's reconstruction of tectonic uplift and nonlinear sea level rise in this region (1988, cf. above; Winslow and Johnson, 1989; Table I) reveals that the shorelines of these two islands have had a complex history of variable uplift and tilting during the Holocene. In particular, the northern ends of these islands have been uplifting at an average rate of 1 mm/year, and the southern ends at an average rate of 3 mm/year, such that the islands as a whole tilt northwest. It must be kept in mind that these average rates were accomplished discontinuously, during major earthquakes, while the sea level rise was continuous at a steadily slowing rate. It is impossible to pinpoint each uplift event, however, so the changes must be presented on the reconstructions as continuous. On the northern ends of the islands, the 6000-B.P. paleoshoreline is found at present sea level, while subsequent paleoshorelines until 1000 B.P. are found above present sea level. On the southern ends, on the other hand, the 9000-B.P. paleoshoreline is found at present sea level, and all subsequent paleoshorelines up until the present time are above present sea level (Figures 1 to 4). The dynamism of the interaction between land and sea in the outer Shumagin Islands indicates that no site's location can be appreciated before its date is known and the shoreline of its time has been reconstructed.

TABLE I. World Sea Level, Uplift on Chernabura and Simeonof Islands, and Island Shorelines

Date (B.P.)	Sea Level	Island Level	Shoreline
A. North Ends of Islands, Uplift Rate 1 MM/Year			
10,000	−37	−10	−27
9,000	−27	−9	−18
8,000	−18	−8	−10
7,000	−11	−7	−4
6,000	−6	−6	0
5,000	−3	−5	2
4,000	−2.5	−4	1.5
3,000	−2	−3	1
2,000	−1.5	−2	0.5
1,000	−1	−1	0
Present	0	0	0
B. South Ends of Islands, Uplift Rate 3 MM/Year			
10,000	−37	−30	−7
9,000	−27	−27	0
8,000	−18	−24	6
7,000	−11	−21	10
6,000	−6	−18	12
5,000	−3	−15	12
4,000	−2.5	−12	9.5
3,000	−2	−9	7
2,000	−1.5	−6	4.5
1,000	−1	−3	2
Present	0	0	0

THE ISLANDS

Simeonof Island (Figure 5), has a peak on its northern peninsula connected to the rest of the island by a sand spit. The main island has one major peak which reaches 438 m and a long, low apron around it. Chernabura (Figure 6), also has a steep, low northern peak connected to the main island by a sand spit. However, the main island is composed of four mountains. The three southern peaks, the highest of which is 491 m, are linked by passes over 150 m in height, while the pass to the northernmost peak is a little over 90 m high. These mountains come down quite steeply to the sea along most of Chernabura's perimeter. The different configurations of these two islands result in quite different shoreline histories.

Today, Simeonof has 18 prehistoric sites, a number of which may be separated remnants of single sites, while Chernabura has 10 (Figures 5 and 6). On Simeonof, the sites along the west shore are almost completely eroded away, as is the site to the east of the spit. On Chernabura, the two sites which bracket the spit are eroding seriously, and the two southern sites do not have good landing spots.

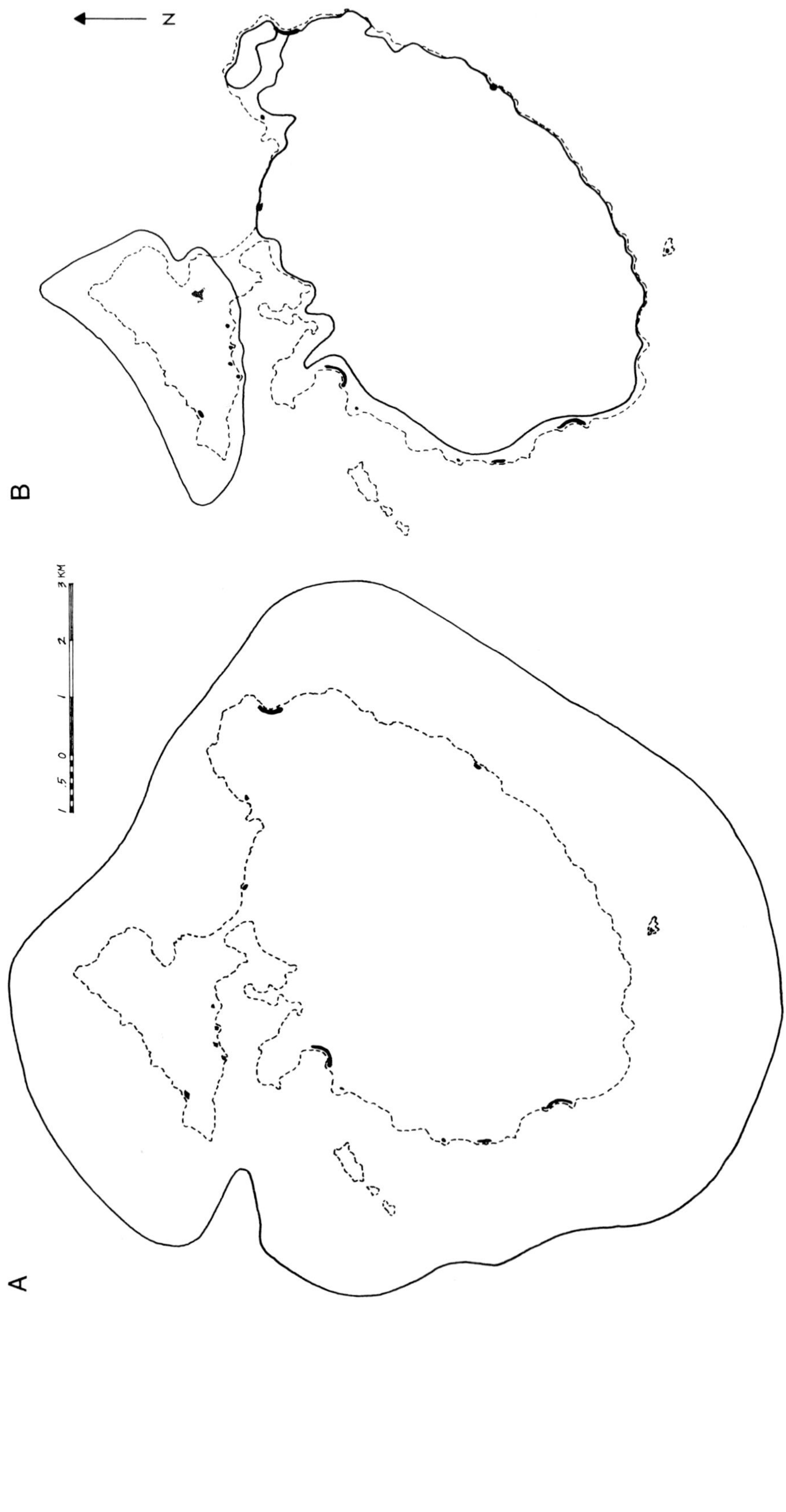

Figure 1. Simeonof Island at 8000 B.P. (A) Shoreline without considering seismicity; (B) shoreline reconstruction, including seismic and sea level change. Dotted lines indicate present coastline; site areas marked.

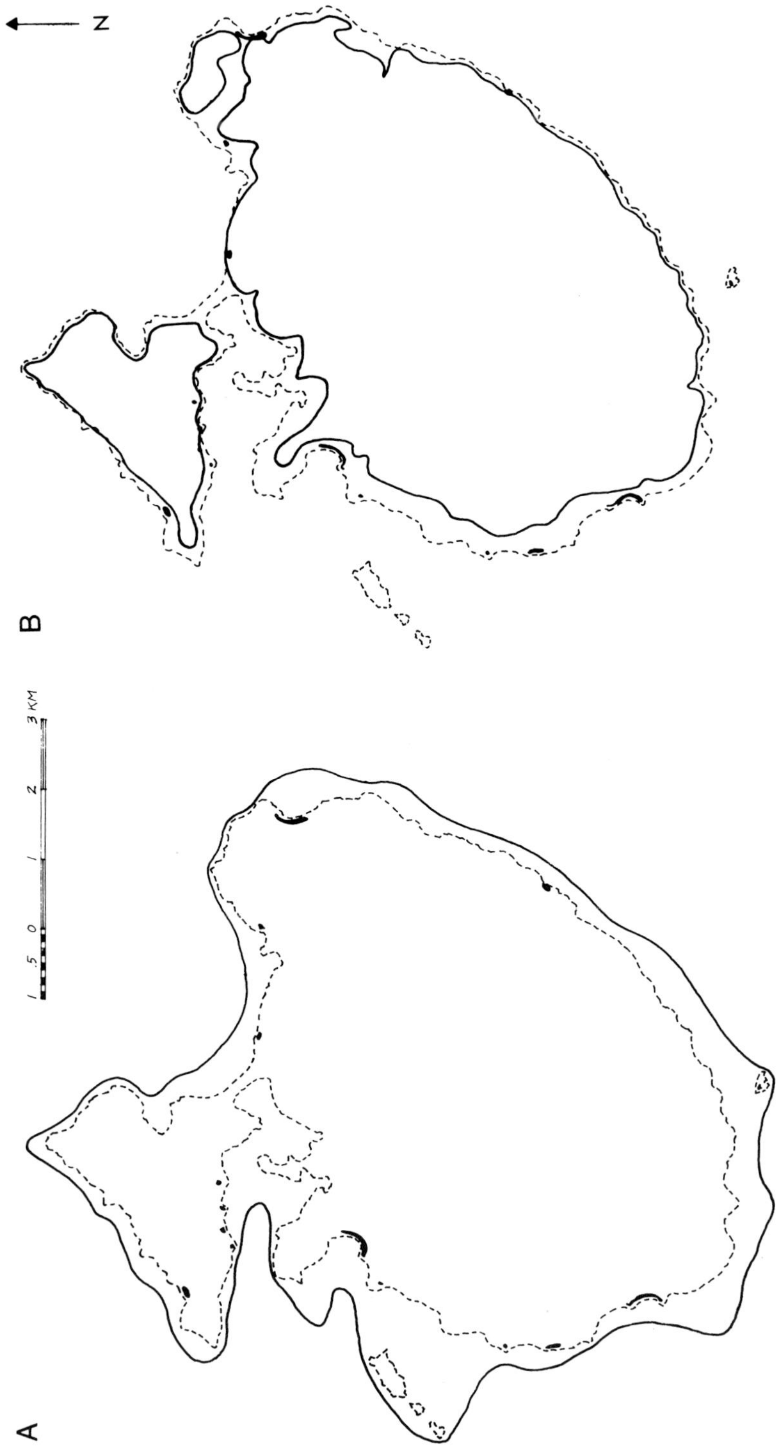

Figure 2. Simeonof Island at 5000 B.P. (A) Shoreline without considering seismicity; (B) shoreline reconstruction, including seismic and sea level change. Dotted lines indicate present coastline; site areas marked.

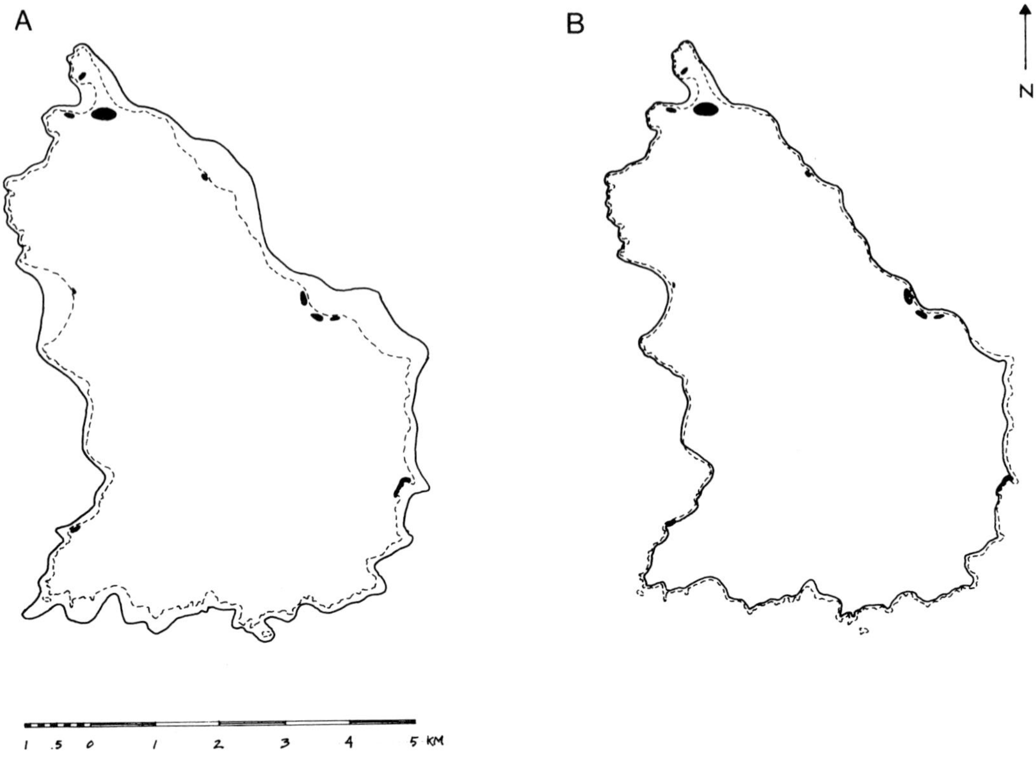

Figure 3. Chernabura Island at 8000 B.P. (A) Shoreline without considering seismicity; (B) shoreline reconstruction, including seismic and sea level change. Dotted lines indicate present coastline; site areas marked.

SIMEONOF

Focusing on Simeonof (Figure 5), Winslow's shoreline reconstructions emphasize the need to combine an understanding of sea level and seismic history to understand island prehistory (1988, cf. above; Winslow and Johnson, 1989; Table I). Although the bathymetry around Simeonof is not completely recorded, enough points are known for a reasonable reconstruction to be made. This shows that at 8000 B.P., if it were not uplifting, Simeonof would have been considerably larger than it is today (Figure 1A). On the other hand, taking uplift into account, it can be seen (Figure 1B) that, while the north end would have been larger than today, the main island would have been smaller. Although the land on which sites XSI-004 and XSI-049 were established may have been above sea level at 8000 B.P., it was too close to the storm tide for the digging of barabaras. Other site locations on the main island were underwater, while those on the small island were inland from the coast. The inland paleoshoreline, which has been walked following eroded cuts along cattle trails, has revealed no evidence that Simeonof was inhabited at this time.

At 5000 B.P., the uncorrected shoreline still would have lain offshore (Figure 2A). However, the true paleoshoreline lay 2 m above present shorelines in the north and 12 m above in the south (Figure 2B). In the next thousand years, during which we have the first evidence of human habitation in the outer Shumagin Islands, the northern paleoshoreline rose 0.5 m, while the south rose 2.5 m. Our previous research (Winslow and Johnson, 1989) has indicated that Shumagin Islanders needed at least 5 m of elevation in order to settle, so that the floors of their semisubterranean houses would be dry. Thus, even though most of the known site locations would have been above sea level, they would not have been high enough to be occupied. In conformation with this hypothesis, none of the known sites on Simeonof date from this millenium.

Sea level change and seismic activity have resulted in the 4000 to 3000-B.P. paleoshoreline of Simeonof Island being located 1 to 7 m above present sea level. While the quality of data on island contours does not permit detailed mapping of the paleoshoreline, elevations of known sites indicate that most locations were not above present sea level at 4000 to 3000 B.P. Of those that were, the north end of the small island, where XSI-001 is located, was not inhabited, while XSI-003 on the eastern edge was inhabited by 3834 B.P. and XSI-014 on the southwest by 3837 B.P. (Table II). As opposed to XSI-001, these sites are located on the ''outer'' edge of the island, which may indicate a focus on deep-sea resources rather than on those of the Bay of Islands. This distribution is particularly noteworthy since the remaining portion of XSI-014, which is seriously eroded, would have been less than 5 m above sea level.

During the last 3000 years, the Simeonof coast has gradually expanded to its present limit. Although more dates are needed for certainty, patterns of habitation on Simeonof

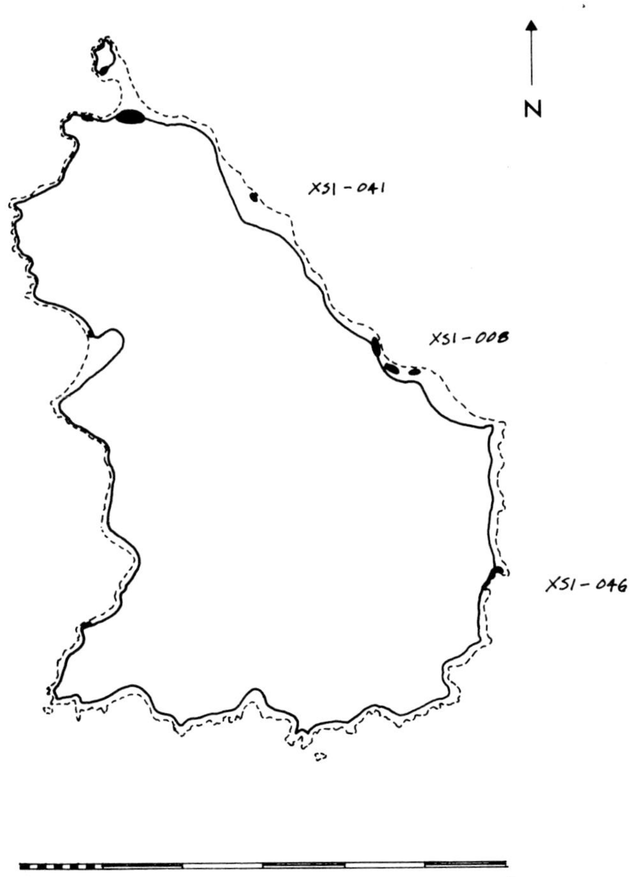

Figure 4. Chernabura Island at 5000 B.P. Shoreline reconstruction, including seismic and sea level change. Dotted lines indicate present coastline; site areas marked.

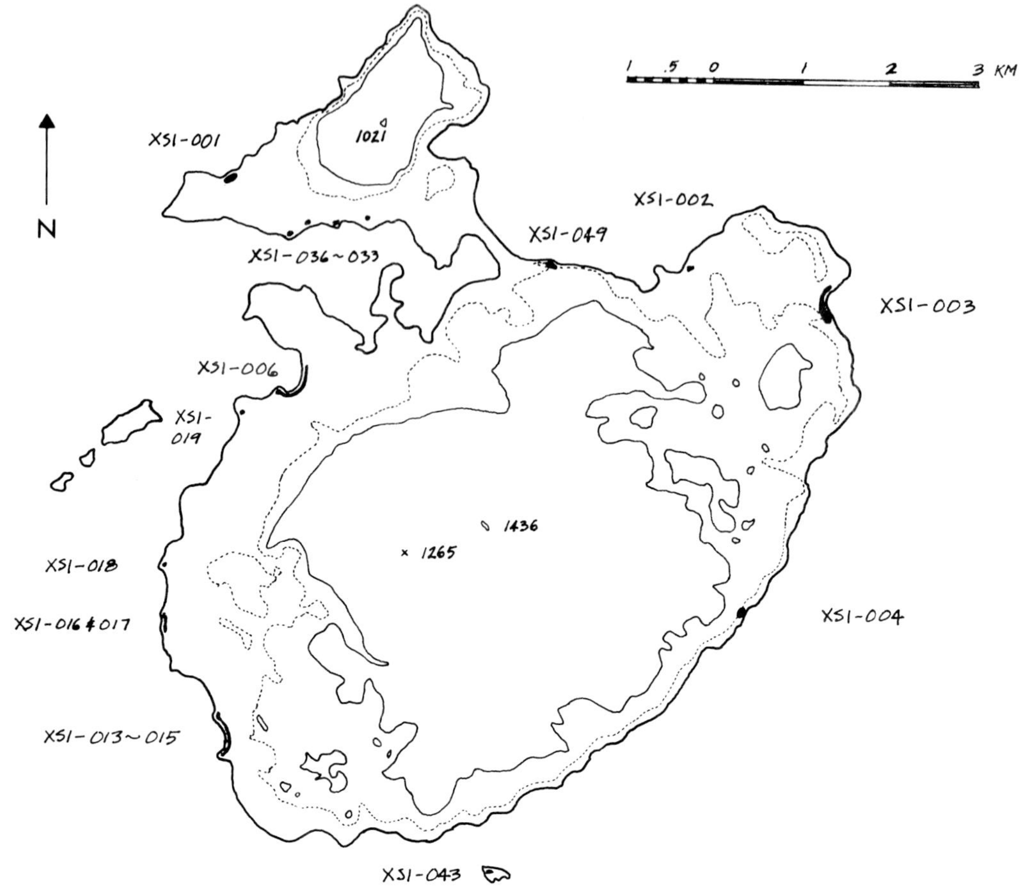

Figure 5. Simeonof Island. Archaeological sites and maximum elevations marked. Dotted line: 50-ft contour; solid line: 100-ft contour.

during this period seem to respond to cultural factors more than to seismic ones. Thus, the low southern end of the island, where both lookouts and redoubts are absent, was abandoned for the steep-sided island to the south of the main island (XSI-043) and the edges of the harbor to the northwest (XSI-001, 006; Figure 6). This would seem to be a reaction to the chronic warfare reported to have existed in this area during the late-prehistoric period (Veniaminov, 1984).

CHERNABURA

On Chernabura Island, the steepness of the island margins leads to a hypothesis of even less relationship between seismicity and settlement patterns than on Simeonof. Unfortunately, there are no offshore contours mapped for Chernabura, making the offshore reconstructions particularly perilous. The reconstructions shown assume that the bathymetry mirrors the above sea level contours. At 8000 B.P., the uncorrected paleoshoreline would have been slightly offshore of the current one (Figure 3A), while the corrected paleoshoreline, taking into account both sea level rise and seismic uplift, is 10 m below the present level on the north end of the island, 6 m above the present level on the south end, and crosses the current shoreline somewhere in the middle of the island (Figure 3B). All known site locations on the southern two thirds of the island would have been available for habitation at this time. The fact that none were inhabited suggests that people had not colonized the outer Shumagin Islands by this time.

The 5000-B.P. paleoshoreline at Chernabura lies entirely above the present sea level (Figure 4), placing all of the site areas south of the spit either underwater or just above mean high water at 5000 B.P. The 5000 to 4000 B.P. shorelines are 9.5 m above present sea level in the south and 1.5 m above present sea level in the north. No known sites were

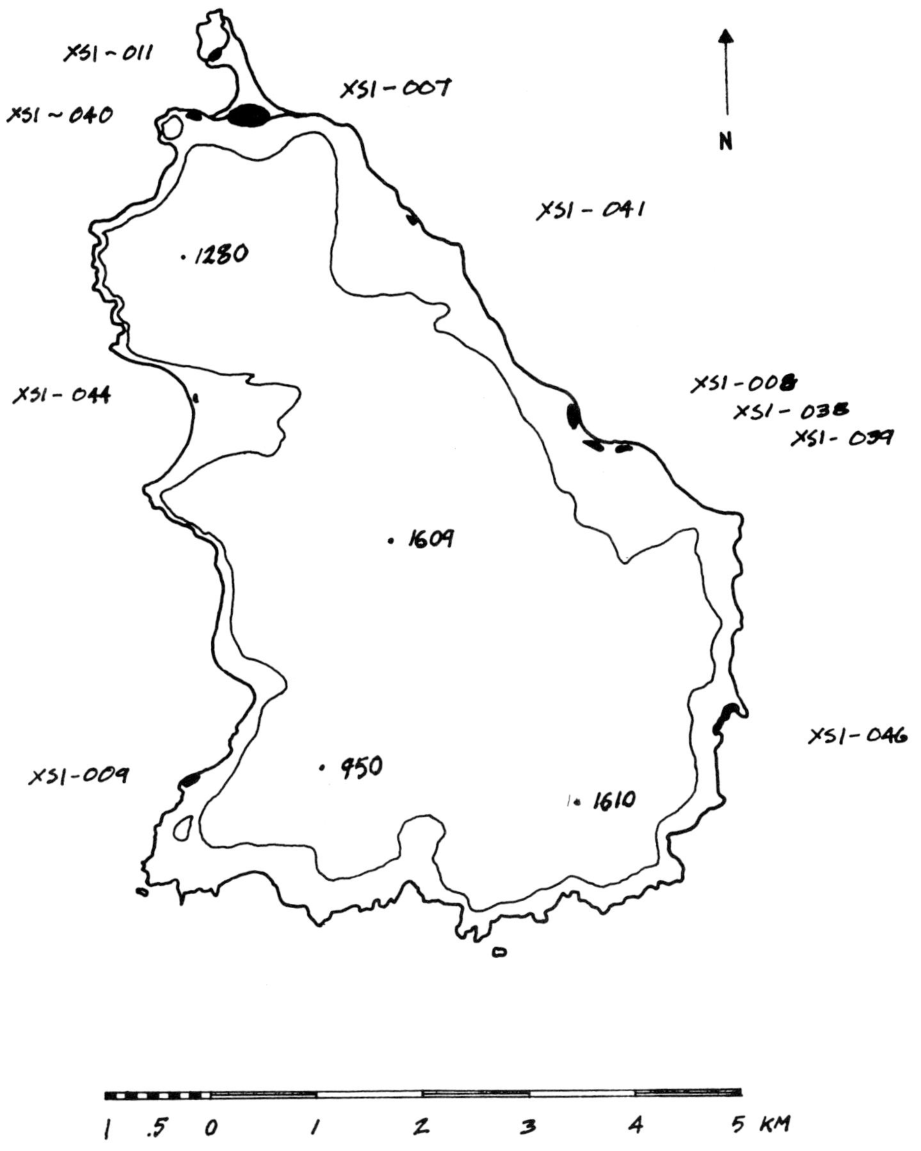

Figure 6. Chernabura Island: archaeological sites and maximum elevations marked. Contour line is at 100 ft.

TABLE II. Chernabura and Simeonof Islands: Sites, Elevations, and Dates

Island	AHRS #	Elevation	C-14 Date B.P.
Chernabura	XSI-007	12 m	2970 ± 270 (Beta-33419)
			2960 ± 140 (Beta-33420)
			2380 ± 130 (RL-2016)
			2090 ± 240 (RL-2015)
			1920 ± 150 (RL-2078)
			1880 ± 110 (RL-2077)
			1420 ± 110 (RL-2039)
	XSI-008	10 m	3520 ± 140 (RL-2025)
	XSI-009	12 m	3010 ± 130 (RL-2026)
	XSI-011	11–19 m	1150 ± 110 (RL-2040)
			1000 ± 110 (RL-2076)
	XSI-038	4–5 m	3120 ± 130 (RL-2014)
	XSI-039	7 m	Not yet dated
	XSI-040	ca. 5–25 m	3510 ± 160 (RL-2079)
			2660 ± 110 (RL-2017)
	XSI-041	9–18 m	3480 ± 140 (RL-2069)
	XSI-044	11 m	Not yet dated
	XSI-046	13 m	3700 ± 180 (RL-2074)
	XSI-047	600–700 m	No sample
Simeonof	XSI-001	7 m	290 ± 100 (RL-2081)
	XSI-002	No record	No sample
	XSI-003N	9 m	3530 ± 200 (RL-2081)
	XSI-003S	To 30 m	Not yet dated
	XSI-004	8 m	Sample too small
	XSI-005	Low	No sample
	XSI-006	7 m	690 ± 110 (RL-2080)
	XSI-012	6.7 m	Sample too small
	XSI-013	ca. 5 m	Not yet dated
	XSI-014	ca. 5 m	3450 ± 140 (RL-2038)
	XSI-015	ca. 3–4 m	No sample
	XSI-016	4–5 m	No sample
	XSI-017	4–5 m	No sample
	XSI-018	ca. 4–5 m	No sample
	XSI-019	Near shore	No sample
	XSI-020	No record	No sample
	XSI-021	No record	No sample
	XSI-033	3–4.5 m	No sample
	XSI-034	4.5 m	No sample
	XSI-035	3 m	No sample
	XSI-036	4.5 m	Sample too small
	XSI-043	14 m	1670 ± 120 (RL-2050)

occupied during this period, with the possible exception of XSI-046 on the southeast edge of the island, which was first occupied between 4082 and 3996 B.P. (Table II).

XSI-046 is particularly interesting in terms of its geological revelations. At 4000 B.P., it would have been located 4.5 m above sea level, probably facing a smooth sand and gravel beach, although it now faces a rocky shore. Today's bedrock shore is overlain by at least 1 m of interbedded sand and gravel beach deposits which are capped by a landslide, possibly related to seismic activity, on top of which the settlement was established. Since landing at this site is next to impossible today, the erosion of the beach probably postdated the site's occupation. As the shoreline receded between 4000 and 3000 B.P., the tidal flow came to hit the bedrock, and the water eroded the beach deposits. This process probably continued until enough of the boulders from the landslide fell onto the bedrock platform to block the wave action from hitting the cliff. There is no way of knowing how much of the site was eroded during this process, but the remaining structures are right at the cliff edge and there is no inland evidence of cultural activity, suggesting that a goodly part of the site eroded.

Although other site areas were available by 4000 B.P., current evidence suggests that they were occupied during the next millennium as the shoreline continued to expand. First occupied were the site areas on the east-central coast to the north of XSI-046, and XSI-040 to the west of the Chernabura Spit, all at about 3800 B.P.; subsequently, at about 3200 B.P., XSI-009 on the southwest shore was inhabited. After this time, there is no evidence of habitation on Chernabura except at the spit itself, the reasons for which appear to be both geological and cultural. The situation of XSI-009 on the southwest coast of the island is very similar to that of XSI-046 on the southeast, and the continuing lowering of the shoreline probably destroyed its beach as well. The east-central coast sites, on the other hand, are still at reasonable levels for habitation and are fronted by decent, if not superior, beaches. However, they do not have nearby lookout points, which probably explains their desertion for sites on the spit, which has excellent beaches and high, easily accessible lookouts.

CHERNABURA SPIT

Moving to the spit area (Figure 7), a most interesting story can be created for the north end of Chernabura, where excavations have allowed multiple dates to be run. Three sites

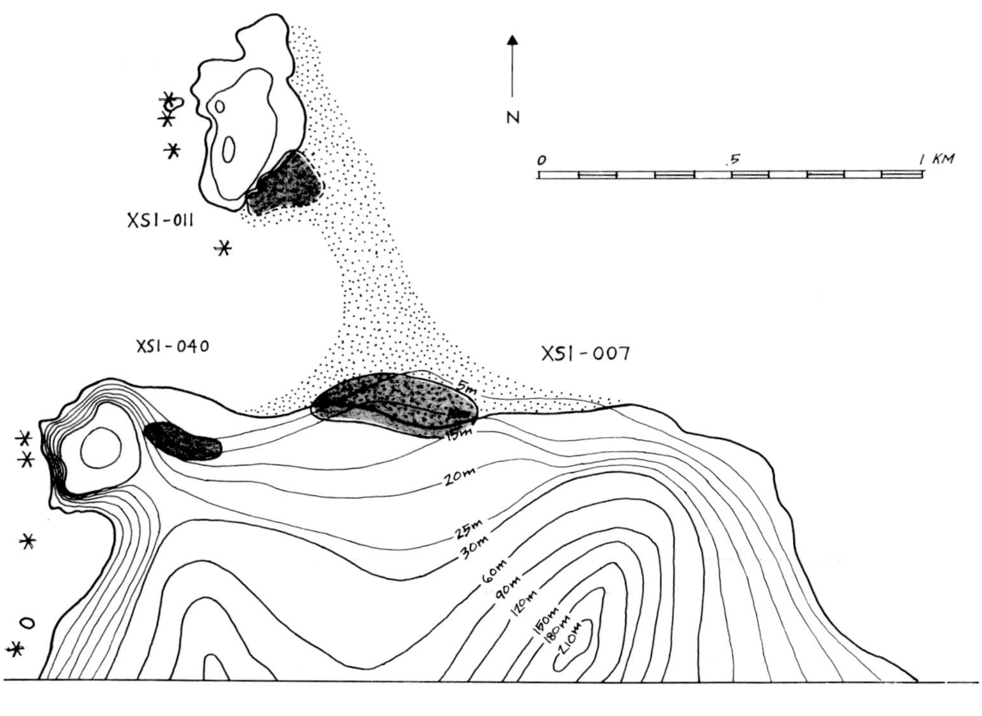

Figure 7. Chernabura Spit. Present shoreline, contours, and site locations.

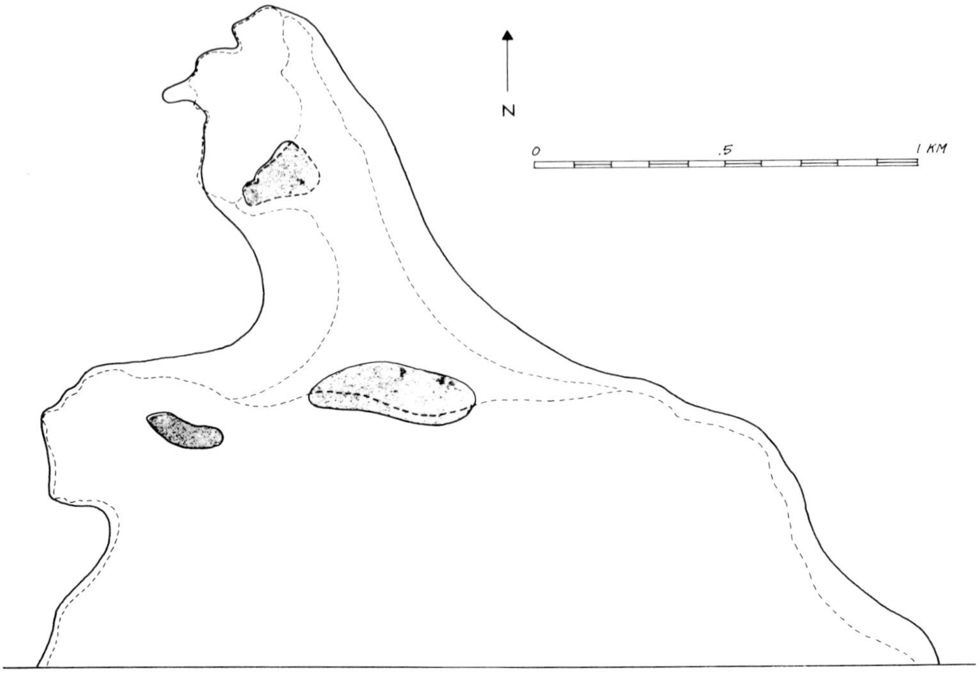

Figure 8. Chernabura Spit at 8000 B.P. Shoreline reconstruction, including seismic and sea level change. Dotted lines indicate present coastline; site areas marked.

have been found around the spit, XSI-040, to the southeast, XSI-007, to the east of XSI-040, and XSI-011, north across the spit from the other two.

At 8000 B.P. (Figure 8), the spit was much wider and the bay on the east side of it was quite open and unprotected. While the west bay looks more propitious for occupation, there are no sites dating to this time period. It should also be considered that the spit may not have existed at all or may have been only in the process of construction. During the period of high relative sea level between 5000 and 4000 years ago (Figure 9), the northern margin of XSI-011 was available for settlement, but was located on top of a vertical sea cliff, along which an active dune was probably growing, making it a difficult place to settle. XSI-040, on the other hand, slopes more gradually, is more protected from blowing sand, and was occupied by 3829 B.P. This early date for XSI-040 comes from within an underground house pit, suggesting a permanent occupation at this time. The site continued to be occupied until at least 2767 B.P.

XSI-007, which is lower than XSI-040, and less protected, shows the effect on the spit of the shifting shorelines. Excavations in the Periwinkle Mound show that dune building on the spit was very active around 3200 B.P. One and one half meters down from the surface of the mound is a thin layer of charcoal which appears to represent a temporary campsite; this dates to between 3204 and 3170 B.P. Sixty centimeters above it is a hearth dating between 3201 and 3165 B.P. The rapid deposit of sand indicated by these dates, even assuming the outside pair of 3204 and 3165, clearly indicates why there was no permanent occupation of the spit at this time. The earliest date on the midden, which begins 15 cm above the hearth, is 2354 B.P., suggesting that the intervening millennium was a time of dune stabilization. The continued regression of the paleoshoreline (Figure 10), coupled with vegetative stabilization of the dune during this period, would have made the

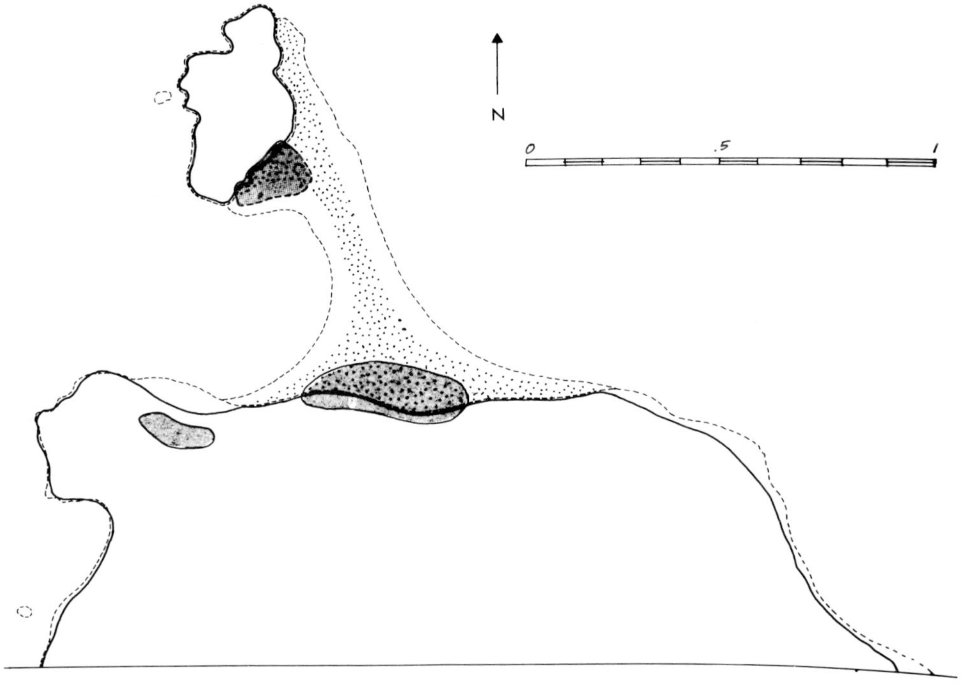

Figure 9. Chernabura Spit at 5000 to 4000 B.P. Shoreline reconstruction, including seismic and sea level change. Dotted lines indicate present coastline; site areas marked.

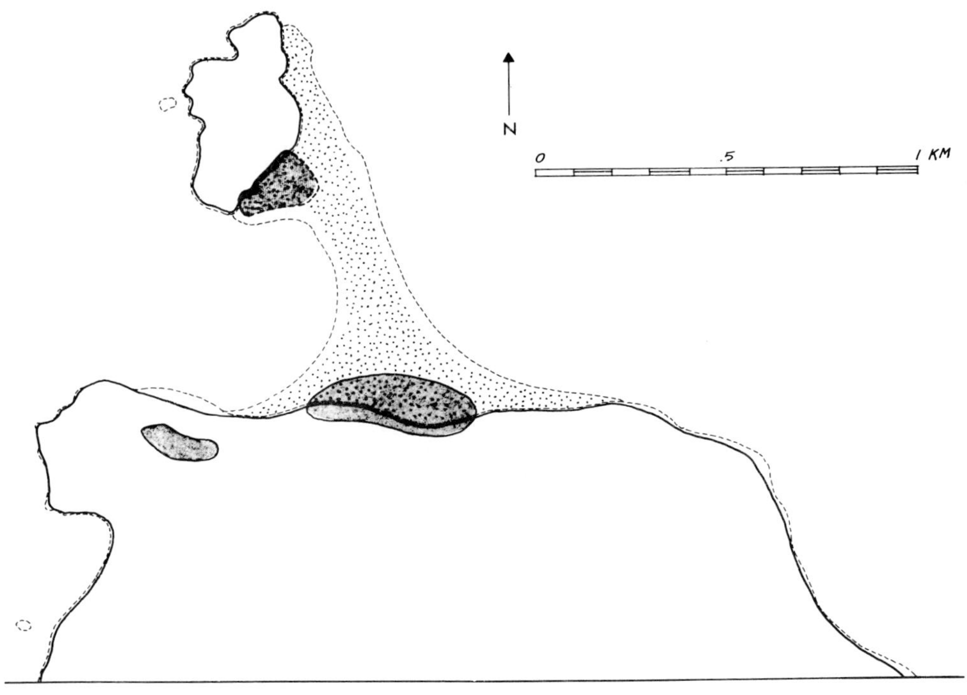

Figure 10. Chernabura Spit at 3000 B.P. Shoreline reconstruction, including seismic and sea level change. Dotted lines indicate present coastline; site areas marked.

area of XSI-007 quite attractive for occupation: it was high enough above sea level for the digging of houses, and not as much of a hike from the beach as was the XSI-040 site. Dense midden deposits and a coherent series of radiocarbon dates suggest that the mound was continuously occupied from 2354 B.P. until at least 1313 B.P. (Table II). The terminal date is uncertain due to erosion. Both XSI-007 and the next occupied site, XSI-011 across the spit, are seriously eroded today, but the precise reason for their condition and the timing of it remain to be fully elucidated. Cultural factors appear to be more important in the sites' current condition than are geological factors.

Andrew Gronholdt, a local Aleut who carried mail to the fox farmer on Chernabura during the 1930s, remembers the spit as grass covered and the farmer's cabin as being well back from the shoreline, rather than right at the vegetation border as it is today. He also does not remember anything like the periwinkle mound, a major feature of today's spit. Two major occupations have affected Chernabura in the past half century which might, independently of or in conjunction with sea level change, have affected the stability of the spit: first, a Navy radio station was established in a cirque on the south of the island with a dug and corduroyed road leading from the spit to it. On the spit itself are oil cans and cable rolls dating to this period. I am currently trying to determine how much disruption of the spit's vegetation was caused by this wartime occupation, but evidence to date suggests that it was not significant.

The second occupation, lasting from the early 1900s until 1985, was by cattle, imported onto the island in the hope of creating a beef industry. Hooves have left deep gouges in the tundra all over the island and they may have been sufficient to destabilize the spit. In 1984, the spit was completely devoid of vegetation while, by 1989, 4 years after the cattle were removed, plants had become reestablished all across the spit. This radical change in the 4 years since the removal of the cattle supports the hypothesis that they were the major agents of erosion.

On the north end of the spit, where XSI-011 is located, erosion has proceeded even farther than on the south end, leaving a tiny remnant of the site and artifacts strewn along the dune face. XSI-011, however, was occupied subsequently to XSI-007, from at least 1130 to 980 B.P., so environmental shifts do not seem to explain the movement of the human population from the former site to the latter. While the deplorable condition of XSI-011 may make it impossible to test, I would hypothesize that the move was caused by the increasing warfare which afflicted the Shumagins in the late prehistoric period.

DISCUSSION

The Shumagin Islanders were caught in a vise between the large eastern Aleutian islands Unalaska and Unimak to the west and the Koniag Eskimo population of Kodiak to the east (Chapter 7, Figure 1). Ivan Veniaminof (1984), the first ethnographer of the Aleut, reports that the Unimak Islanders attacked Kodiak, the Koniag retaliated, and neither was above attacking the Shumagin Islanders en route. By moving to the north end of the Chernabura Spit, with its steep hill immediately at hand, the residents were better able to monitor the surrounding seas for approaching enemies. The hill itself could have served as a refuge from marauders, where residents could hide while peppering enemies with rocks and other deadly missiles.

XSI-011 is the latest site yet dated on Chernabura Island, although Veniaminov reports two villages. To hypothesize the unprovable, one may have been on the now-eroded spit itself. The other may be the "summer" site, XSI-044, found on the west side of the island.

This site shows no barabara remnants, but is adjacent to historic fish-drying terraces and a valley choked with salmon berry bushes. Major occupation in the late-prehistoric period, however, appears to have shifted to Simeonof Island (Figure 5), where two large sites with large, clear barabaras date to 690 ± 110 and 290 ± 100 B.P. The move to Simeonof was probably spurred by the continuing warfare. The site locations are less obvious than the Chernabura Spit, and, while flanking the entrance to Simeonof Harbor, are not visible from it.

SUMMARY

Understanding the changing shorelines of the Shumagin Islands facilitates the understanding of their settlement history. Geology and seismology combine to explain why Simeonof was generally a less propitious island for human habitation and is less explicable to archaeologists than is Chernabura. The whole southern end of Simeonof is low, with few lookout points, which would lessen its value in the eyes of the native inhabitants. In the southwest, which appears to have been moderately popular for settlement, the periodic uplift has not been able to protect sites from the steady sea level rise which has destroyed most of the sites. Remaining evidence suggests that the north end of Simeonof was a more important settlement locale, the northeast being important early in prehistory and the northwest in the late prehistoric period. In the historic period, the harbor itself became the focus of the largest settlement in the outer Shumagin Islands.

Chernabura's overall higher coasts have led to a steadier occupation record, if not to a generally greater occupation. The combination of flat, moderately high platforms with high lookouts suggest that its apparent prehistoric prominence is real. During Chernabura's history, centers of occupation moved from the south end to the north. In the early prehistoric period, the combination of seismic uplift and sea level rise resulted in good beaches in these southern areas, which were closer to the Shumagin Banks' prime codfishing areas and to whale migration routes along the Aleutian Trench. After about 3300 B.P., waves encountered bedrock platforms, resulting in beach erosion, and the southern populations moved north to the protected flats around the developing spit. The population centered on the spit until continual warfare caused it to be deserted for Simeonof.

The above reconstruction of the prehistory of Chernabura and Simeonof Islands is dependent upon detailed knowledge of seismicity and sea level change coupled with a complete archaeological survey of the islands. Without combined geoarchaeological research, past settlement patterns on the islands would be incomprehensible.

REFERENCES

Dall, W.H. (1870) *Alaska and its Resources*. Lee and Shepard, Boston.

Dumond, D.E. (1974) Prehistoric ethnic boundaries on the Alaska peninsula. Anthropological papers of the University of Alaska. Vol 16, 1: 1–6.

Elliott, H.W. (1886) *Our Arctic Province*. Scribners, New York.

Frohlich, B. and Kopjanski, D. (1981) Aleut settlement distribution on Adak, Kagalaska, Buldir and Attu Islands, Aleutian Islands, Alaska. A preliminary report. Laboratory of Biological Anthropology, University of Connecticut, Storrs. offprint

Golder, F.A. (1922) *Bering's Voyages*. American Geographical Society Research Series 1: 1. 2 volumes

Johnson, L.L. (1988) Archeological surveys of the outer Shumagin Islands, Alaska, 1984 and 1986. *Arctic Anthropology* 25:2: 139–170.

Johnson, L.L. (in press) Prehistoric settlement patterns and population in the Shumagin Islands. In: Jordan, R. H., Ed. *Maritime harvesting societies in Southwestern Alaska: the evolution of complex cultural traditions.* In *Aurora, Anthropological Papers of the University of Alaska.*

Martin, J. (1988) Personal Communication. USFWS, Homer, AK 99603. (907) 235–6546.

McCartney, A.W. (1977) Prehistoric occupation of the Rat Islands. In: *The Environment of Amchitka Island, Alaska,* Merritt, M. L. and Fuller, L.G., Eds., pp 59–113. Technical Information Center, ERDA, TID-26712.

Veniaminov, I. (1984(1839)) *Notes on the Islands of the Unalaska District.* tr. Lydia Black and R. H. Geoghegan, Eds. Richard Pierce. Limestone: Ontario.

Winslow, M.A. (1988) Holocene deformation patterns and paleoseismicity in the eastern Shumagin seismic gap: southwestern Alaska. *GSA Abstracts with Program:* 20: 7(Sept).

Winslow, M.A. and Johnson, L.L. (1989) Prehistoric human settlement patterns in a tectonically unstable environment: outer Shumagin Islands, Alaska. *Geoarchaeology* 4: 297–318.

Mainland

Lucille Lewis Johnson
Department of Anthropology
Vassar College
Poughkeepsie, New York

INTRODUCTION

In looking at conditions along active mainland coasts, it is necessary to consider the relationship between the human inhabitants and the coast before proceeding. Clark and Fitzhugh want, like Mason, to use the dates of archaeological sites to establish a time line for geological events, but first they must establish the relationship between shoreline and settlements. Fortunately for our understanding of Labrador sea level changes, the situation is similar to that in Alaska. When the people moved to the coast, they established their camps along the strand. The distinction lies in the Labrador coastal settlements being summer camps, while the prehistoric Alaskans settled on the coast in the winter.

Moseley, Wagner, and Richardson are more concerned with the effect of coastal change on inland peoples, particularly complex agriculturalists. In other papers, Moseley has considered the effect of rising sea level on our understanding of the early adaptation of people to the Peruvian coast, but here he and his co-authors are concerned with a time period after sea level stabilization when essential resources come from irrigation agriculture. Moseley et al. show that in coastal Peru there is a complex interaction between earthquake, rainfall, and coastal currents which determines the effect of each and all on the inhabitants of the area. Individually, earthquakes and El Ninõ events cause great damage; together, they can cause disaster. This synergy had important effects on the development of prehistoric Peruvian civilizations, continues to cause disaster in the present, and should be taken into consideration by the Peruvian government in planning for future agricultural developments along the Peruvian north coast.

Chapter

9

Postglacial Relative Sea Level History of the Labrador Coast and Interpretation of the Archaeological Record

Peter U. Clark
Department of Geosciences
Oregon State University
Corvallis, Oregon

William W. Fitzhugh
Department of Anthropology
Smithsonian Institution
Washington, D.C.

INTRODUCTION

Archaeology, as an integrative discipline, relies heavily on other natural and social sciences for interpreting the context of past human activities. Of all the sciences, geology has contributed the most to archaeological methods and interpretation and has done so since the beginnings of Paleolithic archaeology. Lartet and Christie's (1864) study of artifacts and extinct fauna in cave deposits established the ice age antiquity of man, and their work applying geological principles of stratigraphy to cave deposits provided the first chronological key to human physical and cultural evolution. Equally important to the development of archaeological chronologies was the application of geomorphological principles to the dating of archaeological surfaces such as the Somme River gravels (Boucher de Perthes, 1847).

As archaeological research expanded into glaciated regions of northern Europe, geological dating techniques proved useful for developing chronologies of sites found on raised marine beaches and terraces of northern Norway and Sweden (Nummedal, 1937; Gjessing, 1942; Simonson, 1961; Helskog, 1974). Here, isostatic rebound has outpaced the eustatic rise of sea level for most of the Holocene, preserving archaeological sites and sequences over thousands of years. Later, relative geological dating was applied to the New World Arctic in Greenland (Knuth, 1967), Canada (Meldegaard, 1960; Andrews et al., 1971), and Alaska (Giddings and Anderson, 1986). These studies used curves representing isostatic uplift constructed from dated samples of driftwood, marine shells, or peat to determine the age of archaeological sites associated with former shorelines. The existence of such geological sequences or uplift curves has also assisted archaeological surveys by defining the elevation zones of raised marine formations that could be examined for archaeological sites of a given age. Dating of sites by their altitudinal association to uplift has thus been of critical importance to the development of northern archaeology and is sometimes the only dating method available, especially on sites with few artifacts and no charcoal or other organic remains.

Natural and biological scientists often lament that their association with archaeological projects is a one-way street and that while their contributions are crucial to archaeological reconstructions, such collaboration rarely if ever results in significant advances in their own disciplines. Although that is often the case, this chapter reports a relatively unique exception. Archaeological research has made a significant contribution to our understanding of coastal emergence, ice loading, and the timing and extent of deglaciation for an important sector of glaciated North America. These had proven impossible to research using traditional geological methods because of the absence of datable materials along most of Labrador's 1500-km coast. The implications of this work for understanding paleoclimate and the retreat of the Laurentide Ice Sheet make this unexpected contribution from archaeology of unusual significance.

RELATIVE SEA LEVEL HISTORY AND THE ARCHAEOLOGICAL RECORD

Over the last 18,000 years, global sea level has risen some 120 to 130 m as a result of melting of the late Pleistocene ice sheets (Fairbanks, 1989). It is well known, however, that the relative sea level (RSL) history over this time interval has not been uniform across the globe. Instead, the RSL for various parts of the earth has fluctuated in response to the complex interplay between changing mass distributions across the earth's surface associated with melting of the ice sheets and the eustatic rise of global sea level, and resulting isostatic adjustments of ocean basins and glaciated continental regions (Clark et al., 1978). Therefore, although global sea level has risen, the postglacial RSL history of glaciated coastlines is represented primarily by coastal emergence (sea level regression) as a result of isostatic recovery of glaciated continents during and following the unloading of former ice sheets.

The history of sea level change is important for establishing analytical frameworks in arctic and subarctic archaeology. Because hunters in these areas tend to place their living sites in close physical proximity to the shore, spatial and elevational patterning of archaeological sites on raised beach and terrace sequences can be used to determine relative chronology. As a general rule, old sites are found at higher elevations and younger sites at progressively lower elevations (Figure 1). Thus, within a given region, with careful archaeological and geological evaluation, site elevations can be used as a direct source of chronological control.

Although RSL history is usually determined from geological sources such as marine shells, driftwood, whalebones, or basal peats, in areas where these sources are rare of absent, as in Labrador, archaeological data can serve as an important supplement or substitute for this purpose (Andrews et al., 1971; Walcott, 1972; Fitzhugh, 1972, 1973). As in the case of curves developed with geological data, the accuracy of the reconstructed RSL depends on the degree of association between archaeological sites and the contemporary sea level. This association tends to be quite close (about 1 to 3 m above mean high

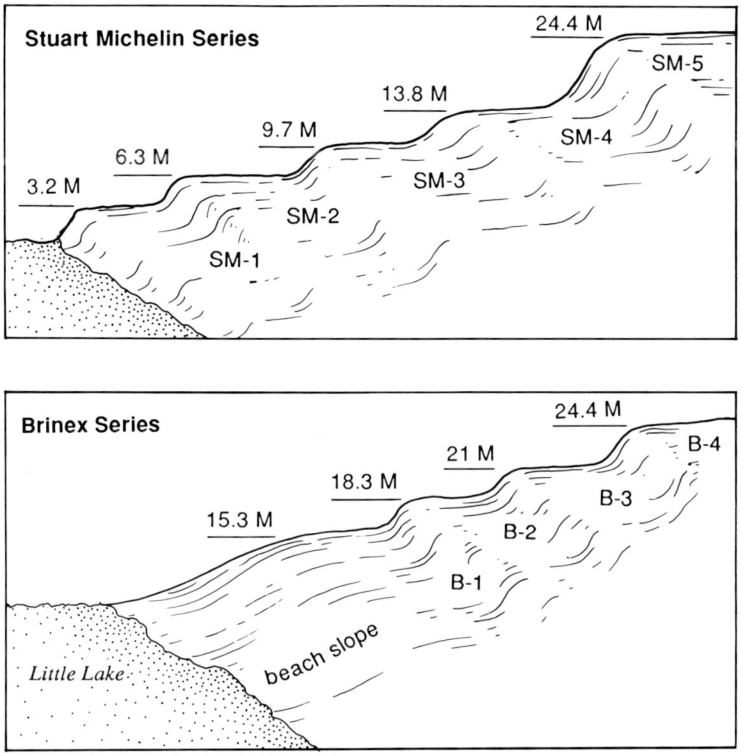

Figure 1. Stuart Michelin and Brinex terrace sequences from northwest River area, Labrador. Elevations are in meters above sea level.

tide) in the case of summer sites of northern coastal hunters. Thus, coastal emergence not only allows the archaeologist to establish a relative chronology directly from sites found on raised marine features, it also facilitates construction of abstracted models (uplift and emergence curves) which can be used for predicting the elevation range in which sites of a given age could be expected to be found and for dating isolated, known sites on the basis of their elevations above sea level (Figure 2). Emergence curves may also indicate when the RSL has risen and possibly destroyed sites by submergence or erosion. This information can provide the archaeologist with a possible explanation for the absence of sites in these regions (e.g., Maine coast, 8000 to 2000 years B.P.; Schnitker, 1974; Speiss et al., 1983).

Because the amount and rate of emergence and submergence varies depending upon the thickness of the former ice sheet, the age of deglaciation, and the rate at which the ice sheet thinned and retreated (Andrews, 1970), sites of the same age may be found at different elevations due to differential geological emergence (Andrews et al., 1971; Fitzhugh, 1972, 1977). Furthermore, because the degree of resolution of cultural sequences varies depending upon the rate of emergence, it is easier to provide chronological and spatial resolution for sites of earlier periods than for later periods, and for areas with greater ice loads than for areas with smaller ice loads.

Another important use of RSL history is to facilitate reconstruction of coastal paleogeography and paleoecology, which will vary with changes in water depth, underwater topography, and coastline position (Figure 3). All of these may have been significantly different from the present, depending on the amount of uplift or submergence. Changes in local and regional geography and ecological conditions almost invariably cause changes in prehistoric settlement patterns. In areas of low topographic relief, found in much of the Canadian Arctic and south-central Labrador, small changes in sea level can have major effects on locations of human settlements, travel routes, animal migrations, and hunting sites. In many cases, former fiords and bays became constricted, lost salinity and biological contact with the sea, and became lakes. The resulting physical and biological changes, which can be monitored by studies of sediment cores, may have had profound impact on resources and sites used by human populations. Thus, reconstructing the RSL history of a

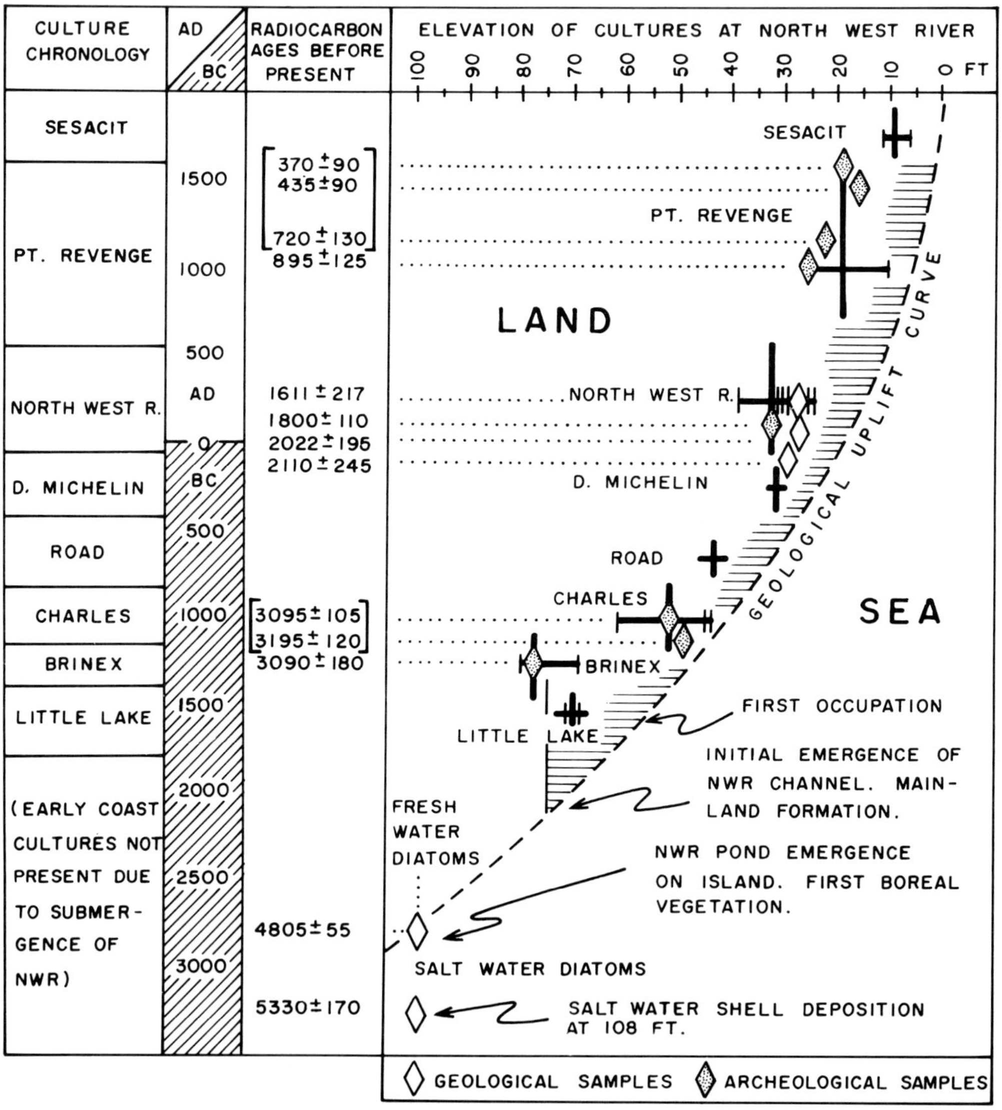

Figure 2. Archaeological sequence at North West River, Labrador, and geological correlation with site elevation (in feet). Marine shell *(Mytilus edulis)* and diatom shift record the emergence of the North West River moraine, whose later history is documented by archaeological site dates. (After Fitzhugh, 1973.)

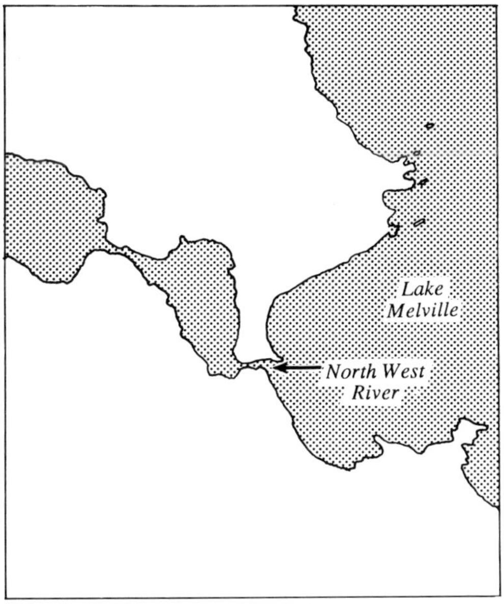

A

North West River Area
Contempory Geography

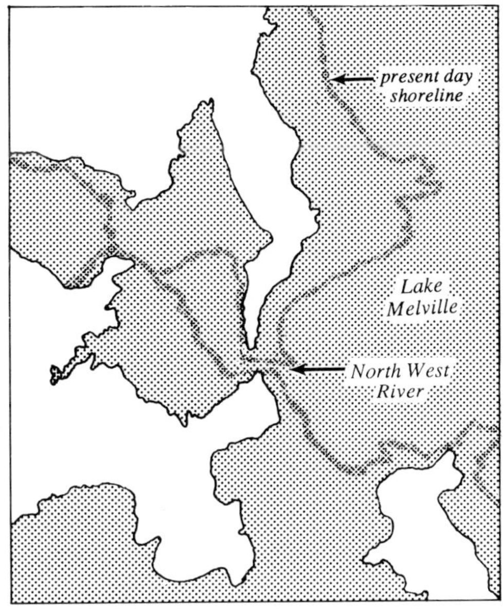

B

Geography, ca. 3000 B.P. (50 ft. contour)

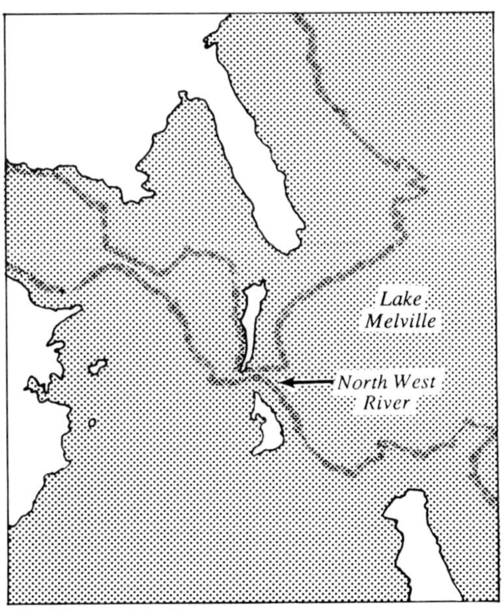

C

Geography, ca. 5000 B.P. (100 ft. contour)

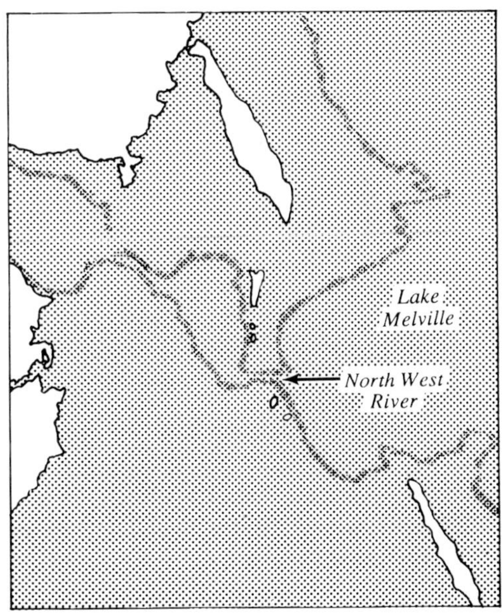

D

Geography, ca. 6500 B.P. (150 ft. contour)

Figure 3. Paleogeographic changes in the North West River area during the past 6500 years as a result of coastal emergence. (A) contemporary geography; (B) ca. 3000 B.P.; (C) ca. 5000 B.P.; (D) ca. 6500 B.P. Scale in miles. (After Fitzhugh, 1972.)

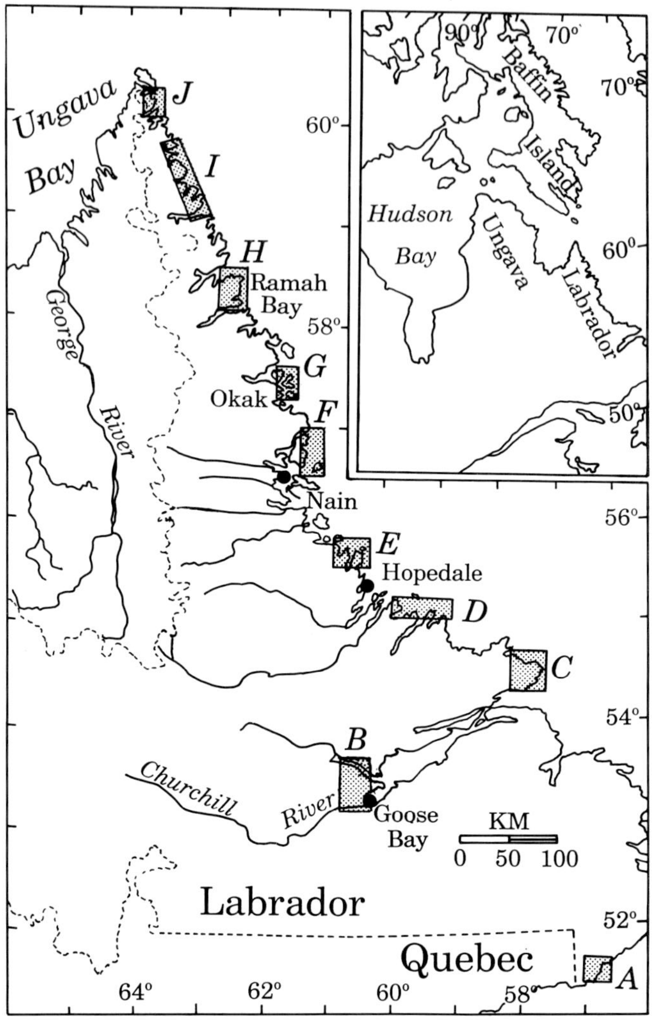

Figure 4. Location map of areas on Labrador coast where RSL curves have been constructed. A = Pinware area, B = Goose Bay area, C = Groswater Bay area, D = Makkovik area, E = Hopedale area, F = Nain area, G = Okak area, H = southern Torngat Mountains, I = central Torngat Mountains, J = Cape Chidley area.

glaciated coast has broad applications for interpreting the paleoecological settings of prehistoric maritime cultures from the archaeological record.

The Labrador coast of eastern Canada (**Figure 4**) was glaciated during the last glacial maximum of the late Wisconsin (Ives, 1978). Evidence of this glacial history is apparent in raised strandlines, marine terraces, and boulder necklaces, in the presence of perched blocks, glacial erratics, and glacial striae, and in observations by 20th century fishermen of continuing shoaling of harbors. Yet, despite the abundant evidence and a history of excellent preradiocarbon era geomorphological research (cf. Tanner, 1944; Wenner, 1947), the dearth of datable fossil materials resulted in declining interest in Labrador during recent decades as research shifted to areas with better dating control. Today, details of Labrador's Holocene geological history are surprisingly poorly known, and this lacuna has increasingly restricted the development of integrative models that define the physical extent, mass balance, dynamics, and history of the eastern edge of the Laurentide Ice Sheet as well as the climatic models built with these data.

A partial solution to this problem has been achieved through use of the archaeological data reported here. Research since 1968 has identified extensive evidence of human settlement of the Labrador coast, much of which can be related to shoreline habitation patterns associated with former sea levels (Fitzhugh, 1972, 1973, 1977). As a result, the postglacial RSL history of the 1500-km coast of Labrador is now as well known as that for any other region of the Northern Hemisphere, and the relationships between RSL history and the archaeological record can begin to be addressed.

The purpose of this chapter is to summarize the most salient results of RSL studies, from both geological and archaeological perspectives, developed over the past 2 decades as a result of extensive archaeological surveys and radiocarbon dating of sites that can be closely associated with former sea levels. We have been able to expand upon the work of Fitzhugh (above) in revising four RSL curves and constructing six additional RSL curves. These define the RSL history for much of the coast and the Goose Bay area (Figure 4) and provide an important added dimension to the archaeological record.

ARCHAEOLOGY

Archaeological studies in Labrador began in the Strait of Belle Isle (Lloyd, 1874) and by the first half of the 20th century had expanded to provide a skeletal framework for the entire Labrador coast from the "Straits" (Harp, 1963; Harp and Hughes, 1968) to Hopedale and Nain (Strong, 1930; Bird, 1945) and the Killinek-Burwell region (Leechman, 1943). Later, systematic regional surveys and interdisciplinary research, begun in Hamilton Inlet in 1968, resulted in a regional culture historical synthesis that incorporated geological, paleoecological, and climatic history (Fitzhugh, 1972). Since then, surveys and excavations have provided modern archaeological information for the entire Labrador coast (McGhee and Tuck, 1975; Cox, 1977, 1978; Jordan, 1978; Fitzhugh, 1980). Concurrently, paleoecological and geological studies were initiated to enhance cultural reconstructions (Jordan, 1975; Short, 1978; Short and Nichols, 1977; Lamb, 1980, 1984, 1985; Fitzhugh and Lamb, 1985).

The Labrador coast is ideally suited for collaborative work in geology and archaeology. Protected with numerous islands and bays, this coast has the advantage of a lee shore (westerly winds predominating) and a rich marine ecology suitable for supporting northern maritime cultures. Human occupation is represented by many distinctive archaeological cultures spanning the last 8000 years. This history has been relatively accessible because isostatic emergence has preserved archaeological sites on raised beaches and marine formations (Figure 1). Coastal archaeological research has been made all the more feasible by a minimum of concealing vegetation, thin soil cover, and lack of site destruction from urbanism and development. Finally, the diversity and complex patterning of archaeological cultures itself has facilitated interdisciplinary research by providing many culturally distinctive features that can be related to a variety of noncultural data sets.

Archaeological surveys begun in Hamilton Inlet in 1968 have since expanded to cover the entire Labrador coast from the Strait of Belle Isle to Killinek at the entrance of Hudson Strait (Figure 4), although coverage is most complete for the central and northern coast. Particular care was taken to standardize data to facilitate integration and comparison. Special attention has been given to gathering radiocarbon samples and site elevation data.

As a result, compared with other arctic and subarctic regions, the archaeology of the

Labrador coast region is quite well known (Figure 5). The Labrador coast extends over a distance of 1500 km from Newfoundland to Hudson Strait, encompassing a large part of the northeastern margin of the Laurentide Ice Sheet. Within the 8000 years of known human habitation, more than 20 culturally distinct groups have been identified from more than 1000 archaeological sites. Elevation data and radiocarbon ages are available for many of these sites. Additional data for defining coastal emergence are provided by radiocarbon ages on geological samples (shell, wood) associated with raised marine deposits and on sediment cores from emergent ponds. These data provide the means for dating the RSL throughout Labrador for most of the postglacial period, for constructing a series of RSL curves, and for preparing culture-ecological syntheses (Fitzhugh and Lamb, 1985), outlined below.

The earliest occupations of Labrador are found in the Strait of Belle Isle region dating to about 8000 B.P. (Tuck, 1975; McGhee and Tuck, 1975), where they occur on beaches at elevations of about 30 m above sea level (m asl) and are associated with the alder-birch shrub tundra zone. Human colonization of the central and northern Labrador coast seems to have followed the northward movement of this shrub zone, which reached Hamilton Inlet about 7500 B.P. (Jordan, 1975), Hopedale about 6000 B.P., and Okak about 4500

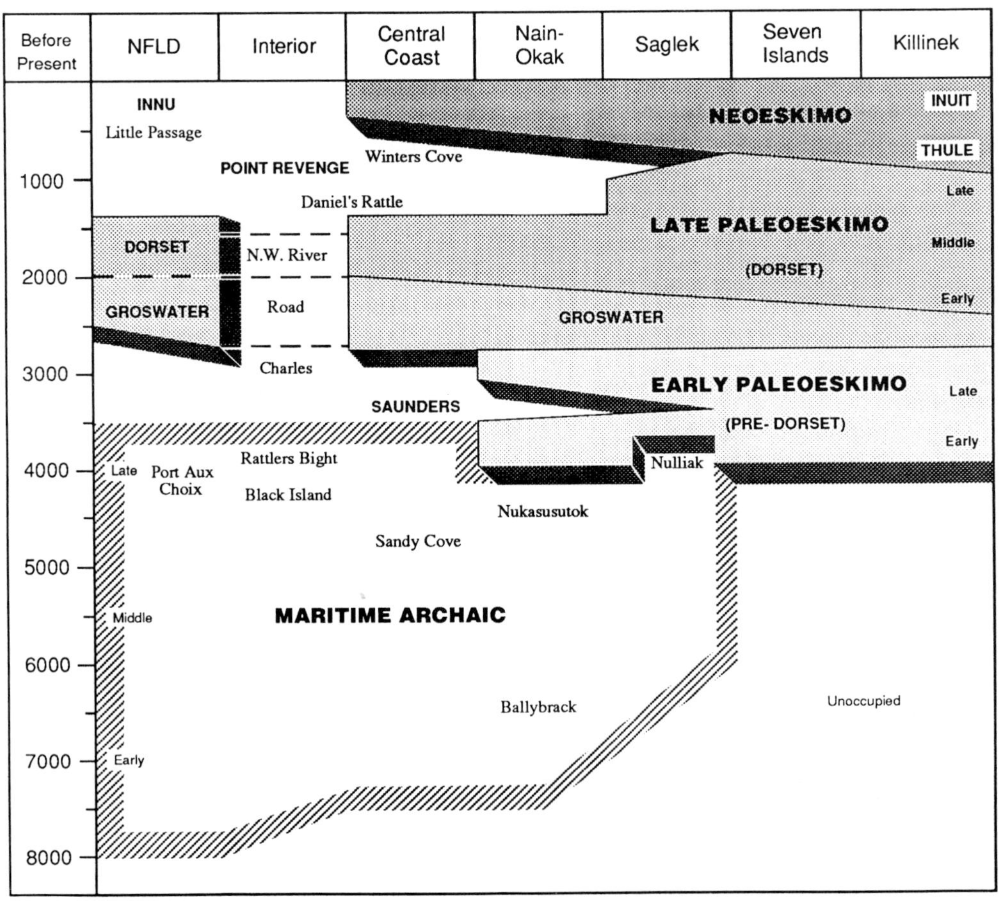

Figure 5. Labrador culture history from Newfoundland to Killinek, northernmost Labrador.

B.P. (Short and Nichols, 1977). This expansion occurred before the arrival of the fir/spruce forest in these regions and is clearly documented in exclusive use of alder and birch in the hearth charcoal of Maritime Archaic sites of this period. The arrival of forest cover, originally thought to have been a precondition for Indian occupation of these northern regions (Harp, 1963), seems to have had relatively little effect on the history and distribution of Maritime Archaic and later populations whose culture history can be traced through a complex series of typological, technological, settlement pattern, and religious changes over the next 4500 years (Fitzhugh, 1978, 1983, 1985). Of these, the most dramatic changes are the development of complex burial ceremonialism, introduction of longhouses and large seasonal settlements, and—most pertinent to the current subject matter—a rapid downward shift in site elevations.

After 4000 years ago, the paleoecological history of Labrador stabilized with the northern forest limit becoming established in the Napartok region north of Okak, although some evidence suggests gradual climatic deterioration to the present (Fitzhugh and Lamb, 1985; Clark et al., 1989). Important cultural changes occur at this time, however, in the form of the appearance of pre-Dorset culture which moved south into Labrador from the central Arctic, having arrived there and in Greenland from the western Arctic only slightly earlier (McGhee, 1969/70; Maxwell, 1985).

During the ensuing 4000 years, a variety of Paleoeskimo (Pre-Dorset, Groswater, and Dorset) and Neoeskimo (Thule and historic Inuit) groups and Indian cultures occupied the Labrador coast and interior. On the interior, Indian groups maintained a continued presence as far north as the forest limit, but rarely occupied the interior tundra regions. On the other hand, Eskimo groups which occupied the coast periodically expanded their range hundreds of miles south of the forest-tundra boundary, as far as Newfoundland and the Gulf of St. Lawrence, while at other times they were restricted to northernmost Labrador. At these times, Indian groups occupied coastal zones as far north as Okak and Ramah Bay, at least seasonally (Fitzhugh, 1980, 1983; Fitzhugh and Lamb, 1985).

This culture history is recorded by spatially discrete and archaeologically distinct cultures whose sites are found consistently in defined elevation ranges and characteristic ecological and geographical settings. However, unlike sites of the earlier Maritime Archaic period, later prehistoric Indian and Eskimo sites (see Fitzhugh, 1987, for discussion of ethnic issues) become more difficult to date by elevation alone because local geomorphologic and cultural factors begin to play a proportionally larger role in site selection than elevation as uplift rates declined.

CONSTRUCTION OF RSL CURVES

RSL curves from glaciated Arctic Canada show similar forms with an initial period of rapid emergence followed by a period of slower emergence or submergence (Andrews, 1970; Walcott, 1972; Clark et al., 1978). Furthermore, the RSL curves of emergence are usually best described by an exponentially decreasing rate of emergence from the marine limit to the present sea level (Andrews, 1968; Ten Brink, 1974), although exceptions to an exponential form have been described (Dyke, 1979; see, however, Quinlan, 1985). The equation for exponential decay is

$$y = e^{-kx}$$

where y is the amount of emergence accomplished in x years, and k is the decay constant that describes the rate of change. High "k" values identify faster rates of emergence. For

TABLE I.

Region	Marine Limit Altitude (m)	Marine Limit Age (yr B.P.)	Decay Constant ("k" value)
Pinware	152	14,000	0.36
Goose Bay	135	7,550	0.66
Groswater Bay	75	11,000	0.39
Makkovik	107	9,700	0.48
Hopedale	119	7,600	0.63
Nain	71	8,500	0.50
Okak	79	9,400	0.47
Southern Torngat	55	9,700	0.41
Central Torngat	56	10,000	0.41
Cape Chidley	16–26	9,000	

comparative purposes, RSL curves are discussed here with respect to the "k" values for each curve (Table I). Similar "k" values indicate that the form of the RSL curves is similar; the greater the difference in "k" values, the more dissimilar the curves will be.

RSL emergence curves represent some combination of three responses. Gravitational attraction of the sea surface by the ice mass and elastic recovery of the lithosphere are represented by the initial period of rapid emergence, and are followed by delayed plastic recovery of the mantle, represented in the later period of slower emergence (Farrell and Clark, 1976). Emergence characterizes those regions that were formerly glaciated (Clark et al., 1978). RSL histories characterized by emergence followed by submergence are found in areas near the former ice margin, and represent migration of a forebulge (Clark et al., 1978).

RSL curves are constructed from dated materials that are tied to the elevation of a feature indicating former sea level. Geological materials commonly dated include shell, bone, and wood that were deposited at or below sea level. In many cases, particularly with shell and to a lesser extent whalebone, these materials are enclosed in sediments that were deposited far below the contemporary sea level. Association of these materials to a former sea level is therefore often uncertain, although it is sometimes possible to trace the sediments to a strandline.

The postglacial RSL history of much of the Labrador coast has previously been poorly documented largely because raised marine deposits, while abundant (Figure 6), are generally barren of dateable geological material. Lacking arctic climates and permafrost, wood, shell, and bone are rarely preserved in the sandy, acidic, unfrozen soils of Labrador beaches north of the Strait of Belle Isle limestone-rich deposits.

Where geological materials are rare or absent, as in Labrador and many other subarctic regions, archaeological sites can provide an alternative means of dating and constructing RSL curves on isostatically emergent coasts. As with geological material, the critical link is the association between site and contemporary sea level. Summer or open-water sites in arctic and subarctic regions have a close shoreline association, whereas winter sites are further removed and more variable due to the greater need for shelter, absence of boats, and other reasons. Historic and modern Labrador Inuit and Indian summer coastal camps are almost all found 1 to 3 m above high tide, whereas winter sites have little reliable shoreline association. Analysis of archaeological structures and assemblages provides a relatively secure method of determining the open-water seasonality of most sites.

Figure 6. Photograph of raised marine beaches near Aillek, Labrador. Several sites have been excavated and radiocarbon dated on these beaches.

As in the case of geological materials, use of archaeological data for constructing emergence curves has its pitfalls, and care needs to be taken to reduce errors introduced by dating or shoreline association. Ideally, water-rolled artifacts would be the best guarantee of a site's shoreline association, but this criterion is almost never encountered, except on submergent coasts where sites are being croded by the sea. Therefore, other factors need to be considered. Site elevations should be set by the lowest elevation at which remains of a component are found, only the youngest ages from a site suite should be used, and consideration must be given to the type of dated material. Twig charcoal from ''nondriftable'' species (ground birch, alder, and small-diameter conifers) is preferred to log samples which could be derived from centuries- or millenia-old arctic driftwood. Wood charcoal is to be preferred over antler, which dates ''young'' in the arctic, while sea mammal bone, ivory, and charred fat and marine shell date 300 to 500 years too old due to the take-up of old carbon from the deep ocean reservoir (McGhee and Tuck, 1976; Arundale, 1981). Finally, local topographic conditions need evaluation. Ideal sites are those found on broad, protected, gently rising raised-beach sequences not exposed to the open sea. Elevations of terrace sites are particularly difficult to determine because there is no way to determine the position of the active shoreline at the terrace front.

In a study comparing RSL history to elevation of archaeological sites of known age from coastal Arctic Canada, Andrews et al. (1971) found a good relationship between the age of any particular site predicted from the RSL history and the known age of the site. Most sites were within a few meters of the elevation predicted by the RSL history. These showed evidence of having been occupied by people actively engaged in using the sea, and were generally summer sites. Those sites found at elevations well above those predicted by the RSL history were either winter occupation sites or muskox and caribou hunting camps located "without any concern for coastal activities" (Andrews et al., 1971), or their anomalous placement was controlled by local topography.

These relationships suggest that prehistoric sites on raised beaches display a close correlation between age and elevation for any given area. Therefore, with a large enough archaeological sample, careful geomorphic analysis, and consideration of dating errors, archaeological samples can be as reliable as geological samples for constructing RSL curves. Results obtained from the construction of RSL curves with radiocarbon-dated samples can be checked and amplified secondarily by plotting elevations of culturally diagnostic sites for which direct radiocarbon-dated cultural materials are not available. While this practice introduces some circularity into the dating/elevation equation, it can help establish confidence intervals for site elevations in cases where a sample includes large numbers of undated sites belonging to well-defined cultural components, as is the case in the Labrador Paleoeskimo period.

In the absence of geological material associated with raised marine deposits from most of the Labrador coast, we have constructed RSL curves based primarily on the large number of radiocarbon-dated archaeological sites that now exist from Labrador. The primary uncertainty in the interpretation of the data used in constructing RSL curves for Labrador is identifying the sea level represented by dated samples. Dated archaeological sites used to constrain the RSL curves represent the lowest of a larger number of summer/open-water season sites of similar age from each area. Interpretation of archaeological structure type and diagnostic assemblages indicate that these sites were summer sites, which are usually closely associated with former sea level (Andrews et al., 1971; Fitzhugh, 1972, 1977). Furthermore, most of the sites occur directly on raised marine beaches and terraces. From these relationships, we infer that the dated sites are closely associated with former sea level and draw a smoothed, exponential curve that falls to the right of archaeological sites. The RSL curve falls to the left of dated geological samples, mostly shell, where they are present.

A further uncertainty in interpreting RSL from radiocarbon ages is differential emergence between dated sites. Differential emergence results from geographic variation in ice loading. For example, the RSL curve for western Lake Melville (Goose Bay area) differs greatly from the RSL curve for Groswater Bay, even though the latter area was in the same path of ice movement. These differences are related to thinner ice over Groswater Bay as well as an earlier retreat from Groswater Bay under ocean-moderated climatic conditions. Surprisingly, a similar differential exists between Groswater Bay and the outer Hopedale coast to the north, discussed further below. For these reasons, the ages used to define each of the ten curves in this study are from areas <200 km^2 and are confined largely to outer coastal regions (Figure 4). Under these conditions, differential emergence within each of the areas is considered to be a relatively insignificant source of error. Nevertheless, differential emergence may explain some of the variation in the elevation of sites above each RSL curve, and the curves should be considered "composite curves" (Dyke, 1979).

MARINE LIMIT

The marine limit is of fundamental importance to the RSL history of a region. It is identified by the maximum level of coastal submergence, as indicated by the upper limit of raised beaches and deltas and washing limits below which perched blocks and other loose glacial deposits do not exist (Bird, 1954, 1967; Andrews, 1986). The age and elevation of the marine limit will vary along any glaciated coastline. Dating the marine limit on emergent coasts dates deglaciation and synchronous exposure of the coast to coastal processes, and thus marks the initiation of recorded RSL. Its elevation is a function of the distance from the ice margin, and thus ice thickness, as well as the rate at which the glacier retreated (Andrews, 1970).

Previous work on the postglacial sea level history of the Labrador coast has focused on identifying the marine limit and describing lower strandline features (Daly, 1902; Tanner, 1944; Wenner, 1947; Andrews, 1961; Loken, 1962, 1964; Johnson, 1969, 1985; Smith, 1969; Fitzhugh, 1972; Fulton and Hodgson, 1979; Clark, 1984; Bell, 1987). These observations of raised strandlines have been supplemented and expanded by our own observations along the coast.

The marine limit is best dated directly by radiocarbon ages on geological material, usually shell, found in deposits at or close to the marine limit. However, along most of the Labrador coast, shells are absent in raised marine deposits, and those that exist and have been dated are generally associated with deposits that have no direct relationship to the marine limit.

We have estimated the age of the marine limit by using a physical model that extrapolates that portion of each RSL curve constrained by radiocarbon ages to the altitude of the marine limit. The basis for this hypothesis is in assuming the similarity of these curves to RSL curves from most of subarctic and arctic Canada and west Greenland, which are characterized by a smoothly decelerating exponential form from the marine limit (Andrews, 1968, 1970; Ten Brink, 1974). This assumption is supported by the forms of RSL curves from the Labrador coast, which are exponential where dated (see below). Thus, we suggest that the curves maintain their exponential form to the marine limit.

Because the forms of most Labrador RSL curves are constrained by radiocarbon ages on archaeological sites, estimated ages of the marine limit for most Labrador RSL curves based on the hypothesis of exponential form for the curves are considered *minimum* estimates. However, various independent estimates for age of deglaciation from several areas of the Labrador coast where we have constructed RSL curves support the estimated ages for marine limit (cf., Awadallah and Batterson, in press).

ISOBASE MAPS

Isobases are lines of equal amounts of emergence (positive) or submergence (negative), with respect to present sea level, that has occurred across an area over the same interval of time. Isobase maps, therefore, illustrate in three dimensions the amount of crustal deformation associated with isostatic loading and unloading of ice sheets.

Isobases may be constructed directly from sea level indicators that have the same age (e.g., Blake, 1970; Dyke, 1984). Such coverage of synchronously dated shorelines, however, is often fortuitous. Therefore, an alternative method that does not rely directly on the stratigraphic record is to use synchronous shoreline elevations as taken from RSL curves (Andrews, 1970).

Isobase maps are of great interest to archaeological studies in showing the elevation of a deformed surface for any particular interval of time. They can thus be used to predict the minimum elevation of any archaeological site of the same age as the reconstructed isobase system (cf., Andrews et al., 1971). Furthermore, the maps identify the possible difference in elevation of widely separated, but contemporaneous, sites.

RSL HISTORY OF THE LABRADOR COAST

The ten RSL curves presented here (Figure 7) provide representative coverage of the RSL history for much of the Labrador coast. Major gaps in coverage are in the area between southernmost Labrador (the Pinware area, 51° N) and Groswater Bay (54° N). With the available coverage, however, isobase maps are constructed which illustrate the RSL history for this area as interpolated from areas of known RSL history.

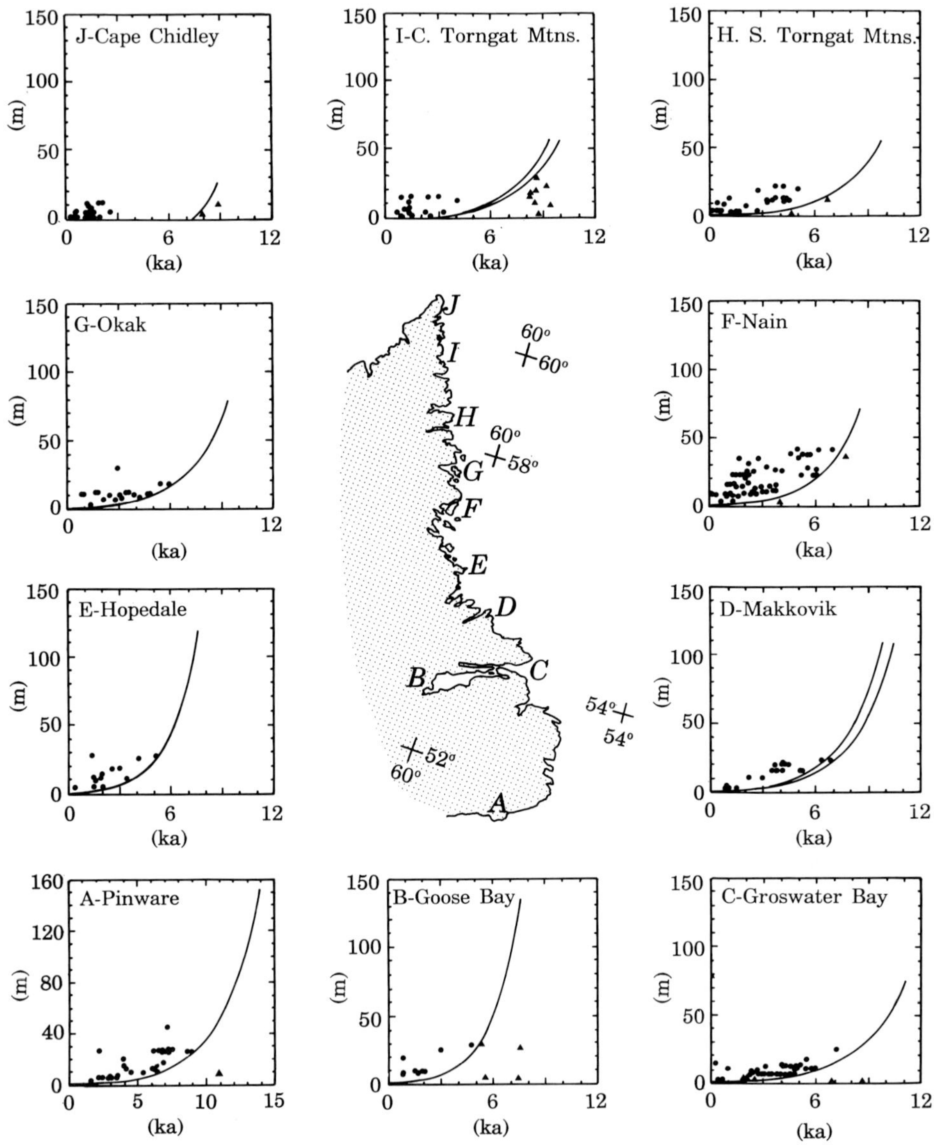

Figure 7. RSL curves from the Labrador coast. Solid circles represent radiocarbon ages on archaeological material, primarily charcoal; solid triangles are radiocarbon ages on fossil shell. Letters correspond to those used on location map (Figure 4).

The estimated minimum age of the marine limit and deglaciation is 14,000 B.P. for the Pinware area on the southernmost Labrador coast (Figure 7A; Table I). The elevation of the marine limit (152 m asl; Prest et al., 1968) is the highest measured on the coast. The estimated minimum age of the marine limit and deglaciation of the Goose Bay area is 7550 B.P. (Figure 7B; Table I), indicating a relatively slow ice retreat from the southern coast to inner Lake Melville.

On the Labrador coast north of Groswater Bay, the estimated minimum age of the marine limit decreases, while its altitude increases, north from Groswater Bay (11,000 B.P., 75 m asl) (Figure 7C), to the Makkovik area (9700 to 10,500 B.P., 107 m asl) (Figure 7D), to the Hopedale area (7600 B.P., 119 m asl) (Figure 7E) (Table I). North of the Hopedale area, the estimated minimum age of the marine limit generally increases, while its altitude decreases. In the coastal Nain area, the marine limit (71 m asl) has an estimated minimum age of 8500 B.P. (Figure 7F). In the Okak area, the marine limit (79 m asl) has an estimated minimum age of 9400 B.P. (Figure 7G). In the southern Torngat Mountains, including the Hebron and Saglek Fiord areas, the marine limit (55 m asl) has an estimated minimum age of 9700 B.P. (Figure 7H). In the central Torngat, including the Nachvak, Kangalaksiorvik, and Eclipse Channel fiord areas, the marine limit (56 m asl) has an estimated age of 9400 to 10,000 B.P. (Figure 7I). Along the northernmost Labrador coast, the marine limit (16 to 26 m asl) has an estimated minimum age of 9000 B.P. (Figure 7J).

Labrador RSL curves record the emergence of the coast to the present time, except for the RSL curves from the central and northern Torngat Mountains, where emergence has been followed by submergence (Figure 7). Rates of emergence have varied along the coast, and the vertical resolution and chronological distinction of archaeological sites as a function of elevation will thus also vary. The degree of vertical resolution expected for sites can be directly compared by evaluating the decay constants (''k'' values) that define the form of each curve (Table I).

The greatest amount of emergence (152 m) has occurred in the Pinware area of the southern Labrador coast. Because deglaciation was relatively early (14,000 years B.P.), however, most emergence was completed early (70% in the first 3000 years), and the RSL at 10,000 B.P. was only 35 m asl, and the RSL at 5000 B.P. was only 5 m asl (Figure 7A). Consequently, resolving the relative age of much of the archaeological record by altitude, particularly for younger sites, will be difficult for this region.

The Goose Bay area has also undergone significant emergence (135 m), but because the estimated minimum age of the marine limit is only 7550 B.P., the rate of emergence has been the fastest of the recorded RSL histories from Labrador, as indicated by the high ''k'' value (Table I). A high degree of vertical separation of sites of differing age is therefore expected for this region, as identified by Fitzhugh (1972).

The Pinware and Goose Bay RSL curves represent the slowest and fastest rates of emergence on the Labrador coast, respectively. The form and timing of the other eight RSL curves can be compared to the Pinware and Goose Bay curves with respect to their ''k'' values; three groups of RSL curves of similar form and decay rates are identified (Table I; Figure 8). RSL curves from the Groswater Bay, southern Torngat, and central Torngat Mountains have similar ''k'' values, and thus have had similar rates of emergence, as the RSL curve from the Pinware area (group III, Figure 8), although the latter part of the RSL from the central Torngat Mountains has been submergent (Figure 7I). The RSL curve from the Hopedale area of the central Labrador coast is similar in form and timing to the RSL curve from the Goose Bay area (group I, Figure 8). The RSL curves from Makkovik, Nain, and Okak are intermediate in form and rate of emergence, with ''k'' values of 0.47 to 0.5

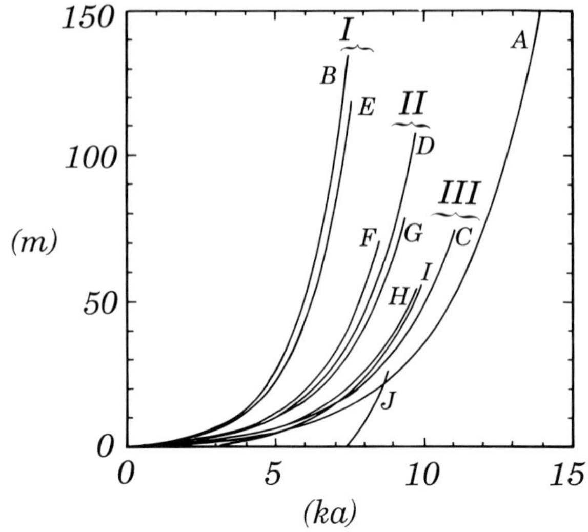

Figure 8. Summary figure of ten RSL curves from Labrador coast. Letters correspond to those from Figure 7. Curves with similar form are grouped as: group I = curves B and E; group II = curves F, D, and G; group III = curves H, I, C, and A.

(Table I; group II, Figure 8). The RSL curve from the Cape Chidley area, northernmost Labrador, is distinguished from all other curves in representing primarily submergence (Figures 7J and 8).

These data indicate that areas with similar RSL histories, as characterized by their decay constants, will have contemporaneous archaeological sites at similar elevations. Those regions with more rapid rates of emergence will have contemporaneous sites at higher elevations than areas which have experienced slower rates of emergence. Furthermore, evaluation of rates of emergence delimits the expected range of vertical resolution of sites from any given area.

RSL curves indicate that submergence has characterized the central Torngat Mountains for the last 2500 years, and for perhaps the last 7300 years on northernmost Labrador (Figures 7I and 7J). As noted previously, this identification of submergence is critical to interpreting the potential consequences to the archaeological record, as submergence may result in the loss of archaeological components. Under these conditions, one expects multicomponent sites as a result of isomorphic habitation and hunting requirements and reuse of earlier winter house sites by later peoples taking advantage of already-excavated house pits and insulating house walls. One also expects to find large-scale site destruction due to submergence and erosion. All of these predictions are upheld, increasingly, along the Torngat coast from Saglek to Killinek. However, good comes with bad; these conditions provide for increased protection of site deposits as early materials are buried and insulated by later deposits, allowing for better organic preservation and physical protection, until the erosion front arrives. On the other hand, such multicomponent sites rarely avoid disturbance as later peoples enlarge earlier house pits for their own use. The consequences of emergent coasts are similarly mixed. Beneficial aspects include enhanced resolution of settlement patterns and lack of component mixing as shoeline displacement results in shifts in settlement location, while negative aspects include the loss of physical and thermal protection of overlying layers.

ISOBASE MAPS

Isobase maps of the Labrador coast (Figure 9) are based on elevation estimates of RSL from the ten RSL curves discussed here and one RSL curve from southeastern Ungava Bay constructed by Allard et al. (1989). The isobase maps identify deformed surfaces at 1000-year intervals from 7000 to 3000 B.P. Surface deformation reflects differential loading and unloading of the earth by the Laurentide Ice Sheet. The deformed surfaces illustrated by the isobase maps are of importance in identifying the spatial and temporal variation in archaeological site elevations.

7000 B.P.

The map at 7000 B.P. illustrates that isobases are generally parallel to the coast (Figure 9). The 0-m isobase is located seaward of the coast except over northernmost Labrador,

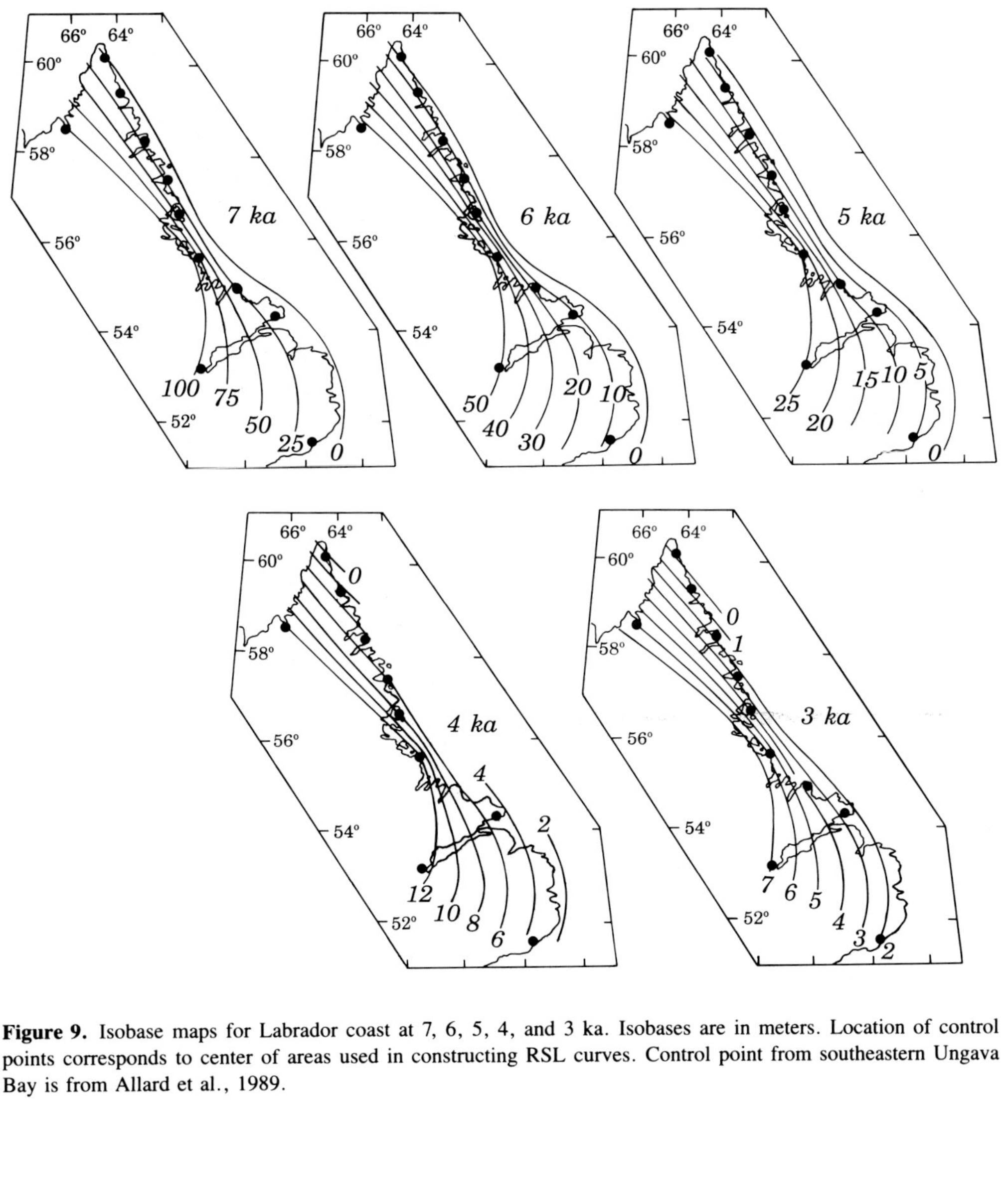

Figure 9. Isobase maps for Labrador coast at 7, 6, 5, 4, and 3 ka. Isobases are in meters. Location of control points corresponds to center of areas used in constructing RSL curves. Control point from southeastern Ungava Bay is from Allard et al., 1989.

where it crosses the peninsula toward Hudson Strait. RSL curves from southern Baffin Island, outer Hudson Strait, also identify the RSL at ≤0 m at this time (Andrews and Miller, 1985; Stravers, 1986), suggesting that the 0-m isobase crosses Hudson Strait to southern Baffin Island.

Rates of emergence have been greatest in the Hopedale area of the central Labrador coast, reflecting the relatively late age of deglaciation and perhaps greater ice thickness over this region. Rates of emergence decrease and age of deglaciation increases north and south of the Hopedale area. Consequently, isobases converge over the central coast and diverge to the north and south. This pattern indicates that the greatest spatial variability in elevation of sites will be along this portion of the central Labrador coast.

The isobase map shows the marked spatial variability in elevation of archaeological sites from the Groswater Bay west along Lake Melville to the Goose Bay area, as identified by Fitzhugh (1972). More specifically, 7000-year-old sites in the Groswater Bay area will be 75 to 100 m lower than contemporaneous sites in the Goose Bay area.

6000 B.P.

The map at 6000 B.P. (Figure 9) shows relative trends in RSL similar to those from 7000 B.P., but by 6000 B.P., absolute values of RSL for any region are lower. Isobases parallel much of the coast, and the 0-m isobase remains offshore except over northernmost Labrador. Isobases converge over the central coast, and the highest sites along the coast north of Groswater Bay will be in the Hopedale area.

In general, 6000-year-old sites along the central and northern Labrador coast which have a direct association with former sea level will be 15 to 20 m lower than 7000-year-old sites. In the Goose Bay and Hopedale areas, however, with their greater ice buildup and later deglaciation, 6000-year-old sites may be as much as 50 m lower than 7000-year-old sites.

Andrews and Tyler (1977) previously constructed a 6000-B.P. isobase map for eastern Canada, including Labrador. Coverage of RSL history for Labrador was restricted to the two preliminary RSL curves constructed by Fitzhugh (1972, 1973) for the Groswater Bay and Goose Bay areas of Lake Melville. Andrews and Tyler (1977) show the 10-m isobase in a position similar to ours along the southern Labrador coast. However, they have drawn their 20-m isobase roughly parallel to our 40-m isobase, except for the central Labrador coastal area, where the difference would be greater.

5000 to 3000 B.P.

The general pattern of isobase orientation remains much the same during this interval, but the altitude of the isobases decreases with each increment (Figure 9). Isobases diverge north and south of the central coast (Hopedale region), and maximum elevation for any given surface is in the Hopedale and Goose Bay regions. The spatial variability in elevation between sites decreases with time, however, such that by 3000 B.P., sites in the Hopedale and Goose Bay regions will only be up to 6 to 7 m above contemporaneous sites elsewhere along the coast.

SUMMARY

We can summarize the spatial and temporal variability in elevation of archaeological sites by representing the RSL data on the Labrador coast north of the Groswater Bay area

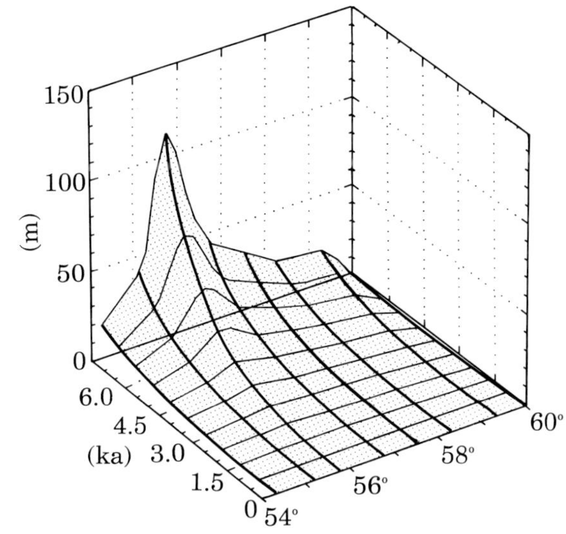

Figure 10. Schematic summary figure showing three-dimensional coastal RSL change for the last 7500 years from latitude 54 to 60° N. Greatest emergence is shown for the Hopedale region (central Labrador coast), with emergence decreasing north and south.

(54° N) as a smoothed, three-dimensional surface (Figure 10). This schematic illustration allows the observer to evaluate the change in RSL with time for one specific area and to identify the possible variation in contemporaneous RSL from one region to another. This perspective clearly identifies the greatest emergence over the Hopedale region (55° 45′), with decreasing rates of emergence to the north and south. Furthermore, it identifies the decreasing variability in site elevation between regions from 7500 B.P. to the present.

RSL HISTORY AND INTERPRETATION OF THE ARCHAEOLOGICAL RECORD

The relationship between RSL history and the archaeological record of the Labrador coast adds to a growing documentation of the relationship between cultural change and environment (Fitzhugh, 1972, 1973; Fitzhugh and Lamb, 1985). Most previous arctic and subarctic archaeological studies of presumed interaction of culture and environment have focused on models of climatic change and its influence on cultural distribution and diversity (e.g., McGhee, 1969/1970; Dekin, 1972; Barry et al., 1978; Fitzhugh, 1972; Fitzhugh and Lamb, 1985). The RSL history of the Labrador coast adds an important dimension, however, to interpreting archaeological events. Furthermore, it provides a fundamental framework for interpreting the spatial and temporal relationships of the former maritime cultures that inhabited coastal Labrador.

Age of deglaciation, as identified by the estimated age of the marine limit, provides a maximum age for potential occupation by prehistoric people of glaciated coasts. The age of deglaciation of the southern Labrador coast from Pinware to the Groswater Bay area was 11,000 to 14,000 B.P., suggesting this region was available to prehistoric people when they first arrived in the area. The earliest ages of human settlement in Labrador (about 8000 B.P.) are from the Strait of Belle Isle region, including the Pinware area (McGhee and Tuck, 1975). Early Maritime Archaic peoples arrived in Groswater Bay by 7500 B.P.

The ice margin, however, remained in western Lake Melville at this time (Jordan, 1975), as early Maritime Archaic people in Groswater Bay must have been aware. The age of marine limit on the central coast (Hopedale region) suggests that deglaciation was 7500 B.P., the same age as in the Goose Bay region. Since this also is the age of early settlement of Groswater Bay and Nain, where deglaciation preceded Hopedale by 1000 years, Clark and Fitzhugh (1990) proposed that an ice barrier may have been a significant factor in the movement of early Maritime Archaic peoples north of the central coast (Figure 11). Previously, the control factor was thought to be related to the northward movement of the shrub-tundra zone to which the Maritime Archaic people had become adapted (Fitzhugh and Lamb, 1985). The coincidence of the appearance of shrub tundra, Maritime Archaic culture, and retreat of ice from the central coast further refines our understanding of the paleoecology of the Labrador coast, and gives a glacier margin-adapted, Greenlandic Eskimo-flavored cast to early Indian adaptations in the far northeast.

The RSL history of the Labrador coast also provides a means for reconstructing coastal paleogeography in some detail. Significant changes in paleogeography occurred along many parts of the Labrador coast which are characterized by low elevations and low relief. Fitzhugh (1972, 1973) reconstructed the paleogeography of the Lake Melville and Groswater Bay areas (e.g., Figure 3), and concluded that paleogeographic changes had a significant impact on local hunting and fishing strategies, migration paths taken by large marine mammals, and human settlement patterns. The greatest impact was in the Maritime Archaic period when sites had to be moved frequently to accommodate rapidly changing local geographic conditions. Probably for this reason, Maritime Archaic sites rarely contain more than a single cultural component and different-age components are widely separated spatially. Furthermore, these sites exhibit a regular pattern of eastward-shifting site location through time as land rise closed off water access to earlier sites and created more advantageous locations for settlement and boat launching in more seaward positions, where marine resources tended also to be more available. Sites of later periods, after 4000 years ago, often contain multicomponent occupations with better organic preservation.

Another implication of our work comes from the fact that many of the earliest and highest Maritime Archaic sites are located on islands of the outer coast. Many of these islands are small and would have been even smaller and more exposed when these sites were occupied from 7500 to 4000 B.P. and RSLs were as much as 20 to 40 m higher than today. Occupation of such exposed locations by early-Maritime Archaic peoples in summer would be surprising, especially as later Maritime Archaic groups generally did not use these seaward locations. Since it is unlikely that the earliest Maritime Archaic groups were better adapted to outer coast life than the more highly developed and more populous late-Maritime Archaic groups, a solution to this problem may lie in settlement-pattern shifts. Probably these sites were not used in summer, but for spring ice-edge hunting, as practiced by ethnographic Inuit peoples.

The RSL history and isobase maps demonstrate that emergence has characterized all of the Labrador coast except for the northernmost area, where submergence has been occurring for up to the last 7000 years. Thus, a complete archaeological sequence should be preserved along all but the northernmost coast, where submergence has resulted in loss of up to 7000 years of potential record. Submergence is the probable cause for the rarity of Paleoeskimo (pre-Dorset and Groswater) sites earlier than 2500 years B.P. in the Killinek region, and for the rarity of early-Paleoeskimo (pre-Dorset) sites between McClelan Strait and Saglek Fiord, except for sites on elevated terraces and other sites not closely associated with former sea level. Where early-Paleoeskimo sites have been found, as at Nunaingok (Fitzhugh, 1980), they are found under active erosion at current sea level and appear to be

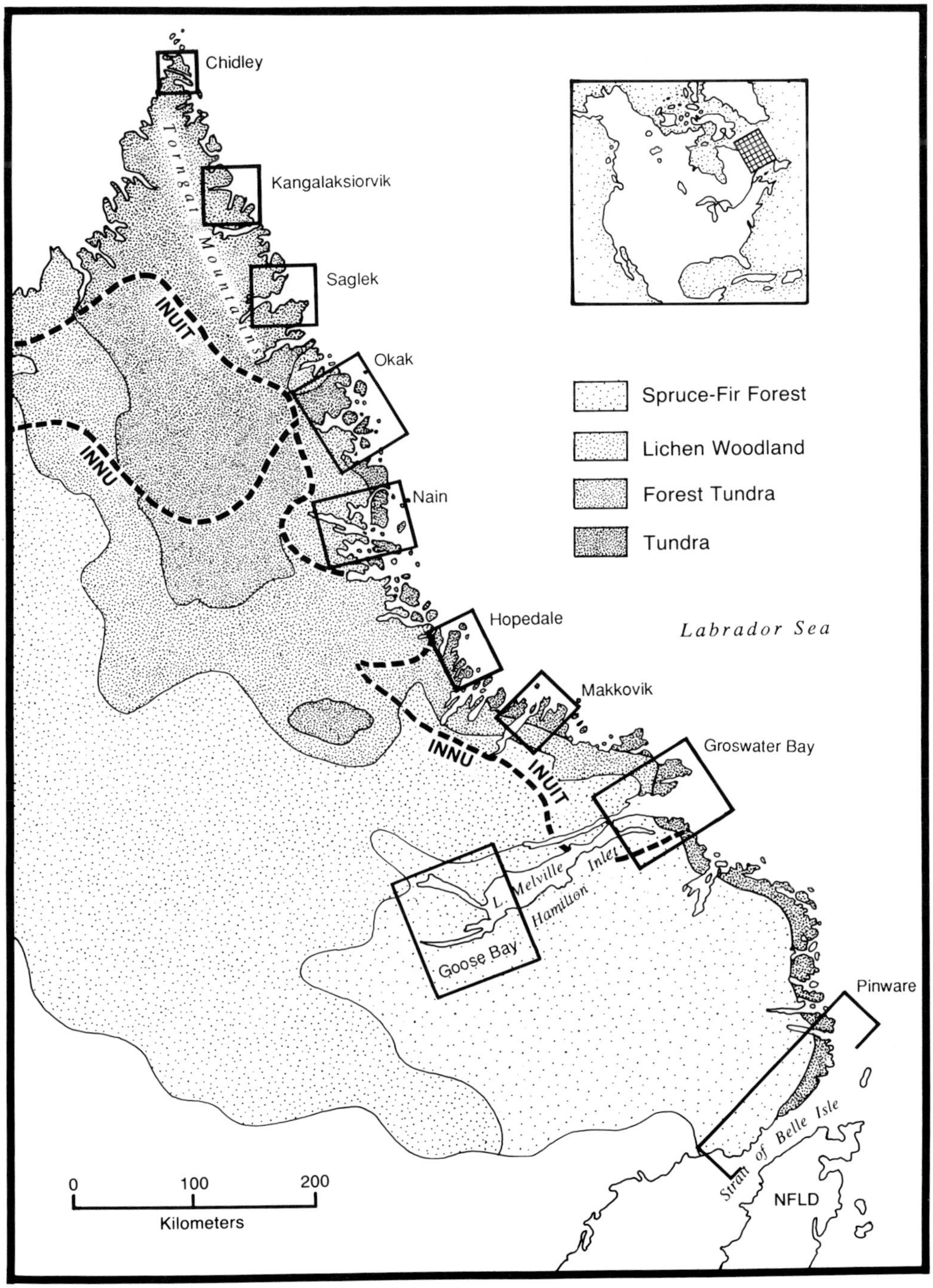

Figure 11. Modern vegetation zones and culture distributions in Labrador-Quebec (after Fitzhugh and Lamb, 1985). Outlined areas correspond to areas where RSL curves have been constructed.

the remnants of much larger sites that have already been lost to submergence-induced erosion, and not simply to erosion of geologically stable shorelines (e.g., Plummet and Gangloff, 1990).

The final aspect of the RSL history of the Labrador coast that has important implications for interpreting the archaeological record is the spatial and temporal variability of archaeological sites with respect to elevation above present sea level. Because the prehistoric Labrador people were active seafarers, identification of the height of former shorelines serves as an important first-order guide for the field archaeologist searching for sites. As a result of our studies, it is now possible to predict elevation ranges for all of the Labrador cultures now known, corrected for differential uplift by region. We have identified those regions where the greatest emergence has occurred over the time interval of interest to the archaeologist (≤8000 B.P.) as the Goose Bay and Hopedale regions. Archaeological sites of any given age will be found at the highest elevations in these areas in comparison to other regions of the coast. While better definition will always be possible for the earlier sites wherever they occur, these highly and rapidly emergent areas contain the greatest potential for vertical separation of archaeological components of all ages.

ACKNOWLEDGMENTS

Grateful acknowledgment is given to the support of the Smithsonian Institution, the Newfoundland Historic Resources Division/Newfoundland Museum, the National Science Foundation (Grant DPP-8922205 to Clark), and the Labrador people. We thank Steve Cox, Jim Tuck, and Callum Thomson for providing unpublished radiocarbon ages and site information, and Tore Vorren for helpful discussions. Special thanks are due to Robert Stuckenrath, who dated many of the samples on which this paper is based, Dosia Laeyendecker, who identified and processed many of the charcoal samples, and Julie Permutter who provided Figures 1 and 5.

REFERENCES

Allard, M., Fournier, A., Gahe, E., and Seguin, M.K. (1989) Le Quaternaire de la cote sudest de la Baie d'Ungava, Quebec nordique. *Geogr. Phys. Quat.* 43: 325–336.

Andrews, J.T. (1961) The glacial geomorphology of the northern Nain-Okak section of Labrador. M.S. thesis, McGill University, Montreal, Quebec.

Andrews, J.T. (1968) Postglacial rebound in Arctic Canada: similarity and prediction of uplift curves. *Can. J. Earth Sci.* 5: 39–47.

Andrews, J.T. (1970) *A Geomorphological Study of Post-Glacial Uplift, with Particular Reference to Arctic Canada.* Spec. Publ. 2, Institute of British Geographers.

Andrews, J.T. and Miller, G.H. (1985) Holocene sea level variations within Frobisher Bay. In: *Quaternary Environments: Eastern Canadian Arctic, Baffin Bay and Western Greenland,* pp. 585–607. Andrews, J.T., Ed. Allen and Unwin, Boston.

Andrews, J.T. and Tyler, K. (1977) The observed postglacial recovery of Quebec and nouveau-Quebec since 12,000 BP. *Geogr. Phys. Quat.* 31: 389–400.

Andrews, J.T., McGhee, R., and McKenzie-Pollack, L. (1971) Comparisons of elevations of archaeological sites and calculated sea levels in arctic Canada. *Arctic.* 24: 210–228.

Arundale, W. (1981) Radiocarbon dating in eastern Arctic Archeology: a flexible approach. *Am. Antiq.* 46: 244–271.

Awadallah, S.A. and Batterson, J.J. (in press) Comment on "Late deglaciation of the central Labrador coast: implications for the age of glacial lakes Naskaupi and McLean and prehistory". *Quat. Res.*

Barry, R.G., Arundale, W.T., Bradley, R., Andrews, J.T., and Nichols, H. (1978) Environmental change and cultural change in eastern Canada Arctic during the last 5,000 years. *Arct. Alp. Res.* 9: 193–210.

Bell, T.J. (1987) Quaternary geomorphology, glacial history and relative sea level change in outer Nachvak Fiord, northern Labrador. M.Sc. thesis, Memorial University, St. John's.

Bird, J. (1945) Archaeology of the Hopedale area, Labador. *Anthropol. Pap. Am. Mus. Nat. Hist.* 39(2).

Bird, J.B. (1954) Postglacial marine submergence in central Arctic Canada. *Geol. Soc. Am. Bull.* 65: 457–464.

Bird, J.B. (1969) *The Physiography of Arctic Canada.* Johns Hopkins University Press, Baltimore.

Blake, W., Jr. (1970) Studies of glacial histories in Arctic Canada. I. Pumice, radiocarbon dates, and differential postglacial uplift in the eastern Queen Elizabeth Islands. *Can. J. Earth Sci.* 7: 634–664.

Clark, J.A., Farrell, W.E., and Peltier, W.R. (1978) Global changes in postglacial sea level: a numerical calculation. *Quat. Res.* 9: 265–287.

Clark, P.U. (1984) *Glacial Geology of the Kangalaksiorvik-Abloviak Region, Northern Labrador, Canada.* Ph.D. thesis, University of Colorado, Boulder.

Clark, P.U., Short, S.K., Williams, K.M., and Andrews, J.T. (1989) Late Quaternary chronology and environments of Square Lake, Torngat Mountains, Labrador. *Can. J. Earth Sci.* 26: 2130–2144.

Clark, P.U. and Fitzhugh, W.W. (1990) Late deglaciation of the central Labrador coast: implications for the age of glacial lakes Naskaupi and McLean and prehistory. *Quat. Res.* 34: 296–305.

Cox, S.L. (1977) Prehistoric Settlement and Culture Change at Okak, Labrador. Ph.D. thesis, Department of Anthropology, Harvard University, Cambridge. MA.

Cox, S.L. (1978) Paleo-Eskimo occupations of the north Labrador coast. *Arct. Anthropol.* 15: 96–118.

Daly, R.A. (1902) The geology of the northeast coast of Labrador. *Bull. Mus. Comp. Zool. Harv. Univ.* 38:205–269.

Dekin, A. (1972) Climatic change and cultural change: a correlative study from eastern Arctic prehistory. *Polar Notes.* 12: 11–31.

Dyke, A.S. (1979) Radiocarbon-Dated Holocene Emergence of Somerset Island, Central Canadian Arctic, pp. 307–318. Geol. Surv. Can. Pap. 79–1B.

Dyke, A.S. (1984) Quaternary Geology of Boothia Peninsula and Northern District of Keewatin, Central Canadian Arctic. Geol. Surv. Can. Mem. 407.

Fairbanks, R.G. (1989) A 17,000 year glacio-eustatic sea level record: influence of glacial melting rates on the Younger Dryas event and deep-ocean circulation. *Nature.* 343: 637–642.

Farrell, W.E. and Clark, J.A. (1976) On postglacial sea level. *Geophys. J. R. Astron. Soc.* 46: 647–667.

Fitzhugh, W.W. (1972) *Environmental Archeology and Cultural Systems in Hamilton Inlet, Labrador.* Smithsonian Contrib. Archaeol. 16. Washington, D.C.

Fitzhugh, W.W. (1973) Environmental approaches to the prehistory of the north. *J. Wash. Acad. Sci.* 63: 39–53.

Fitzhugh, W.W. (1977) Population movement and culture change on the central Labrador coast. *Ann. N.Y. Acad. Sci.* 288: 481–497.

Fitzhugh, W.W. (1978) Maritime Archaic cultures of the central and northern Labrador coast. *Arct. Anthropol.* 15: 61–95.

Fitzhugh, W.W. (1980) Preliminary report on the Torngat archeological project. *Arctic.* 33: 585–606.

Fitzhugh, W.W. (1983) Residence pattern development in the Labrador Maritime Archaic. In: *Archeology in Newfoundland and Labrador.* Vol. 4: 6–47. Thomson, J.S. and Thomson, C., Eds. Government of Newfoundland and Labrador, Historic Resources Division.

Fitzhugh, W.W. (1985) The Nulliak pendants and their relation to spiritual traditions in northeastern prehistory. *Arct. Anthropol.* 22: 87–109.

Fitzhugh, W.W. (1987) Archaeological ethnicity and the prehistory of the Labrador coast. In: *Ethnicity and Culture,* pp. 141–153. Auger, R., Glass, M.F., MacEachern, S., and McCartney, P., Eds. Archaeological Association, University of Calgary.

Fitzhugh, W.W. and Lamb, H. (1985) Vegetation history and culture change in Labrador prehistory. *Arct. Alp. Res.* 17: 357–370.

Fulton, R.J. and Hodgson, D.A. (1979) Wisconsin Glacial Retreat, Southern Labrador, pp. 17–21. Geol. Surv. Can. Pap. 79–1C.

Giddings, L. and Anderson, D. (1986) Beachridge Archeology of Cape Krusenstern. Prehistoric Settlements Around Kotzebue Sound, Alaska. Publ. Archaeol. 20, U.S. Department of the Interior, National Park Service, Washington, D.C.

Gjessing, G. (1942) *Yngre steinalder i Nord-Norge.* Instituttet for Sammenlignende Kulturforskning, Ser. B, 39. Oslo.

Harp, E., Jr. (1963) Evidence of Boreal Archaic culture in southern Labrador and Newfoundland. *Natl. Mus. Can. Bull.* 193: 184–261.

Harp, E., Jr. and Hughes, D. (1968) Five prehistoric burials from Port aux Choix, Newfoundland. *Polar Notes.* 8: 1–47.

Helskog, K. (1974) Two tests of the prehistoric cultural chronology of Varanger, north Norway. *Norw. Archaeol. Rev.* 7(2): 97–103.

Ives, J.D. (1978) The maximum extent of the Laurentide ice sheet along the east coast of North America during the last glaciation. *Arctic.* 31: 24–53.

Johnson, J.P. (1969) Deglaciation of the central Nain-Okak Bay section of Labrador. *Arctic.* 22: 373–394.

Johnson, J.P. (1985) Geomorphic evidence and the upper marine limit in Northern Labrador. In: *Our Geographic Mosaic: Research Essays in Honour of G.C. Merrill,* pp. 70–91. Knight, D.B., Ed. Carleton University Press, Ottawa.

Jordan, R.H. (1975) Pollen diagrams from Hamilton Inlet, Labrador, and their environmental implications for the northern Maritime Archaic. *Arct. Anthropol.* 12(2): 92–116.

Jordan, R.H., Ed. (1978) Selected papers from a symposium on central Labrador archeology. *Arct. Anthropol.* 15(2).

Knuth, E. (1967) *Archaeology of the Musk-ox Way.* Ecole Pratique des Hautes Etudes. Contributions du Centre d'Etudes Arctiques et Finno-Scandinaves 5, Paris.

Lamb, H. (1980) Late Quaternary vegetational history of southeastern Labrador. *Arct. Alp. Res.* 12: 117–135.

Lamb, H. (1984) Modern pollen spectra from Labrador and their use in reconstructing Holocene vegetation history. *J. Ecol.* 72: 37–59.

Lamb, H. (1985) Palynological evidence for post-glacial change in the position of the tree limit in Labrador, eastern Canada. *Ecol. Monogr.* 55: 241–258.

Leechman, D. (1943) Two new Cape Dorset sites. *Am. Antiq.* 8(4): 363–375.

Lloyd, T.G.B. (1874) Notes on Indian remains found on the coast of Labrador. *J. R. Anthropol. Instit. G. B. Ir.* 4: 39–44.

Loken, O.H. (1962) The late-glacial and postglacial emergence and the deglaciation of northernmost Labrador. *Geogr. Bull.* 17: 23–56.

Loken, O.H. (1964) A Study of the Late and Postglacial Changes of Sea Level in Northern Labrador. Unpubl. rep., Arctic Institute of North America.

McGhee, R. (1969/1970) Speculations on climatic change and Thule culture development. *Folk.* 11/12: 173–184.

McGhee, R. and Tuck, J.A., Jr. (1975) An Archaic Sequence from the Strait of Belle Isle, Labrador. Archaeol. Surv. Can. Pap. (Mercury Ser.) 34.

McGhee, R. and Tuck, J.A., Jr. (1976) Undating the Canadian Arctic. *Am. Antiq. Mem.* 31: 6–14.

Maxwell, M. (1985) *Prehistory of the Eastern Arctic.* Academic Press, Orlando.

Nummedal, A. (1937) Yngre stenaldersfund fra Nyelv og Karlebotn I. In: *Universitetets Odlsaksamlings Arbok 1935–1936.* Oslo.

Perthes, B. de (1847) *Antiquites Celtiques et Antediluviennes.* Paris.

Plumet, P. and Gangloff, P. (1990) Contribution a l'Archeologie et l'Ethnohistoire de l'Ungava Oriental. Paleo-Quebec 19, Les Presses de l'Universite du Quebec, Quebec.

Prest, V.K., Grant, D.R., and Rampton, V.N. (1968) Glacial Map of Canada: Geol. Surv. Can. Map 1253A.

Quinlan, G. (1985) A numerical model of postglacial relative sea level change near Baffin Island. In: *Quaternary Environments: Eastern Canadian Arctic, Baffin Bay and Western Greenland,* pp. 560–584. Andrews, J.T., Ed. Allen and Unwin, Boston.

Schnitker, D. (1974) Postglacial emergence of the Gulf of Maine. *Geol. Soc. Am. Bull.* 85: 491–494.

Short, S.K. (1978) Palynology: a Holocene environmental perspective for archeology in Labrador-Ungava. *Arct. Anthropol.* 15(2): 9–35.

Short, S.K. and Nichols, H. (1977) Holocene pollen diagrams from subarctic Labrador-Ungava: vegetational history and climatic change. *Arct. Alp. Res.* 9: 265–290.

Smith, P.A.W. (1969) *Glacial Geomorphology of the Saglek Fiord Area of Northeastern Labrador,* pp. 115–123. McGill Sub-Arctic Res. Stn. Res. Pap. 24, McGill University, Montreal, Quebec.

Spiess, A., Bourque, B.J., and Gramly, R.M. (1983) Early and middle Archaic site distribution in western Maine. *North Am. Archaeol.* 4(3): 225–244.

Stravers, J.A. (1986) Glacial Geology of Outer Meta Incognita Peninsula, Southern Baffin Island, Arctic Canada. Ph.D. thesis, University of Colorado, Boulder.

Strong, W.D. (1930) A stone culture from northern Labrador and its relation to Eskimo-like cultures of the northeast. *Arct. Anthropol.* 32: 126–144.

Tanner, C. (1944) Outlines of the geography, life, and customs of Newfoundland-Labrador. *Acta Geogr.* 8: 1–906.

Ten Brink, N.W. (1974) Glacio-isostasy: new data from west Greenland and geophysical implications. *Geol. Soc. Am. Bull.* 85: 219–228.

Tuck, J.A., Jr. (1975) The northeastern maritime continuum. 8000 years of cultural development in the far northeast. *Arct. Anthropol.* 12(2): 139–147.

Walcott, R.I. (1972) Late Quaternary vertical movements in eastern North America. *Rev. Geophys. Space Phys.* 10: 849–884.

Wenner, C.G. (1947) Pollen diagrams from Labrador. *Geogr. Ann.* 29(3–4): 137–373.

Chapter

10

Space Shuttle Imagery of Recent Catastrophic Change Along the Arid Andean Coast

Michael E. Moseley
Department of Anthropology
University of Florida
Gainesville, Florida

David Wagner
Institute for Technology Development
Space Remote Sensing Center
NSTL Station, Mississippi

James B. Richardson, III
Division of Anthropology
Carnegie Museum of Natural History
Pittsburgh, Pennsylvania

INTRODUCTION

This paper discusses the use of space shuttle and other high-altitude imagery of the Santa River region of the Andean coast to monitor 4 decades of change in the littoral zone. The environment of the desert landscape is often regarded as extremely stable. This stability is reflected in minimal yearly temperature variation and in hyperarid conditions along the lower 45% + of the Pacific watershed. Here, there is normally no rainfall below elevations of 1500 to 2000 m. The barren landscape of coastal northern Peru is composed of eolian sand and unconsolidated alluvium banked against the igneous bedrock escarpment of the Cordillera (Figure 1). Drainages discharging seasonal runoff from high elevations transect the escarpment and create oases valleys irrigated by large canals that have supported agrarian life for more than 3 millennia.

Monitoring the Santa River littoral zone allows us to recognize modern analogs for two types of geological structures—beach ridges and dune fields. The Santa beach ridges have long attracted attention because they are thought to originate under catastrophic circumstances. Suggested causal agencies have included major tectonic events, severe El

Figure 1. A LFC photograph of the arid Peruvian coast crossed by irrigated river valleys beginning in the north (top) with the Moche Valley and descending south through the Viru, Chao, Santa, Lacramarca, and Nepeña Valleys. Spanning more than 160 km, this section of coast represents approximately half the area covered by a single LFC photograph.

Niño-Southern Oscillation (ENSO) events, or a combination of the two (Sandweiss, 1986). Our high-altitude imagery indicates that a new ridge arose in conjunction with a strong 1972 ENSO event, which followed a devastating 1970 earthquake. We are further able to show that beach-ridge building is one aspect of episodic coastal progradation that also gives rise to dune fields. Understanding the origin and movement of dunes is important because they can have disastrous cultural consequences. After summarizing an archaeological case of sand inundation, we will turn to the Santa River region to see how beach ridges and dune fields have arisen in recent times. The cultural implications of the formation processes will then be assessed.

ARCHAEOLOGICAL PROBLEMS

As archaeologists, our interest in aeolian and littoral processes results from several decades of encountering prehistoric remains in physical contexts reflecting geomorphic change. Change in ancient dune conditions was uncovered during 1972 excavations at Huaca del Sol and Huaca de la Luna located on the south side of the Moche Valley some 9 km inland from the sea. This is the northernmost valley in Figure 1. Bedrock appears dark in the photograph, whereas the unconsolidated sediment is lighter, and the lightest surfaces are the youngest or most active dune areas. Strong daily winds off the ocean transport sand inland until it comes to repose against the escarpment. At the time, we did not have the high-altitude imagery to recognize that eolian formations and active dunes are most prominent between drainages and on the south sides of coastal valleys. Southern valley sides, such as where Huacas Sol and Luna are situated, can now be identified as settings likely to experience sand incursions.

The Sol and Luna platform mounds, or huacas, have long been considered the capital of the ancient Moche state or polity that controlled much of the northern coast from about 100 B.C. to 600 A.D. Huaca del Sol is the biggest platform mound of solid mud brick ever erected in the continent, and its Luna counterpart is also of impressive size. These enormous mounds tower above a flat, featureless plain that lacks standing architecture, and which is largely devoid of cultural structures. The apparent dearth of ancillary architecture contributed to the impression that the platforms were isolated ceremonial structures built and used by vast numbers of people who resided elsewhere. This impression proved to be the product of sand having buried the surrounding city. Archaeological deposits, reaching 7 m in depth, were encountered in the plain between the two Huacas during the 1972 excavations. Deep cuts indicated that the area was settled by at least 100 B.C. Excavations revealed a very high density of elite buildings and residential structures of adobe, as well as high-status cemeteries. The remains reflect a lengthy occupation of an urban nature (Topic, 1982).

Between 500 and 600 A.D., when the prosperous metropolis was at its height, it experienced an extremely calamitous rainfall and large-scale flooding attributable to El Niño precipitation. Flood waters stripped as much as 4 m of deposit from some occupation surfaces. The scale of erosion was unprecedented in the long occupational and postoccupational history of the site. It is unclear whether the unusual magnitude of destruction reflects more than one El Niño event, or if the severity of erosion was exacerbated by prior tectonic activity. Whatever the case, the surviving population rebuilt the capital. Yet, abandonment ensued shortly due to sand encroachment. As dunes migrated into the city, people removed the roofs and the valued contents from the buildings in front of the advancing drifts. The adobe structures are in otherwise pristine condition due to their burial by yellow sands. By 600 A.D., all standing architecture was interred, except for the two enormous Huacas. Nearby canal systems and agricultural lands were also engulfed. Much of the lower valley south of the river was deserted as populations resettled inland and across the river to avoid the eolian drifts.

The ancient Moche capital is the most calamitous case of dune disaster that we have encountered. Yet it is not the only case of dunes burying prehistoric sites. Several kilometers upstream and upwind from Sol and Luna, a long low-bedrock ridge projects two km outward to the southern bank of the Moche River. The summit, flanks, and general region of the ridge, called Cerro Arena, supported a very extensive settlement of scattered buildings (Brennan, 1978). This settlement was abandoned more than 2 centuries before the Huacas were built. On the lee side of the ridge, there are several massive linear dunes. Masonry buildings near the dunes exhibit exceptionally well-preserved adobe mortar, plaster, and floors. It is not clear if sand incursion caused their abandonment. However, it is clear that all standing buildings with preserved plaster and mortar were interred by sand prior to the episode of severe ENSO erosion that struck Sol and Luna. Without the protection of entombment, the adobe would not have survived. Today, the dunes at Cerro Arena are in an active state, and the yellow sand is moving off of the former settlement, exposing once-buried remains. We do not know what triggers the reactivation of older eolian deposits. However, the reactivation complicates our understanding of the numbers of independent sand sweeps that repose against the escarpment of the Cordillera. Resting against the bedrock, there is a very old generation of stabilized drifts. These undated, unconsolidated deposits are heavily gullied and eroded. They are often covered with scree and rocky detritus from adjacent hills, and they comprise the darkest eolian deposits visible in Figure 1. In the Moche area, this old generation of dunes is in part overlain by inactive and active drifts of yellow sand that may be products of several separate incursions.

By late prehistoric times, the sands covering the settlement at Huacas Sol and Luna had stabilized, leveled, and had been lightly compacted by El Niño precipitation or ephemeral irrigation. The plain and the adjacent desert surfaces are now being crossed by occasional barcan dunes composed of grey sand. Traced upwind, the barcans arise as a moderate-sized dune field immediately behind the littoral on the south side of the valley mouth. Here the field of barcans overrides an earlier, much larger formation of stabilized linear dunes. Scattered Moche-phase ceramics occur on deflated surfaces between the old linears, but we cannot tell if the formation overrode the occupation, or if people resided between the dunes when they were relatively inactive. The linear dune field is composed of yellow sands similar in color and composition to the eolian deposits at Huacas Sol and Luna. Because this formation is crossed by grey barcans, we concluded that dune fields arose episodically along the shoreline.

Lacking modern analogs for the formation process, we have long been puzzled by the shoreline conditions that give rise to sand sweeps. If dune fields do not form under normal conditions, then catastrophic processes are seemingly implicated. If the dune fields arise behind the littoral zone, then the same processes that gave rise to the beach ridges should be involved. Yet, the origins of Andean beach ridges have been debated because they, too, lack analogs.

METHODS OF RESOLUTION

Observing beach ridges and dune fields during their course of formation would do much to clarify the origins of their ancient counterparts. In 1987, we inadvertently discovered that this was possible while working with the Figure 1 negative. Testing enlargement capabilities above 20 amplifications revealed a new beach ridge in the Santa River formation, which Moseley and Richardson then field verified in July. Initially, we thought the new structure was the product of the 1982–1983 ENSO event; however, it proved to be present in an aerial photograph predating the event, and this led us to develop a diachronic coverage for the region.

Space Shuttle Perspective

As Figure 1 reveals, the space shuttle offers a truly encompassing overview of the arid coast and the watershed. This perspective allows the movement of sediment to be viewed on a regional basis and elucidates long-distance relationships that are often obscure on the ground. The figure is a large-format camera (LFC) photograph taken in October 1984 on the camera's maiden voyage aboard the 13th mission of the space shuttle Challenger. Designed for precise stereophotography and photogrammetric mapping, the LFC has a unique lens. It combines a high resolution for small objects 20 to 30 m in length with a wide range of view that captures an area larger than the State of Massachusetts. A single 9 × 18 in. negative covers a very long section of the Peruvian coastline; yet many archaeological and geoarchaeological features are discernable, and this led to our involvement with the imagery. Excellent photographic quality allows for the sizable enlargements that accompany this chapter.

Diachronic Perspective

The 1984 LFC photographs revealed a new beach ridge, but we needed earlier coverage to determine when the structure arose. Although an annual succession of images was not available, Wagner developed a multitemporal coverage by digitizing and overlaying a sequence of eight high-altitude images of the Santa area that span 40 years, from 1944 to 1984. The sequence includes (1) aerial photographs from the Servicio Aerofotografico Nacional (SAN) of the Peruvian Air Force for the years 1944, 1955, 1961, 1970, and 1982, (2) Landsat MSS imagery for 1975 and 1978, and (3) LFC photography for 1984. Partial cloud cover and limited resolution made the two Landsat images the most difficult with which to deal.

The images were converted to a digital format at the National Space Technology Laboratory utilizing an Eikonix Eikonixscan 78/99 device composed of both a camera and an electronic scanner. The latter employs a moving linear array of photodiodes spaced 15 cm apart that registers an amount of voltage proportional to the amount of reflected light from the section of the photograph that a diode is viewing. Voltage registrations are then converted into an 8-bit digital matrix yielding gray scale values that range from 0 (black) to 255 (white). Digitized black and white LFC coverage was converted to false color images to enhance features of interest.

The eight digitized images were then overlaid and coregistered using modules of the Earth Resources Laboratory Analysis Software (ELAS). Coregistration employed ELAS georeferencing modules that allowed images of different scale to be rescaled to a common size. All images were overlaid to the 1955 aerial photograph because of its clarity and its previous use in the Santa beach-ridge literature (Sandweiss, 1986). Once overlain, a set of identifiable ground points, common to each image, was used to establish scale. Pixel resolution was determined by comparing the 1955 image with the Instituto Geografico Militar 1:2,500,000 topographic map of the region.

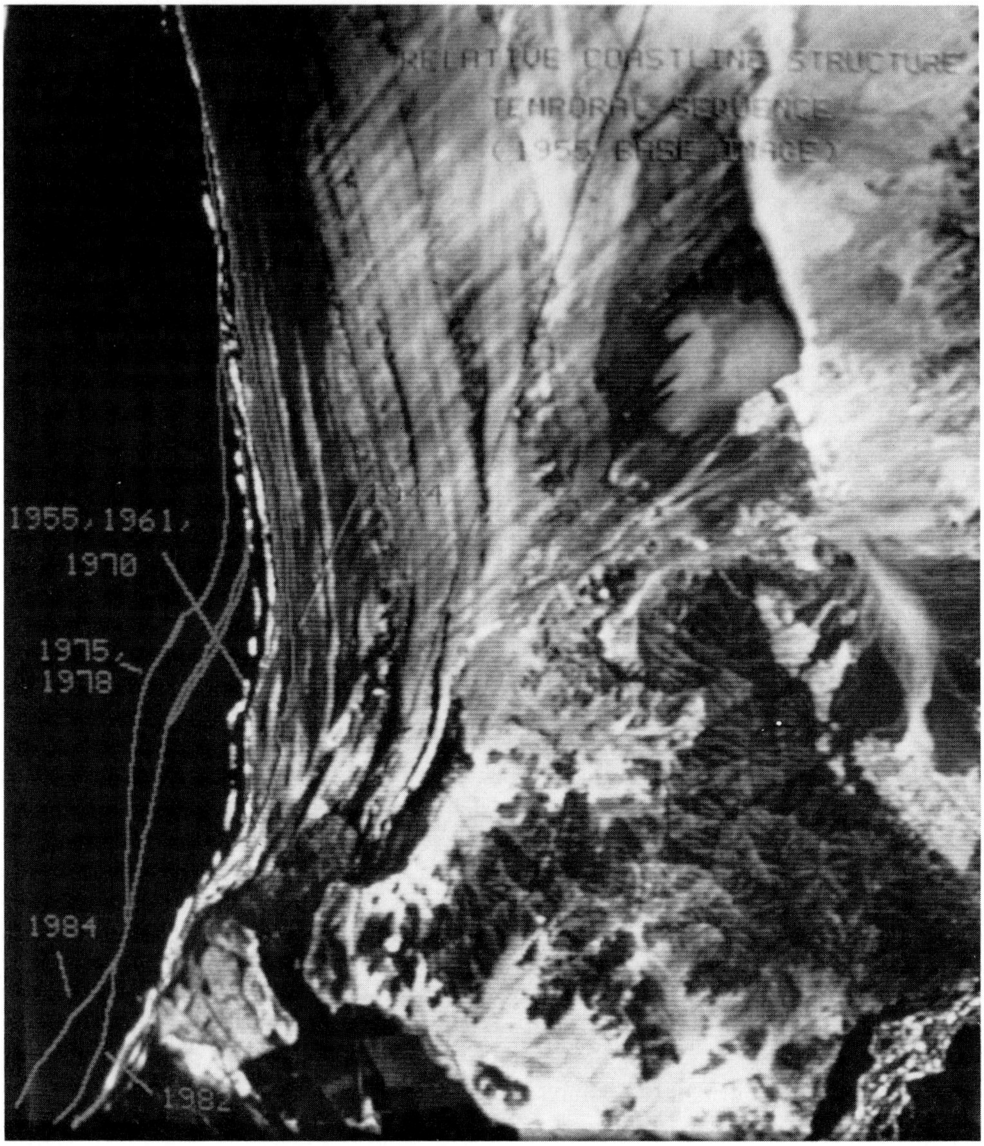

Figure 2. The 1955 SAN air photograph with overlay of seven high-altitude images of the Santa beach-ridge formation beginning in 1944 and ending in 1984.

This map was also used to scale coastal progradation revealed in the composite overlay of Figures 2 and 3. The maximum point of progradation is on a pronounced seaward bow of the new ridge. Perpendicular or east of this, an inland bedrock outcrop was used as a visual control point for measurement. From this point, the distance to the shoreline was calculated at 1906 m in 1955, and 2928 m in 1975. The scale of the topographic map makes these distances approximations. However, maximum progradation between 1955 and 1975 was on the order of 1 km.

AN OVERVIEW

LFC coverage allows sediment movement to be pictured on a regional scale. This overview reveals large erosional and depositional features that define the pathways along which material moves. Different types of pathways involve distinct mechanisms of transport. It is necessary to identify the major agents of transport in order to interpret the geomorphic change evident in the diachronic imagery. We can recognize five mechanisms. Three operate continuously under normal conditions, and two are stochastic.

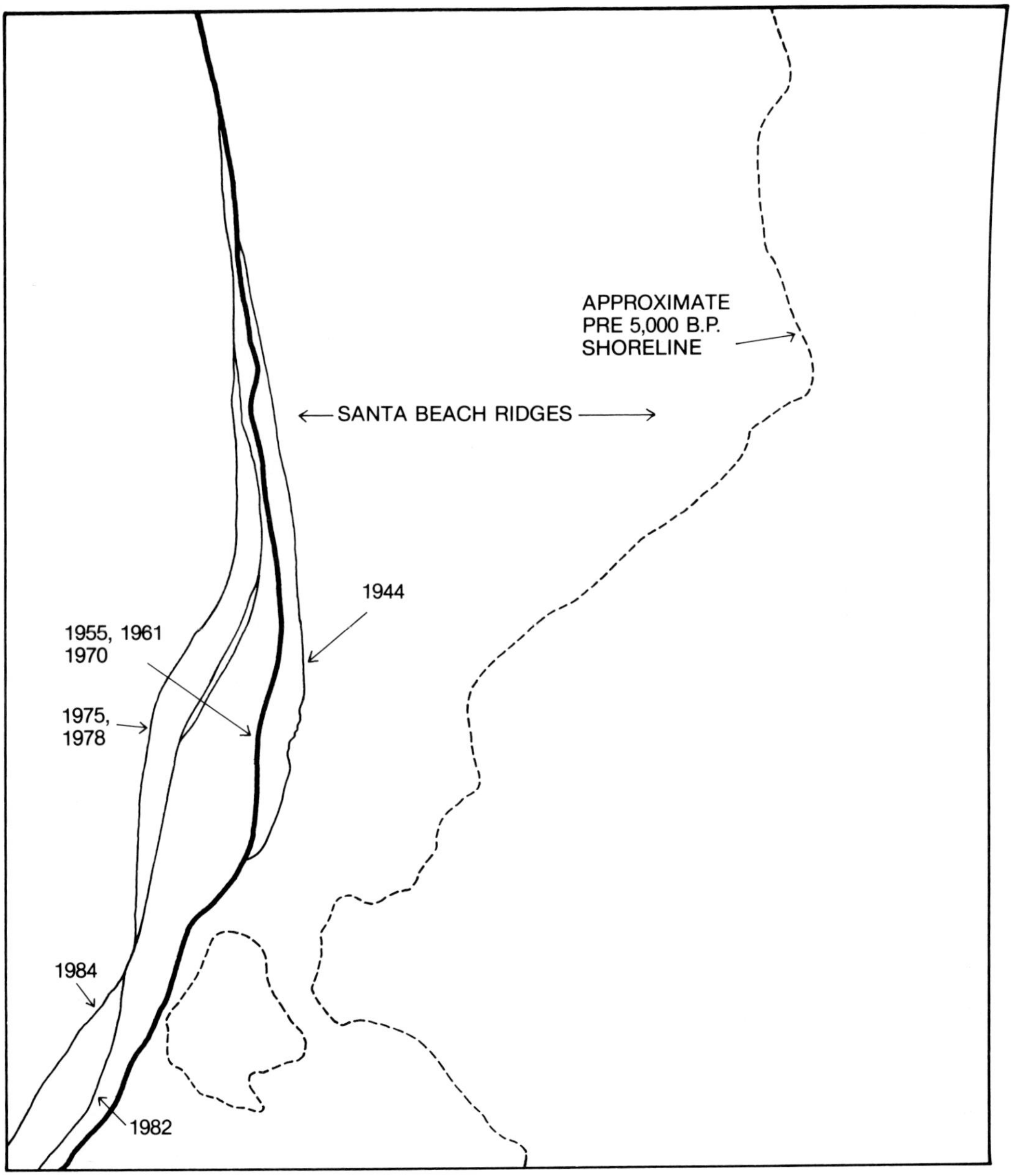

Figure 3. A diagrammatic guide to Figure 2 with overlay of eight high-altitude images of the Santa beach-ridge formation.

Pathways of sediment movement are dramatic in the Santa region because an exceptional quantity of material experiences transport. The Santa River basin is the largest and longest drainage of the entire arid Andean watershed. Consequently, stream flow and discharged sediment are some 15 to 40 times greater than those of adjacent rivers. The river disgorges a unique load of very large clasts, which contributes to the most massive beach ridges on the Pacific coast. Cobble ridges arise near the river mouth and extend north for 20 km along the littoral. Behind the northern littoral, there are immense inland deposits of eolian sand (Figure 4). These features are aspects of a regional system of sediment transport that entails fluvial, marine, and eolian components as normal mechanisms of transport. The system, diagramed in Figure 5, begins with the Santa River and then reaches northward along the coast and inland, where it extends well past the Chao River.

Fluvial Component

Most coastal streams and rivers are only supplied by seasonal precipitation in the upper 55% of the watershed. Therefore, smaller, shorter basins, such as the Chao and the Viru, are much drier than longer ones. The Santa is the only Pacific river to break through the western Cordillera Negra and to extend affluents deep into the Andes. LFC coverage suggests that the river transects the western range along a bedrock structure that coincides with a major change in coastal physiography. North of the Santa, the littoral shifts from rocky to sandy, bays are J-shaped, and the coastal plain and continental shelf widen significantly (Richardson, 1981). To the south, the coast is rocky, bays are C-shaped, and the coastal plain disappears as the continental shelf narrows. In the high mountains, the Santa turns south and runs along the western flanks of Cordillera Blanca, draining a major intercordillera basin known as the Callejon de Huaylas. This extended reach gives the river unusual length and discharge volume as well as a very distinctive sediment load. The load is unusually large, and the Santa disgorges exceptional quantities of cobbles because it drains an oversteep landscape, and it has the ability to transport large clasts. The sediment load is also compositionally distinct because it includes materials from the intercordillera basin that are not found in shorter adjacent basins, which only drain the Cordillera Negra. The river disgorges fine sediment and sands that are grey in color, whereas adjacent drainages release sediment of lighter tone and yellow sands.

Marine Component

The coast is prograding and building out from a wave-cut sea cliff (Figure 5, #3). The sea cliff is a prominent feature along much of the coast, and it can be traced northward past the Chao, Viru, and Moche Rivers. The cliff marks the maximum marine transgression, and it is thought to have formed with the Holocene onset of relative sea level stabilization. Where the littoral is sandy, the modern shoreline often lies well to the west of the sea cliff. Westward coastal progradation out from the sea cliff is due to sediment disgorged by streams and rivers being deposited along the littoral by marine processes. These processes are dominated by very strong longshore currents that sweep northward paralleling the littoral. The currents act against the formation of a well-developed, protruding delta by the Santa despite the river's massive sediment load. A short bedrock peninsula juts seaward several kilometers south of the river mouth. Grey beach sands, but not cobbles, of Santa origin extend down to this eminence. However, south of it, beach sands are yellow, as are the sands of stabilized dunes that arise immediately behind these beaches.

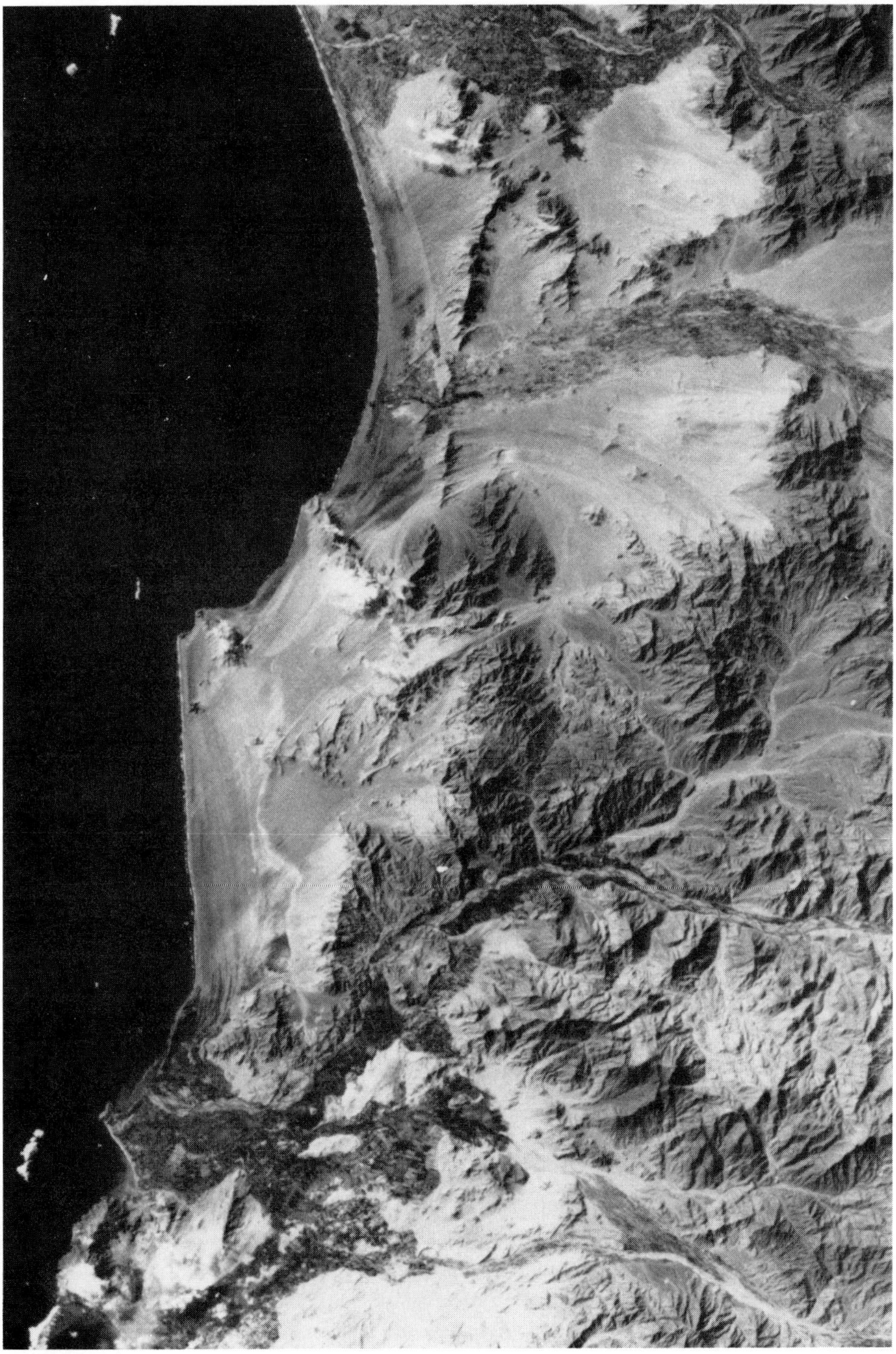

Figure 4. An enlarged section of Figure 1, showing, from top to bottom, the Viru, Chao, and Santa Valleys.

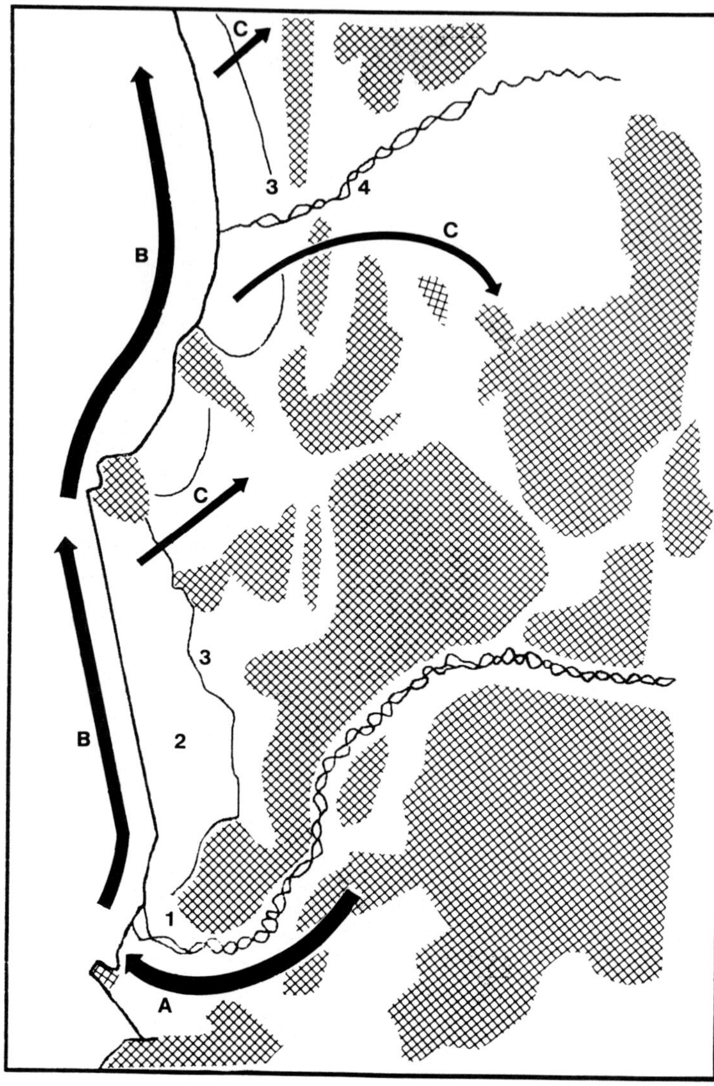

Figure 5. A diagrammatic guide to Figure 4. Cross-hatching indicates rocky hills. Arrows indicate direction of sediment movement by fluvial (A), marine (B), and aeolian (C) transport. Numbers indicate the Santa River (1), the Santa beach-ridge formation (2), the paleoshoreline demarcated by a wave-cut sea cliff (3), and the Chao River (4).

Beach ridges are not present south of the river mouth. Located west of the wave-cut sea cliff, these features only occur north of the river, where they are associated with grey sands. The Santa beach-ridge formation is comprised of nine major cobble-clad ridges that parallel one another and represent a time-transgressive sequence. The earliest structure, C-14 dated at 4236 ± 115 B.P., is located inland to the east, and later structures are situated coastward to the west (Sandweiss et al., 1983). The beach ridges arise as wide, linear mounds 2 to 4 m high that are covered by large clasts and cobbles, and the Santa River is the only source for these materials. The well-rounded and well-sorted condition of the clasts covering the ridges indicates that they were transported up the littoral and deposited by marine action after being disgorged by the river. The ninth ridge, which formed recently, arises north of the river, then bows seaward to a maximum point 1 km west of the 1955 shoreline before bending back to parallel the earlier ridges. Cobble clad, it is compositionally similar to the earlier ridges, although somewhat shorter.

Some 20 km north of the river, the beach-ridge formation ends against a large, rocky promontory. At this point, individual ridges have lost most of their original height, and their covering materials have graded down to pebbles and smaller clasts. Grey sands extend north along the shore past the Chao River to the southern reaches of Viru Valley, if not beyond. The grey sand flats, between the old sea cliff and the modern shoreline, are quite wide in the Chao region, indicating the Santa sediment had contributed significantly to coastal progradation in front of this small drainage. Beginning at the Santa and continuing past Viru, LFC imagery reveals a south-to-north decrease in the overall width of the coastal sand flats. Presumably, this decreased width reflects a decrease in the amount of sediment available for deposition as transport distance from the Santa mouth increases. North of the Chao River, coastal rivers become progressively larger. Therefore, as the Santa contributions to coastal progradation decline, the contributions from other basins enlarge. Yet, the overall width of the sand flats decreases because the northern rivers are all much smaller than the Santa.

Aeolian Component

Sand and fine sediment deposited along the shoreline by marine action are subject to inland transport by strong daily winds blowing off the ocean out of the southwest. The courses of inland transport and deposition are readily visible in LFC imagery (Figure 4). Very large, linear coppice dunes have accumulated on the lee sides of coastal hills. At their distal ends, the coppice formations often break into barcan fields. The barcans march inland and then spread out as sand sheets that are banked against the lower escarpment of the Cordillera. Some eolian formations have stabilized, but others have not. Marching barcans continually cross the Pan American highway at many points. Presumably, it requires many years for sand leaving the beach to reach final repose along the escarpment.

Large-scale, long-distance aeolian transport is particularly graphic in the Chao drainage. Here, much of the southern basin is buried by thick deposits of sand that have swept more than 18 km inland and now repose along the escarpment of the Cordillera Negra. Moving around or over hills, sand has invaded the Chao by a number of routes that are visible in Figure 4, and that have been diagramed in Figure 5. The coastal flats in front of the valley are the source of sand that migrates directly inland. Some of this material settles on the north side of the Chao, but much of it crosses the intervalley divide and swamps the southern portion of the Viru drainage. As Figure 1 indicates, eolian formations invading the south regions of valleys do not cross active river channels and continue north. Even small, intermittent rivers such as the Chao and the Viru have sufficient competence to erode encroaching dunes and to return the sand to the sea. Thus, annual stream action tends to mitigate eolian incursion by restricting dramatic deposition to the southern sides of coastal valleys.

El Niño Component

In addition to wind, water is a major agent that has sculpted the hyperarid landscape. Figures 1 and 2 reveal numerous dry drainages and land surfaces that have been shaped by rainfall and by runoff. Many of these features are associated with archaeological sites, roads, and canals that exhibit erosional damage from water. In historic times, significant desert rainfall has only occurred during El Niño events, and a causal association between ENSO events and desert precipitation is inferred for the Holocene (Wells, 1988). El Niño perturbations of normal marine and meteorological conditions are accompanied by a southward influx of warm tropical waters that override the cold Peru current. Marked by high sea-surface temperatures, strong events last about 18 months. Longshore currents become erratic, abate, and often cease entirely. Winds off the ocean also become erratic or cease.

The influx of tropical waters breaks down the normal temperature inversion that inhibits coastal precipitation. Weather changes from cool and dry to hot and humid. Storms arise at sea, then move inland, bringing torrential deluges to the desert. Generally confined to a few months, heavy rainfall tends to be highly sporadic and variable, but calamitous. In the Santa River drainage, ENSO precipitation is not abnormally greater in the upper, inter-Cordillera basin. However, it is catastrophic along the lower half of the Cordillera Negra (Waylen and Caviedes, 1986; Waylen, personal communication). El Niño rainfall is the erosional and depositional agent responsible for the transport of Holocene alluvium. ENSO runoff disgorges substantial amounts of material into the sea. This produces rapid, but generally temporary, progradation due to the temporary relaxing of strong longshore currents during El Niño conditions.

Tectonic Component

Earth movements affect the Santa basin, and it has a long history of seismic shocks (Ferro, 1978). There is also evidence suggesting at least one episode of coastal uplift during the latter half of the Holocene (Sandweiss, 1986). Uplift events or strong seismic events result in mass wasting along the steep watershed. An inordinate amount of material may be moved, but only over short distances. The tectonic component acts mainly to loosen and to liberate material that is subsequently transported by other agencies. Debris in the upper basin undergoes normal weathering from rainfall and from runoff, but this is not the case along the lower arid watershed. Here debris languishes on the landscape. This results in a sedimentary imbalance that is only adjusted for during rare, but recurrent El Niño precipitation. If earth movement alters ground slope, then the slope of normally dry drainages will remain out of equilibrium until ENSO runoff fosters erosional adjustment.

AN EXPLANATION

Our data indicate that the ninth Santa beach ridge formed in association with exceedingly rapid coastal progradation due to a surge of sediment released by El Niño runoff which carried debris from an earlier earthquake. Put in sequential order, the condition contributing to beach-ridge and dune-field formation are (1) mass wasting and release of material by a strong seismic shock, (2) abrupt subsequent transport of the debris to the river and to the sea by ENSO rainfall and runoff, (3) rapid coastal deposition in front of the river during the absence of longshore currents, (4) northward transport, reworking, and shoreline deposition of material with the resumption of normal marine conditions, and (5) inland movement of the new sand supply by strong daily winds.

Diachronic Constraints

The eight high-altitude images monitoring the Santa shoreline are overlain in Figures 2 and 3. The first time frame, 1944 to 1955, identifies an episode of moderate progradation with a slight seaward bow at the south end of the beach-ridge formation. The second and third windows of 1955 to 1961, and 1961 to 1970, reflect a long period of shoreline stasis. The fourth frame from 1970 to 1975 documents dramatic progradation and formation of the ninth beach ridge with a southern bow. The cobble bow protruded some 1022 m west of the 1955–1970 shoreline, creating a large lagoon between the new ridge and earlier ridges. The new coastline is largely maintained in the fifth window, 1975 to 1978. However, the sixth frame, 1978 to 1982, reveals limited (about 288 m) landward contraction of the most protruding section of the bow, as material was reworked and transported northward. The bow is then stable in the concluding frame, 1982 to 1984, but south of the bow, this frame reveals renewed progradation of substantial scope near the mouth of the river.

The imagery depicts two discrete incidences of shoreline building separated by 15 years of relative shoreline stability. We interpret the era of stasis between 1955 and 1970 as a time when the quantity of material disgorged by the Santa River was similar to the quantity of sediment removed by longshore currents and other marine processes. The shoreline remained stable because depositional and erosional processes were in equilibrium. Therefore, incidences of progradation should reflect episodes of imbalance when deposition exceeds erosion. If the sediment transport system operates in a balanced manner under normal conditions, then disequilibrium must result from abnormal conditions.

Environmental Events

During the 4 decades spanned by the imagery, the Santa region was affected by natural disasters of ENSO and seismic origin. The dates and magnitude of significant events are summarized in Table I. The imagery indicates that the ninth Santa beach ridge arose between 1970 and 1975. During this time, there were two natural disasters. The first was the 7.7-Richter scale earthquake of 1970. The second was the strong 1972–1973 ENSO event. We propose that these events were linked in a synergistic causal association. The quake generated an exceptional load of debris that the El Niño deposited at the river mouth, creating a sedimentary overload when longshore currents were inoperative. We will discuss the El Niño and seismic components of this recent episode of progradation before addressing the 1944–1955 incident of shoreline building.

The 1972–1973 El Niño

During strong ENSO events, coastal currents are perturbed and marine sediment transport breaks down, while flash floods carry substantial material to the sea. During the 1982–1983 event, we observed significant progradation at the mouths of many rivers. This included a sediment build up more than 1 km wide in front of the Jequetepeque River, which is among the larger northern drainages. Comprised largely of sand and fine sediment, these formations persisted until the resumption of strong longshore currents initiated erosion that gradually removed the deposits, which disappeared first in front of small rivers and a year or more later from large rivers.

The latest Santa beach ridge was well developed 2 years after the 1972–1973 ENSO events. We postulate that El Niño precipitation flushed material from the lower, normally dry watershed into the Santa River, that the expanded sediment load discharged by the river exceeded marine transport capabilities, and that this contributed to coastal progradation and beach-ridge building. Yet it is evident that every El Niño event does not produce a new

TABLE I. El Niño and Seismic Events Affecting Santa Study Area

Event Type	Date	Magnitude
Niño	1940–1941	S
Niño	1943	M+
Seismic	1946, Nov. 8.3/77.8	7.25
Niño	1951	NM
Niño	1953	M+
Niño	1957–1958	S
Niño	1965–1966	M+
Seismic	1962 9.9/78.0	6.75
Seismic	1970, May 9.2/78.8	7.7
Niño	1972–1973	S
Niño	1976–1977	M
Niño	1982–1983	VS

Note: NM = near moderate, M = moderate, S = strong, VS = very strong. Magnitude = Richter scale; approximate epicenter = Lat. S/Lon. W

El Niño magnitude ratings from Quinn et al., 1986; seismic events from Ferro, 1978.

ridge. The ''strong'' magnitude rating of the 1972–1973 perturbation is noteworthy because it was preceded by a comparable strong event in 1957–1958 that did not generate a new beach ridge. Nor was one generated by the ''very strong'' event of 1982–1983, which is generally considered the most severe El Niño of the century. Therefore, there is no evident association between ENSO magnitude and beach-ridge origin. Furthermore, the 1972–1973 event did not appear to generate new ridges elsewhere along the coast. The beach ridges north and south of the Chira River (Richardson, 1983) should be studied using the same photographic overlay techniques we applied to the Santa ridges. The SAN air photos (5/5/83 and 7/15/83) of the effects of the 1982–1983 El Niño show a massive sediment buildup and plume at the mouth of the Chira River, but a ground inspection in 1987 did not show a new ridge formation. The Colán ridges (Woodman and Polia, 1974; Ortlieb et al., 1989) are south of the mouth of the Chira River and, again, they did not exhibit any new ridge formations or progradation when inspected in 1987 and 1990.

The 1970 Earthquake

On the basis of the foregoing observations, it appears that while there is an association between the 1972–1973 El Niño and beach-ridge formation at Santa, another variable must also be involved. An additional contributory agency is implicated by the lack of correspondence between ENSO-event magnitude and beach-ridge origin and the fact that the 1972–1973 ENSO generating a new structure at one locale but not at others. For the ninth Santa ridge, we postulate that the contributory agency was the 1970 earthquake. Epicentered some 30 km seaward of the river mouth, the quake had a very severe impact upon the entire Santa basin, adversely affecting lateral drainages as well. The oversteep watershed experienced thousands of landslides, and mass wasting transpired on a vast scale. In the Callejon de Huaylas, this included the most devastating debris avalanche in the history of the hemisphere. Enormous quantities of detritus from this slide crossed and temporarily dammed the Santa River before the obstruction was breached by the river (Ericksen et al., 1970; Plafker et al., 1971). We propose that the material contributing to coastal progradation, beach-ridge formation, and dune building originated with the 1970 earthquake.

1944–1955 Progradation

An episode of moderate coastal progradation is framed by our 1944 and 1955 aerial photographs. We presume that this episode was triggered by the 7.25-Richter scale earthquake of November 10, 1946. This quake differed from the 1970 event because it was epicentered high in the Atlantic watershed, roughly 20 km east of the Santa River headwaters (Ferro, 1978). Therefore, mass wasting within the Santa Basin was of lesser magnitude than in 1970, and less pronounced along the lower, arid watershed. If the majority of seismic debris resided in the upper basin, then it was flushed out to sea gradually by highland rainfall and runoff. Material in the lower basin would have languished until subsequent transport was initiated by the moderate El Niño of 1951.

VARIABLES

Graphic monitoring of the Santa shoreline illustrates some of the processes contributing to recent coastal progradation. However, the monitoring also raises unresolved issues about the release and reworking of large sediment loads that require mention.

Tectonic Triggers

If the Santa has but nine beach ridges, then every incident of earth movement cannot lead to significant progradation because the basis has experienced many seismic events. For example, the imagery reflects no coastal change attributable to the 6.75-Richter scale quake of 1962. To serve as a triggering mechanism, tectonic events must exceed a certain threshold. Once the threshold is crossed, there may be substantial variability in the amount of debris generated by a tectonic event. Where a seismic event is epicentered is certainly important, and there must be significant differences between the high, wet and the low, arid divisions of the watershed. A high mountain epicenter, as in the 1946 case, will produce debris subject to relatively gradual disgorgement via annual rainfall and runoff. Given transport distance and reworking, the upper basin material reaching the coast will be relatively fine grained. Alternatively, an event epicentered close to the dry watershed, such as in the 1970 case, will produce substantial debris that can only be discharged by ENSO runoff, which increases the ability of the Santa River to disgorge large clasts from the lower watershed.

El Niño Events

Episodes of progradation should have two phases. Shortly after a tectonic event, the first phase will entail debris from the upper basin where rainfall and runoff move material into the river. When the volume and rate of sediment discharge are sufficient to override normal marine processes, shoreline deposition will occur. Without additional satellite data for the early 1970s, we cannot gauge the extent of this pre-ENSO phase of change. The second, or El Niño phase will entail a surge of sediment off the lower watershed. Yet it may require more than sediment to build a beach ridge. Along nonarid coasts, beach-ridge building is associated with violent storms. Turbulent sea storms accompany strong ENSO events, and they may be critical contributors to beach-ridge formation. In this vein, Sandweiss et al. (1983) report that a gravel pit cut into the second-oldest ridge reveals a core structure of unsorted, unrounded large clasts indicative of minimal marine transport. The investigators propose that the material was disgorged by the river under ENSO conditions and then thrown up on the shore by accompanying storm activity to create the ridge core. A thick deposit of well-rounded and well-sorted cobbles covers the core. This armor is attributed to deposition by normal wave action accompanying the return of strong longshore currents. If other ridges have similar cores, then the El Niño phase of progradation is critical to beach-ridge building as well as important in supplying sediment at a time when normal marine processes have broken down.

Synergistic Interactions of Tectonic Events and ENSO Events

There should be a relationship between the severity of a seismic or uplift event and the magnitude of the next El Niño that will generate coastal progradation and beach-ridge building. Weak seismic triggers followed by very strong ENSO events or powerful triggers followed by a moderate El Niño might have similar consequences. Depending upon the nature and severity of the tectonic agency, coastal progradation could continue through more than one ENSO episode. The new Santa beach ridge and shoreline should not stabilize until the river basin adjusts to the seismic mass wasting of 1970. Adjustment will proceed faster in the upper watershed than in its lower, arid regions where transport depends upon ENSO precipitation. Our 1982 and 1984 images bracketed renewed progradation in the river delta area. This progradation is attributable to the strong El Niño of 1982–1983. If debris from the 1970 quake contributed to this buildup, as seems probable, then adjustment to mass wasting takes a very long time, because many years pass between significant incidences of ENSO precipitation. Progradation between 1982 and 1984 was substantial, but relatively modest in comparison to the 1972 ENSO event. This suggests that the lower basin is nearing equilibrium, in which case the next El Niño will release less material. Even if the next ENSO event flushes out the last of the 1970 debris, it will require considerable time for the marine and aeolian processes to rework and to redeposit the material.

Marine and Aeolian Processes

One of the curious features of the new cobble ridge is the marked seaward bow of the southern end of the structure. The bow created an enormous lagoon between the new structure and the previous ridges. Evidence of littoral uplift was not observed in association with the 1970 seismic event (Ericksen et al., 1970; Plafker et al., 1971). If the quake did not result in seafloor or groundslope alteration, then the bow must be a product of marine processes that accompanied or followed the 1972–1973 ENSO event. The imagery indicates that by 1975 dramatic quantities of debris had accumulated both in front of the river mouth and along the coast to the north. Remaining more or less in place through 1978, the beach-ridge bow then eroded back several hundred meters before assuming its present configuration in 1982. Loss of material in the bow area should be associated with northward sediment movement and shoreline deposition of sand by the marine transport system. This, in turn would sustain an ongoing imbalance in the eolian component of the regional system of sediment movement. When abnormal quantities of fine material cease to be deposited along the shore, it will still require many years for wind-borne sand that has left the beach to achieve inland repose along the flanks of the Cordillera. Once this happens, the cycle of landscape alteration that began in May of 1970 will have run its course.

Counterpart Formations

If earlier ridges formed under conditions similar to the new one, then the Santa formation reflects rare, but recurrent incidents of tectonic crises followed by severe ENSO erosion, rapid coastal progradation, and episodic sand-dune incursion. Because the Santa disgorges an enormous sediment load that includes large clasts, the signatures of these recurrent processes are dramatic. If episodic progradation is a consequence of tectonic activity and subsequent El Niño perturbations, then other desert drainages must have experienced similar event combinations, although they need not have been coeval with those affecting the Santa. There seems to be evidence for this in aerial imagery of coastal sand flats adjacent to other northern streams and rivers. The many flats do not appear to be homogeneous. Rather, in the imagery, they are characterized by sets of linear features paralleling the shore. On the ground, these low-relief features are difficult to detect because changes in elevation or composition are generally subtle. We suggest that these linear features were formed by processes analogous to those of the Santa ridges. However, smaller basins release less sediment and finer clasts. Therefore, geomorphic changes in the littoral zone are less pronounced.

IMPLICATIONS

Identifying the origin and nature of recent change in the Santa coastline carries cultural implications, because beach ridges are symptomatic of abnormally stressful conditions. These conditions have occurred periodically in the past, and they will certainly recur in the future. We will touch upon the archaeological implications before concluding with what the future may hold.

The Past

During the latter half of the Holocene, nine major beach ridges have formed at the mouth of the Santa River. If all beach ridges, dune fields, and incidences of progradation arise from the synergistic consequences of major tectonic events followed by erosive ENSO ordeals, then coastal societies experience rare, but recurrent episodes of exceptional stress. The stress is prolonged because a threefold sequence of natural disasters must be played out before the desert landscape regains equilibrium (Moseley et al., 1981). The archaeological record should reveal the impact of the rare, but recurrent catastrophic processes that give rise to coastal progradation. However, all processes will not be equally detectable. Dune incursions should be discernable. Abnormally severe El Niño erosion prior to eolian activity should also be recognizable. However, it will be difficult to discover evidence of tectonic events that did not result in uplift. Problems with not being able to detect seismic events will greatly complicate geoarchaeological interpretations of the magnitude of El Niño floods. Following the 1970 seismic shock, the 1972–1973 ENSO event resulted in much more dramatic erosion and deposition than did the significantly stronger event of 1982–1983. If the quake went unrecognized, then the magnitude of the lesser El Niño would appear to be the greater of the two events.

Probable seismic damage to ancient architecture (Marcus, 1987) and possible tectonic slope alteration of ancient canals (Ortloff et al., 1982) have been reported. However, this type of evidence survives because people did not rebuild or repair the damaged structures. When reconstruction transpired, documentation will be difficult. This is not because rebuilding is obscure. Rather, it is because archaeological monuments on the coast were built in multiple stages interspersed with long periods of use. We began this chapter with the case of dune incursion which followed severe El Niño erosion at Huacas Sol and Luna. Both monuments assumed their final form after many separate building phases. In the case of Sol, we documented eight construction stages separated by periods of use. In hindsight, we should have been alert for evidence that one or more of these stages followed a tectonic event. Yet at the time, we thought they were all due to purely cultural considerations. Now that there is a modern analog for the events that generate episodic sand movement, archaeologists can be more attentive to detecting evidence of seismic disturbance. This is very important since tectonic events produce the highest mortalities of the three types of natural disasters that are associated with beach ridges. The 1970 seismic event did more than generate an episode of shoreline progradation. It killed in excess of 60,000 individuals. In terms of death and property damage, this was the most devastating New World earthquake to be historically documented (Plafker et al., 1971). It brought profound cultural changes to the affected region (Oliver-Smith, 1986). Ranging from where people resettled to new economic realities, many types of change should be discernable in the archaeological record. Yet their cause would be misconstrued if the quake went undocumented.

Of the three types of disasters associated with beach ridges, dune incursions are only as disastrous as people make them, and from the Santa River north they can be very calamitous. This is because the coastal plain widens to afford more arable land, and rivers are progressively bigger, thereby favoring large-scale irrigation. More than one third of all land farmed along the entire arid Andean coast has traditionally lain between the Santa and Olmos Rivers. For millennia, pre-Hispanic populations in the region built and operated the largest canal systems on the continent, and at least 30% more land was reclaimed in the past than is under cultivation today. Thus, sand incursions across the coastal plain have long occurred where their impact could be disastrous to large constructs of economic importance. Irrigation works are easily perturbed by events of tectonic and ENSO origin and, where studied, abandoned prehistoric systems provide sensitive records of past environmental events. Historically and prehistorically, irrigation systems tend to be reconstructed and reused after most disasters. However, this is probably not true for dune incursions. The reasons for this, to judge from the case of Huaca del Sol, is that there is no way to turn back an encroaching sea of sand; movement of aeolian material persists for a long time, and after it is over, the landscape is not the same as it was when formerly farmed.

Space shuttle LFC imagery indicates that irrigation systems on the southern sides of oasis valleys are particularly susceptible to stress and disruption from dune inundation. The problem with southern land tracts is their allure in times of landscape stability. When sand is not moving about, southern land tracts invite reclamation, albeit at the cost of constructing large canals. The risk is twofold. First, reclaimed land must remain in production for generations if yields are to meet and to exceed the costs of constructing the irrigation systems. Second, when dunes do inundate southern irrigation systems, there is often little latitude north of the rivers to reexpand irrigation and to compensate for the long enduring loss of arable land. It remains for archaeologists to determine how often agriculture has expanded southward during times of landscape stasis, only then to be lost to cycles of renewed sediment movement. In the Moche and Jequetepeque Valleys, this seems to have transpired at least twice in pre-Hispanic times (Ortloff et al., 1982; Eling, 1987). Ultimately, southern land tracts could prove to be energy sumps in the evolution of irrigation agriculture on the north coast.

The Present and Future

In terms of the present, the Peruvian government is engaged in building ever larger reclamation works to open more of the desert to cultivation. Between the Santa and Omos Rivers, these undertakings reactivate a great deal of land that was farmed in pre-Hispanic times, and some projects lie entirely within the confines of ancient canal networks. Unfortunately, the projects do not proceed from any prior study of why the prehistoric irrigation systems may have failed in the first place. This reflects an unstated premise that past loss of irrigated land is attributable to conditions that will not reoccur. The most ambitious northern reclamation project was launched shortly after LFC coverage of the region was flown. The undertaking is designed to carry abundant Santa River waters up through the Chao, Viru, and Moche drainages to the Chicama Valley and to transform the sandy coastal plain into a vast greenhouse that will feed untold generations to come. At the time we field checked the new beach ridge, the course of the Canal Madre that will make all this happen was being cut across the more stable sand deposits banked against the southern escarpment of the Chao basin.

In terms of the future, this ambitious project could represent a great disaster in the making if the cycle of sediment movement that opened in 1970 has yet to run its course. If marine currents are still reworking and redepositing the debris that contributed to beach-ridge building, then there is a continuing source of sand for dune incursions. In this situation, stress from saltating sands could come to exceed agrarian design expectations and diminish the performance of the irrigation project. Potential stress on the reclamation system will increase with proximity to its water supply source because the Santa River is also the source of sediment that will interdict the irrigation system. In a worst-case scenario, a sea of sand will seek repose upon the southern escarpment of the Chao basin. If the Canal Madre is a trough of open-channel design in this section of the basin, end of story! A duct of enclosed channel design could mitigate water transport problems. Yet fields and planting surfaces would present difficulties hard to alleviate. This is because sand travels across much of the landscape that is to be irrigated, and is incompatible with farming when moving en masse. To picture this point, we conclude with a false color image (Plate 1)* of the Chao basin that illustrates the truly vast sweep of eolian material. Two decades of satellite and high-altitude imagery are available for detailed, time-lapse modeling of Santa sediment movement. Therefore, impending human consequences could be forecasted with precision if interest in the agrarian future was more than a footnote to agricultural problems of the past.

*Plate 1 follows text.

ACKNOWLEDGMENTS

A.J.S. Guggenheim Fellowship supported the senior author's contributions to this report. Grants to Richardson from the Center for Latin American Studies, University of Pittsburgh, and the M. Graham Netting Fund of the Carnegie Museum of Natural History supported field verification of the ninth Santa beach ridge in 1987. We are indebted to the Stennis Space Center for the LFC photographs and for the facilities to digitize and coregister the imagery employed in this article. Many people at Stennis provided invaluable assistance and special acknowledgement is due to Thomas Sever, Harry Hoff, and Scott Madry.

REFERENCES

Brennan, C.T. (1978) Investigations of Cerro Arena, Peru: Incipient Urbanism on the Peruvian North Coast. Ph.D. thesis, Department of Anthropology, University of Arizona, Tucson.

Cossio, A. and Jaen, H. (1967) Geologia de los Cuadrangulos de Puemape, Chocope, Otuzco, Trujillo, Salaverry, y Santa. *Bol. Serv. Geol. Min.* 17: 1.

Craig, A.K. and Shimada, I. (1986) El Niño flood deposits at Batan Grande. *Geoarchaeology.* 1: 29.

Eling, H.H., Jr. (1987) The Role of Irrigation Networks in Emerging Societal Complexity During Late Prehispanic Times. Jequetapegue Valley, North Coast, Peru. Ph.D. thesis, Department of Anthropology, University of Texas, Austin.

Ericksen, G.E., Plafker, G., and Concha, J.F. (1970) Preliminary Report on the Geological Events Associated with the May 31, 1970 Peru Earthquake. U.S. Geol. Surv. Circ.

Ferro, E.S. (1978) Historia de Los Sismos Mas Notables Ocurridós en El Peru (1513–1947). *Inst. Geol. Min.* 3: 7.

Marcus, J. (1987) Late Intermediate Occupation at Cerro Azul, Peru: A Preliminary Report. Tech. Rep. 20, University of Michigan Museum of Anthropology, Ann Arbor.

Moseley, M.E., Feldman, R.A., and Ortloff, C.R. (1981) Living with Crises: Human Perception of Process and Time. In: *Biotic Crises in Ecological and Evolutionary Time,* pp. 321–367. Nietcki, M., Ed. Academic Press, New York.

Nials, F.L., Deeds, E.R., Moseley, M.E., Pozorski, S.G., Pozorski, T., and Feldman, R.A. (1979) El Niño: The catastrophic flooding of coastal Peru. *Field Mus. Nat. Hist. Bull.* 50 (Part 1,2): 4.

Oliver-Smith, A. (1986) *The Martyred City: Death and Rebirth in the Andes,* pp. 1–280. University of New Mexico Press, Albuquerque.

Ortlieb, L., Machare, J., Fournier, M., and Woodman, R. (1989) La Secuencia de Cordones Litorales de Colan (Piura): Un Registro del Fenomeno "El Niño" en el Holoceno Superior. *Bol. Soc. Geol. Peru.* 80: 107–121.

Ortloff, C.R., Moseley, M.E., and Feldman, R.A. (1982) The Chicama-Moche intervalley canal: social explanations and physical paradigms. *Am. Antiq.* 48: 375.

Plafker, G., Ericksen, G.E., and Concha, J.F. (1971) Geological Aspects of the May 31, 1970 Peru Earthquake. *Bull. Seismol. Soc. Am.* 61: 543.

Quinn W.H., Neal, V.T., and Antunez de Mayolo, S.E. (1986) Preliminary report on El Nino occurrences over the past four and a half centuries. Oregon State University, Corvallis.

Richardson, J.B., III (1981) Modeling the development of sedentary maritime economies on the coast of Peru: a preliminary statement. *Ann. Carnegie Mus.* 50: 39.

Richardson, J.B., III (1983) The Chira beach ridges, sea level change, and the origins of maritime economies on the Peruvian coast. *Ann. Carnegie Mus.* 52: 265.

Sandweiss, D.H. (1986) The beach ridges at Santa, Peru: El Niño, uplift, and prehistory. *Geoarchaeology.* 1: 17.

Sandweiss, D.H., Rollins, R.B., and Richardson, J.B., III (1983) Landscape alteration and prehistoric human occupation on the north coast of Peru. *Ann. Carnegie Mus.* 52: 277.

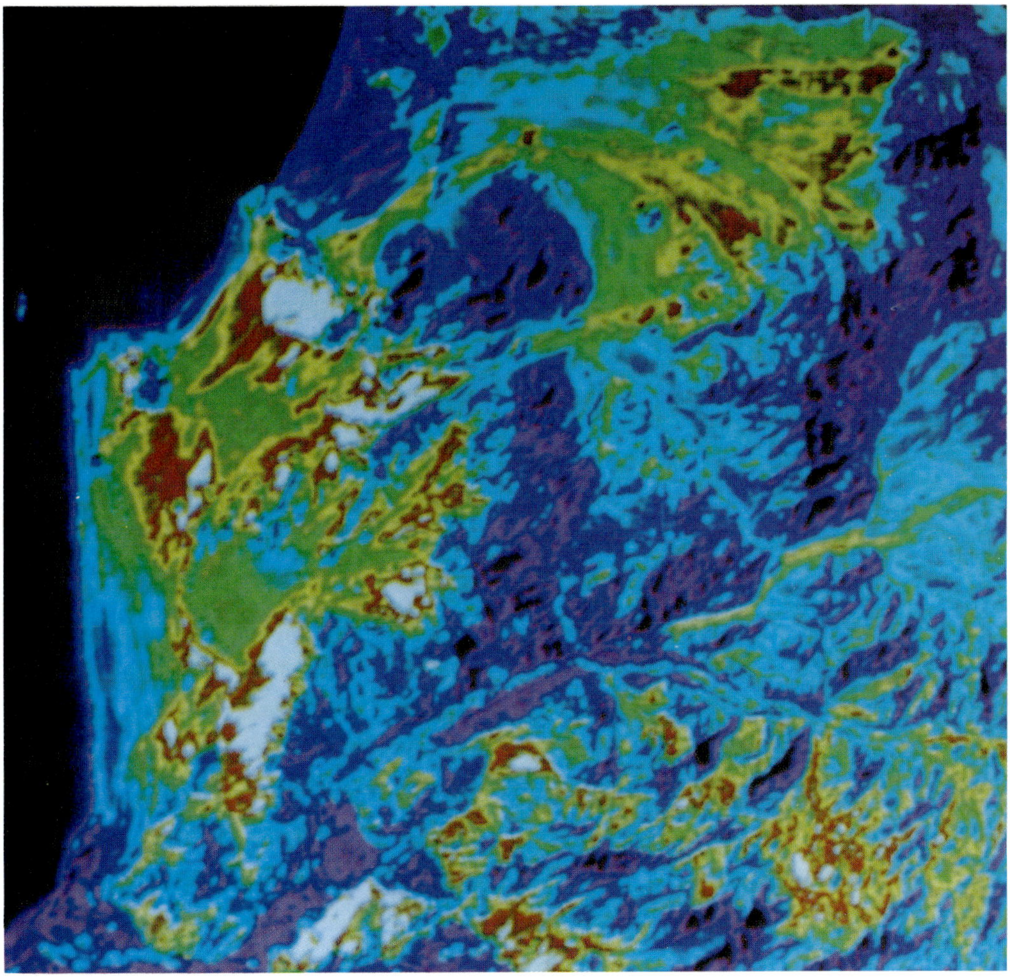

Plate 1. A false color LFC image showing aeolian sands in the Chao River region.

Topic, T.L. (1982) The early intermediate period and its legacy. In: *Chan Chan: Andean Desert City.* Moseley, M.E. and Day, K.C., Eds. School of American Research, University of New Mexico Press, Albuquerque.

Waylen, P.R. and Caviedes, C.N. (1986) El Niño and annual floods on the north Peruvian littoral. *J. Hydrol.* 89: 141.

Wells, L.E. (1988) Holocene Fluvial and Shoreline History as a Function of Human and Geologic Factors in Arid Northern Peru. Ph.D. thesis, Department of Geology, Stanford University, Palo Alto, CA.

Woodman, R. and Polia, M. (1974) Evidencia Arqueologica del Levantamiento Continental al Norte del Peru en los Ultimos 4 mil Años. *Bol. Soc. Geogr. Lima.* 43: 63–66.

Conclusion

Lucille Lewis Johnson
Department of Anthropology
Vassar College
Poughkeepsie, New York

We have shown some of the ways in which geologists and archaeologists working together have approached paleoshorelines and the interpretation of past geological and cultural events along them. Other scientists are also needed for us to fully understand past coastal adaptations and change. Particularly critical, and missing from most of our studies, is an understanding of marine fauna, its life condition, and how these are affected by changes in coastal contours. Watters, Donahue, and Stuckenrath, for example, will need to consult a marine biologist in order to resolve the *Strombus* line problem. The sea-grass beds west and south of the island are the kinds of environment in which *Strombus* live. A marine biologist might be able to determine where such environments would have been during the Archaic and early Ceramic periods, when the landscape and seascape were quite different. The nature of the deposit might also be obvious to an experienced marine biologist, while it is opaque to archaeologists and geologists. It might also be useful for Watters and Donahue to adopt some of Mason's methods to study the Barbudan beach ridges.

Dunbar, Webb, and Faught have shown that underwater investigations can reveal unexpected details of prehistoric occupations, and others should follow their lead. This would be particularly easy in the Caribbean, where conditions are ideal for scuba diving. Despite the high-energy marine environment in this hurricane-prone region, sites that were inundated during periods of rapid sea level rise may still survive. Keegan rightly deplores the havoc caused by hurricanes on the Bahamian archaeological record, but he might be able to find preserved sites in inundated paleobays. With Florida occupied by 10,000 B.P. at least and the Antilles by 5000 B.P., it seems improbable that occupation of the Bahama banks occurred so late in prehistory. Watters, Donahue, and Stuckenrath might be able to discover through sub-marine exploration whether River was a single isolated Archaic site, or is only the last Archaic site left, others having been flooded out. Another major question about River, which sub-marine mapping might answer, concerns where it was located relative to the shoreline at the time it was occupied.

Although research problems are never entirely solved, some of our studies have reached a point of conclusion in that a particular set of questions has been answered, while other studies have been presented midway in ongoing research. The mainland situations, despite the greater cultural complexity of mainland occupations, are most resolved geologically. Owen Mason has the geomorphology of Kotzebue Sound well mapped, while, to the east, Clark and Fitzhugh's relative sea level maps cover the zone of concern both spatially and temporally. In South America, Suguio et al. have a strong understanding of the Brazilian coastal sequence and Moseley et al. have presented a strong argument for the combined effects of tectonism, El Niño-Southern Oscillation currents, and wind in shaping the north Peruvian coast. All of these scholars have created methods of analysis which have solved their local problems and which can be used by them and others in further investigations of comparable situations.

Elsewhere, the underwater teams in the Gulf of Mexico need to find the Clovis shoreline, create models for its occupation, and discover the sites predicted by the models. All of the island studies presented here are also in progress, and require further fieldwork and analysis to resolve the problems which interest the authors. As discussed above, both Caribbean studies need surveys of the islands' shallow shelves to determine the nature and extent of early occupation, and Watters et al. have outlined necessary steps for resolving the *Strombus* line problem. In the Shumagin Islands, it is essential that Johnson and Winslow complete their survey of the inner Shumagins to determine whether the raised shorelines of these islands contain sites of the dates predicted by Winslow's model.

To conclude this book as I began it: the studies which have been presented here indicate both how much we have learned about paleoshorelines and human adaptations to them and how much we still have to learn. The research is proceeding and I am sure that, within a decade, the methods and results presented here will have been amplified by exciting additional studies.

INDEX

A

B

C

D

E

F

G

H

I

J

K

L

M

N

O

P

Q

R

S

T

U

V

W